D1702678

Gunther Kolb
Fuel Processing

Further Reading

K. Sundmacher, A. Kienle, H. J. Pesch, J. F. Berndt, G. Huppmann (Eds.)

Molten Carbonate Fuel Cells

Modeling, Analysis, Simulation, and Control

2007
ISBN: 978-3-527-31474-4

W. Vielstich, A. Lamm, H. Gasteiger (Eds.)

Handbook of Fuel Cells - Fundamentals, Technology, Applications

4 volume set

2003
ISBN: 978-0-471-49926-8

B. Elvers (Ed.)

Handbook of Fuels

Energy Sources for Transportation

2007
ISBN: 978-3-527-30740-1

A. Züttel, A. Borgschulte, L. Schlapbach (Eds.)

Hydrogen as Future Energy Carrier

2008
ISBN: 978-3-527-30817-0

H.-W. Häring (Ed.)

Industrial Gases Processing

2008
ISBN: 978-3-527-31685-4

Gunther Kolb

Fuel Processing

for Fuel Cells

WILEY-VCH Verlag GmbH & Co. KGaA

The Author

Dr. Gunther Kolb
IMM - Institut für Mikrotechnik Mainz GmbH
Carl-Zeiss-Str. 18 - 20
55129 Mainz
Germany

Cover Illustration:
Photograph courtesy of Nuvera.
The APU model was developed
by Tenneco within the European
project Hytran ("Hydrogen and
Fuel Cell Technologies for Road Transport"),
contract no. TIP3-CT-2003-502577
co-ordinated by Volvo Technology
Corporation.

All books published by Wiley-VCH are carefully produced. Nevertheless, authors, editors, and publisher do not warrant the information contained in these books, including this book, to be free of errors. Readers are advised to keep in mind that statements, data, illustrations, procedural details or other items may inadvertently be inaccurate.

Library of Congress Card No.: applied for

British Library Cataloguing-in-Publication Data
A catalogue record for this book is available from the British Library.

Bibliographic information published by the Deutsche Nationalbibliothek
Die Deutsche Nationalbibliothek lists this publication in the Deutsche Nationalbibliografie; detailed bibliographic data are available on the Internet at <http://dnb.d-nb.de>.

© 2008 WILEY-VCH Verlag GmbH & Co. KGaA, Weinheim

All rights reserved (including those of translation into other languages). No part of this book may be reproduced in any form – by photoprinting, microfilm, or any other means – nor transmitted or translated into a machine language without written permission from the publishers. Registered names, trademarks, etc. used in this book, even when not specifically marked as such, are not to be considered unprotected by law.

Typesetting Thomson Digital, Noida, India
Printing Strauss GmbH, Mörlenbach
Binding Litges & Dopf GmbH, Heppenheim
Cover Grafik-Design Schulz, Fußgönheim

Printed in the Federal Republic of Germany
Printed on acid-free paper

ISBN: 978-3-527-31581-9

Contents

Acknowledgement *IX*

1 Introduction and Outline *1*

2 Fundamentals *3*
2.1 Common Fossil Fuels *3*
2.2 Basic Definitions, Calculations and Legislation *6*
2.3 The Various Types of Fuel Cells and the Requirements of the Fuel Processor *12*
2.3.1 PEM Fuel Cells *12*
2.3.2 High Temperature Fuel Cells *15*

3 The Chemistry of Fuel Processing *17*
3.1 Steam Reforming *17*
3.2 Partial Oxidation *22*
3.3 Oxidative Steam Reforming or Autothermal Reforming *29*
3.4 Catalytic Cracking of Hydrocarbons *38*
3.5 Pre-Reforming of Higher Hydrocarbons *39*
3.6 Homogeneous Plasma Reforming of Higher Hydrocarbons *43*
3.7 Aqueous Reforming of Bio-Fuels *44*
3.8 Processing of Alternative Fuels *44*
3.8.1 Dimethyl Ether *44*
3.8.2 Methylcyclohexane *45*
3.8.3 Sodium Borohydride *45*
3.8.4 Ammonia *46*
3.9 Desulfurisation *46*
3.10 Carbon Monoxide Clean-Up *48*
3.10.1 Water–Gas Shift *48*
3.10.2 Preferential Oxidation of Carbon Monoxide *49*
3.10.3 Methanation *51*

Fuel Processing for Fuel Cells. Gunther Kolb
Copyright © 2008 WILEY-VCH Verlag GmbH & Co. KGaA, Weinheim
ISBN: 978-3-527-31581-9

3.11	Catalytic Combustion	52
3.12	Coke Formation on Metal Surfaces	52
4	**Catalyst Technology for Distributed Fuel Processing Applications**	**57**
4.1	A Brief Introduction to Catalyst Technology and Evaluation	57
4.1.1	Catalyst Activity	58
4.1.2	Catalyst Stability	60
4.1.3	Catalyst Coating Techniques	61
4.1.4	Specific Features Required for Fuel Processing Catalysts in Smaller Scale Applications	68
4.2	Reforming Catalysts	69
4.2.1	Catalysts for Methanol Reforming	71
4.2.2	Catalysts for Ethanol Reforming	77
4.2.3	Overview of Catalysts for Hydrocarbon Reforming	80
4.2.4	Catalysts for Natural Gas/Methane Reforming	81
4.2.5	Catalysts for Reforming of LPG	84
4.2.6	Catalysts for Pre-Reforming of Hydrocarbons	86
4.2.7	Catalysts for Gasoline Reforming	88
4.2.8	Catalysts for Diesel and Kerosene Reforming	92
4.2.9	Cracking Catalysts	96
4.2.10	Deactivation of Reforming Catalysts by Sintering	98
4.2.11	Deactivation of Reforming Catalysts by Coke Formation	98
4.2.12	Deactivation of Reforming Catalysts by Sulfur Poisoning	101
4.3	Catalysts for Hydrogen Generation from Alternative Fuels	105
4.3.1	Dimethyl Ether	105
4.3.2	Methylcyclohexane	106
4.3.3	Sodium Borohydride	107
4.3.4	Ammonia	107
4.4	Desulfurisation Catalysts/Adsorbents	108
4.5	Carbon Monoxide Clean-Up Catalysts	111
4.5.1	Catalysts for Water–Gas Shift	111
4.5.2	Catalysts for the Preferential Oxidation of Carbon Monoxide	116
4.5.3	Methanation Catalysts	123
4.6	Combustion Catalysts	124
5	**Fuel Processor Design Concepts**	**129**
5.1	Design of the Reforming Process	129
5.1.1	Steam Reforming	129
5.1.2	Partial Oxidation	146
5.1.3	Autothermal Reforming	149
5.1.4	Catalytic Cracking	154
5.1.5	Pre-Reforming	155
5.2	Design of the Carbon Monoxide Clean-Up Devices	155
5.2.1	Water–Gas Shift	155
5.2.2	Preferential Oxidation of Carbon Monoxide	161

5.2.3	Selective Methanation of Carbon Monoxide	*164*
5.2.4	Membrane Separation	*164*
5.2.5	Pressure Swing Adsorption	*174*
5.3	Aspects of Catalytic Combustion	*176*
5.4	Design of the Overall Fuel Processor	*181*
5.4.1	Overall Heat Balance of the Fuel Processor	*181*
5.4.2	Interplay of the Different Fuel Processor or Components	*188*
5.4.3	Overall Water Balance of the Fuel Processor	*190*
5.4.4	Overall Basic Engineering of the Fuel Processor	*192*
5.4.5	Dynamic Simulation of the Fuel Processor	*205*
5.4.6	Control Strategies for Fuel Processors	*213*
5.5	Comparison with Conventional Energy Supply Systems	*215*

6 Types of Fuel Processing Reactors *217*
6.1 Fixed-Bed Reactors *217*
6.2 Monolithic Reactors *217*
6.3 Plate Heat-Exchanger Reactors *221*
6.3.1 Conventional Plate Heat-Exchanger Reactors *223*
6.3.2 Microstructured Plate Heat-Exchanger Reactors *225*

7 Application of Fuel Processing Reactors *227*
7.1 Reforming Reactors *227*
7.1.1 Reforming in Fixed-Bed Reactors *227*
7.1.2 Reforming in Monolithic Reactors *230*
7.1.3 Reforming in Plate Heat-Exchanger Reactors *240*
7.1.4 Reforming in Membrane Reactors *254*
7.1.5 Reforming in Chip-Like Microreactors *260*
7.1.6 Plasmatron Reformers *264*
7.2 Water–Gas Shift Reactors *269*
7.2.1 Water–Gas Shift in Monolithic Reactors *269*
7.2.2 Water–Gas Shift in Plate Heat-Exchanger Reactors *270*
7.2.3 Water–Gas Shift in Membrane Reactors *272*
7.3 Catalytic Carbon Monoxide Fine Clean-Up *272*
7.3.1 Carbon Monoxide Fine Clean-Up in Fixed-Bed Reactors *272*
7.3.2 Carbon Monoxide Fine Clean-Up in Monolithic Reactors *273*
7.3.3 Carbon Monoxide Fine Clean-Up in Plate Heat-Exchanger Reactors *275*
7.3.4 Carbon Monoxide Fine Clean-Up in Membrane Reactors *282*
7.4 Membrane Separation Devices *283*
7.5 Catalytic Burners *285*

8 Balance-of-Plant Components *289*
8.1 Heat-Exchangers *289*
8.2 Liquid Pumps *290*
8.3 Blowers and Compressors *290*

8.4	Feed Injection System	292
8.5	Insulation Materials	293
9	**Complete Fuel Processor Systems**	**295**
9.1	Methanol Fuel Processors	295
9.2	Ethanol Fuel Processors	316
9.3	Natural Gas Fuel Processors	317
9.4	Fuel Processors for LPG	327
9.5	Gasoline Fuel Processors	332
9.6	Diesel and Kerosine Fuel Processors	344
9.7	Multi-Fuel Processors	348
9.8	Fuel Processors Based on Alternative Fuels	350
10	**Introduction of Fuel Processors Into the Market Place – Cost and Production Issues**	**355**
10.1	Factors Affecting the Cost of Fuel Processors	355
10.2	Production Techniques for Fuel Processors	359
10.2.1	Fabrication of Ceramic and Metallic Monoliths	359
10.2.2	Fabrication of Plate Heat-Exchangers/Reactors	361
10.2.3	Fabrication of Microchannels	365
10.2.4	Fabrication of Chip-Like Microreactors	367
10.2.5	Fabrication of Membranes for Hydrogen Separation	369
10.2.6	Automated Catalyst Coating	370

References 373

Index 409

Acknowledgement

I would like to cordially thank my colleagues at IMM, in particular Dr. Karl-Peter Schelhaas for fruitful discussions and input in the fields of calculations and material properties, Dr. Hermann Ehwald for input in the field of desulfurization catalysts, Tobias Hang for dealing with the figures, Carola Mohrmann and Christina Miesch-Schmidt for dealing with the tables, Dr. Athanassios Ziogas and Martin O'Connell for dealing with the literature ordering and Sibylle for dealing with me when I was "hacking" through weekends and nights.

Gunther Kolb

1
Introduction and Outline

Mankind's energy demand is increasing exponentially. Between 1900 and 1997, the world's population more than tripled and the average energy demand per human being has also more than tripled, resulting in greater than thirteen times higher overall global emissions [1]. Thus the carbon dioxide concentration rose from 295 parts per million in 1900 to 364 parts per million in 1997 [1]. In 1997 almost all European countries committed to reducing greenhouse gas emissions to an amount 8% below the emissions of 1990 in the period from 2008 to 2012. With this scenario, fuel cell technology is attracting increasing attention nowadays, because it offers the potential to lower these emissions, owing to a potentially superior efficiency compared with combustion engines. Fuel cells require hydrogen for their operation and consequently numerous technologies are under investigation worldwide for the storage of hydrogen, aimed at distribution, and mobile and portable applications.

The lack of a hydrogen infrastructure in the short term, along with the highly attractive energy density of liquid fossil and regenerative fuels, has created widespread research efforts in the field of distribution and on-board hydrogen generation from various fuels. This complex chemical process, generally termed fuel processing, is the subject of this book.

The electrical power output equivalent of the fuel processors that are currently under development world wide covers a wide range, from less than a watt to several megawatts. Portable and small scale mobile fuel cell systems promise to be the first commercial market for fuel cells, according to a market study of *Fuel Cell Today* in July 2003 [2]. According to the same report, the number of systems built has increased dramatically to up to more than 3000 in 2003. To date, most of these systems have used Proton Exchange Membrane (PEM) fuel cells.

Low power fuel processors (1–250 W) compete with both conventional storage equipment, such as batteries, and simpler fuel cell systems, such as Direct Methanol Fuel Cells (DMFC).

Fuel cell systems for residential applications are typically developed for the generation of power and heat, which increases their overall efficiency considerably, because even low temperature off-heat may be utilised for hot water generation, which reduces energy losses considerably.

Fuel Processing for Fuel Cells. Gunther Kolb
Copyright © 2008 WILEY-VCH Verlag GmbH & Co. KGaA, Weinheim
ISBN: 978-3-527-31581-9

For mobile applications, systems designed to move a vehicle need to be distinguished from the Auxiliary Power Unit (APU), which either creates extra energy for the vehicle (e.g., the air conditioning and refrigerator system of a truck) or works as a stand alone system for the electrical power supply.

This book provides a general overview of the field of fuel processing for fuel cell applications. Its focus is on mobile, portable and residential applications, but the technology required for the smaller stationary scale is also discussed.

In the second chapter fundamental definitions and the basic knowledge of fuel cell technology are provided, as far as is required to gain an insight into the interplay between the fuel cell and its hydrogen supply unit – the fuel processor.

The third chapter deals with the reforming chemistry of conventional and alternative fuels, and with the chemistry of catalytic carbon monoxide clean-up, sulfur removal and catalytic combustion.

An overview of catalyst technology for fuel processing applications is provided in Chapter 4, covering all the processes described in Chapter 3.

The design of the individual components of the fuel processor is the subject of Chapter 5. Design concepts and numerical simulations presented in the open literature are discussed for reforming, catalytic carbon monoxide clean-up and physical clean-up strategies, such as membrane separation and pressure swing adsorption. In addition, fuel processor concepts are then presented and the interplay between the various fuel processor components is explained. Details of the basic engineering of fuel processors and dynamic simulations are discussed, covering start-up and control strategies. Some tips and the basic knowledge required to perform such calculations are provided.

There are three basic types of fuel processing reactors, namely fixed catalyst beds, monoliths and plate heat-exchangers, which are explained in Chapter 6.

Chapter 7 then shows the practical applications of such reactors, as published in the literature.

In Chapter 8 some important aspects of balance-of-plant components are discussed, and Chapter 9 presents complete fuel processors for all types of fuels, while cost and production issues are the subject of Chapter 10.

2
Fundamentals

This chapter provides information about common fossil fuels, necessary definitions in the field of fuel processing and the basic knowledge from the wide field of fuel cell technology. It is by no means comprehensive and is not a substitute for the dedicated literature in these fields. Rather, it provides a brief summary for readers who wish to gain an overview of the topic of fuel processing without the need to use too much additional literature.

2.1
Common Fossil Fuels

Fuels are solid, liquid or gaseous energy carriers. To date, practically all of the fuels available on the market are based upon fossil sources and thus contain hydrocarbons of varying composition. However, alternative fuels such as alcohols and hydrides may serve as future energy carriers. Table 2.1 provides an overview of the conventional fuels and of the most important alternative fuels, which may act as future hydrogen source for fuel cells along with their key properties.

A comparison of the gravimetric and volumetric density of various hydrogen carriers shows that liquid hydrocarbons have – apart from borohydrides – by far the best combined properties (see Figure 2.1).

Table 2.2 shows the maximum amount of work that can be converted into electricity from various fuels, in theory. Compared with the gravimetric and volumetric energy density of 1 MJ kg^{-1} or <2 MJ L^{-1} of lithium-ion and zinc-air batteries, these values are considerably higher.

The composition of fossil hydrocarbon fuels may vary widely depending on the source of the crude oil that is processed in the refinery.

The composition of natural gas is predominantly methane, and also contains several percent ethane and propane. In addition, minor amounts of butane and higher hydrocarbons are present, plus carbon dioxide and nitrogen.

Table 2.3 shows the composition of natural gas from various sources [5]. Natural gas also contains sulfur compounds at the ppm-level, such as hydrogen sulfide and

Fuel Processing for Fuel Cells. Gunther Kolb
Copyright © 2008 WILEY-VCH Verlag GmbH & Co. KGaA, Weinheim
ISBN: 978-3-527-31581-9

Table 2.1 Overview of important fuels for fuel processing.

Fuel	Formula	Sulphur content [wt. ppm] [only commercially available fuels]	Lower heating value [kJ/mol]	Flammability limits lower, higher [Vol.%]	Density [kg/m³]	Boiling point or boiling range [°C]	Heat of vaporization [kJ/mol]	Heat capacity [J/(mol K)] At 20 °C
Hydrogen	H_2	–	240	4.1, 74	0.09	–252.7	0.92	28.6
Methanol	CH_3OH	–	643	7.3, 36	794 (l)	64.6	35.2	49.0 (g)
Ethanol	C_2H_5OH	–	1240	4.3, 19	790 (l)	78.3	38.9	77.3
Methane	CH_4	–	802	–	0.72	–161.4	8.2	35.5
Natural gas	$C_{1.07}H_{4.1}$	7–25	797	5.3, 15	0.77	–	8.2	–
Propane	C_3H_8	–	2015	2.2, 9.5	1.96	–42	73.5	73.5
Liquified Petroleum Gas [LPG]	C_3H_8/C_4H_{10}	50–200	2024	1.5, 11	540 (l) (at 8 bar)	–42––0.5	–	–
Iso-octane	C_8H_{18}	–	4731	–	–	99.3	34.8	–
Gasoline	$C_{7.1}H_{14.3}$	150 50 (European Regulation 2005)	4,720	0.8, 8	720–770	30–200	33.5	180
Dodecane	$C_{12}H_{26}$	–	7,392	–	750	216.4	40.6	270.2
Hexadecane	$C_{16}H_{34}$	–	9,792	–	770	286.8	51.3	386.5
Diesel	$C_{13.6}H_{27.1}$	50 (European regulation 2005) Heating oil 2,000	8,080	1, 6	0.856	120–430	47.0	340
Bio-diesel	$C_{18.7}H_{34.5}O_2$	<3,000	10,800	–	–	–	67.8	–

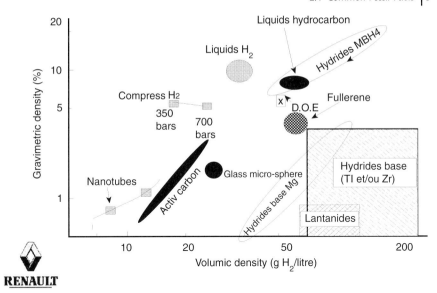

Figure 2.1 Comparison of gravimetric and volumetric storage densities as provided by Heurtaux et al. [3].

diethyl sulfide, and mercaptanes, such as ethyl mercaptane [$(C_2H_5)CHS$] and tertiary butyl mercaptane [$(CH_3)_3CHS$].

Amongst all the fossil fuels, propane contains the highest amount of hydrogen on a gravimetric basis, which even exceeds liquefied hydrogen, when the weight of the storage tanks is taken into consideration [6]. Propane is usually marketed as liquefied petroleum gas, which is a mixture of propane and butane in various ratios.

For gasoline, only approximate characterization parameters are provided, such as the octane number, the boiling point distribution, and the saturated hydrocarbons (alkanes), unsaturated hydrocarbons (olefins) and aromatics content. The content of contaminants, such as sulfur, is important.

Table 2.2 Energy density of various fuels related to different properties [4].

Fuel	Maximum amount of work				MJ/Mol H_2 via reforming
	MJ/Mol fuel	MJ/Kg fuel	MJ/L fuel	MJ/Mol C in fuel	
Methanol	−0.69	−22	−17[a]	−0.69	−0.23
Ethanol	−1.31	−28	−22[a]	−0.65	−0.22
n-Octane	−5.23	−46	−32[a]	−0.65	−0.21
Ammonia	−0.33	−19	−10[a]		−0.22
Methane	−0.80	−50	−3.9[b]	−0.80	−0.20
Hydrogen	−0.23	−113	−0.89[b]		−0.23

[a] density of the liquid fuels calculated at 298K and 1 bar, for ammonia at 10 bar.
[b] density of the gaseous fuels calculated at 298K and 100 bar.

2 Fundamentals

Table 2.3 Composition of natural gas from various sources [5].

Component	North Sea	Qatar	Netherlands	Pakistan	Ekofisk
CH_4 (Vol.%)	94.86	76.6	81.4	93.48	85.5
C_2H_6 (Vol.%)	3.90	12.59	2.9	0.24	8.36
C_3H_8 (Vol.%)		2.38	0.4	0.24	2.85
$i\text{-}C_4H_{10}$ (Vol.%)	0.15	0.11		0.04	0.86
$n\text{-}C_4H_{10}$ (Vol.%)		0.21	0.1	0.06	
C_5^+ (Vol.%)		0.02		0.41	0.22
N_2 (Vol.%)	0.79	0.24	14.2	4.02	0.43
S (ppm)	4	1.02	1	N/A	30

Regular gasoline, at least according to German standards, is well represented by the overall formula C_7H_{12} [7].

A standard jet fuel that is frequently cited is the American JP-8 fuel. It contains about 1 000 ppm sulfur and up to 1.5 vol.% non-volatile hydrocarbons [8, 9]. Jet fuels widely used in the world are Jet fuel A and A1 [10] with a boiling range between 150 and 300 °C.

Diesel fuels contain mainly iso-paraffins, but also n-paraffins, mono-, di-, tri-, tetra-cycloparaffins, alkylbenzenes, naphthalenes and phenanthrenes and even pyrenes [11].

2.2
Basic Definitions, Calculations and Legislation

Fuel processing is the conversion of hydrocarbons, alcohol fuels and other alternative energy carriers into hydrogen containing gas mixtures. The chemical conversion is achieved in most instances in the gaseous phase, normally heterogeneously catalysed in the presence of a solid catalyst and less frequently homogeneously at high temperature without a catalyst.

The first step of the conversion procedure is generally termed reforming, and has been well established in large scale industrial processes for many decades. The industrial applications most commonly (about 76% [12]) use natural gas as feedstock. The purpose of this process is the production of synthesis gas, a mixture of carbon monoxide and hydrogen, which is then used for numerous processes in large scale chemical production, which are not subject of this book.

Rather, the focus of this book is the technology that provides a hydrogen containing gas mixture, termed the reformate, which is suitable for feeding into a fuel cell. The fuel cell then converts hydrogen into electrical energy. Carbon monoxide may also be converted, which depends on the fuel cell type (see Section 2.3.2).

The lower heating value of a chemical substance is defined as its standard enthalpy of formation. The lower heating value of any fuel $C_xH_yO_z$ is easily determined by the following formula [13]:

$$LHV\left[\text{kJ mol}^{-1}\right] = \left(\frac{y}{2} + 2x - z\right)198.8 + 25.4 \quad (2.1)$$

The performance of a fuel processor is measured by its overall efficiency, which is commonly defined as the ratio between the Lower Heating Value (LHV) of the hydrogen and carbon monoxide that are produced to the LHV of the fuel consumed:

$$\eta_{\text{Fuel processor}} = \frac{LHV(H_2)\, n_{H_2} + LHV(CO)\, n_{CO}}{LHV(\text{Fuel})\, n_{\text{Fuel}}} \quad (2.2)$$

n are the molar flows and the lower heating value is in units of kJ mol^{-1}. The efficiency of the reformer may be calculated by a simplified version of Eq. (2.2):

$$\eta_{\text{Reformer}} = \frac{LHV(H_2)\, n_{H_2}}{LHV(\text{Fuel})\, n_{\text{Fuel}}} \quad (2.3)$$

A certain portion of the hydrogen produced by the fuel processor is frequently fed back to it, because it is not completely consumed by the fuel cell (see Section 2.3). The curious situation may then arise where the fuel processor efficiency exceeds 100%. In particular, this is the situation for steam reforming, where substantially more heat is required to run the process compared with partial oxidation and autothermal reforming (see Section 3). A fuel processor running on steam reforming may reach up to 120% efficiency according to the Eqs. (2.2) and (2.3).

The carbon monoxide content of the reformate obviously needs to be minimised for low temperature proton exchange membrane fuel cells, but other fuel cells may well utilize it as a fuel (see Section 2.3.2). The same applies for methane in certain fuel cells. Therefore, the heating value of the hydrogen alone does not provide the appropriate number for the calculation of efficiency in this instance.

A modified definition of the fuel processor efficiency provides a more realistic value than Eqs. (2.2) and (2.3) [14]:

$$\eta_{\text{Fuel processor}} = \frac{\begin{array}{c}LHV(H_2)n_{H_2} + LHV(CO)n_{CO} + LHV(CH_4)n_{CH_4} \\ - [LHV(H_2)n_{H_2} + LHV(CO)n_{CO} + LHV(CH_4)\, n_{CH_4}]_{\text{recirculated}}\end{array}}{LHV(\text{Fuel})\, n_{\text{Fuel}}} \quad (2.4)$$

In addition to the formula provided by Lutz et al. [14], it takes into consideration the release of unconverted methane and the formation of methane by the reforming process (see Section 3). Unconverted methane is commonly re-circulated to the fuel processor, along with unconverted carbon monoxide, in particular for high temperature fuel cells.

However, for PEM fuel cells methane and carbon monoxide could be excluded from efficiency calculations, because they are not converted in the fuel cell.

The following definition of efficiency was proposed by Feitelberg [15]. It was modified to also take methane and carbon monoxide fed to the fuel cell into consideration as discussed above:

$$\eta_{\text{Fuel processor}} = \frac{LHV(H_2)n_{H_2} + LHV(CO)n_{CO} + LHV(CH_4)n_{CH_4}}{LHV(\text{Fuel})n_{\text{Fuel}} + [LHV(H_2)\, n_{H_2} + LHV(CO)n_{CO} + LHV(CH_4)\, n_{CH_4}]_{\text{recirculated}}} \quad (2.5)$$

This definition seems to be the most realistic, because it takes all products into consideration in the numerator and all feed entering the fuel processor in the denominator, which is in agreement with the rules for energy balancing.

Hagh [13] has derived a general formula for the fuel processor efficiency. Thus, by describing the fuel processing reactions with the following general and simplified formula (by-products such as methane are not taken into consideration, the same applies for unconverted methane in case of methane fuel processing):

$$C_xH_yO_z + a\,O_2 + b\,H_2O \rightarrow d\,CO + e\,CO_2 + f\,H_2 \tag{2.6}$$

defining the stoichiometric ratio (SR) as the ratio of oxygen to oxygen required for complete combustion:

$$SR = \frac{2a}{\frac{y}{2} + 2x - z} \tag{2.7}$$

and by further defining the hydrogen ratio (HR) as the ratio of moles of hydrogen produced to the moles of fuel reformed:

$$HR = \frac{\frac{y}{2} + \left(\frac{3\frac{d}{e} + 4}{\frac{d}{e} + 1} - 2\right)x - z}{\frac{y}{2} + 2x - z} - SR \tag{2.8}$$

The fuel processor efficiency is expressed by the simple term:

$$\eta_{\text{Fuel processor}} = \frac{HR}{1 - SR} \tag{2.9}$$

However, these calculations still assume ideal conditions, namely complete conversion of the fuel, absence of methane formation and negligible heat losses and they do not consider the water excess fed to practical systems (see Section 3.1).

Unconverted fuel is generally not desirable for most fuel processors, because the fuel molecules might well contaminate the gas purification devices and the fuel cell itself, which is especially critical for higher hydrocarbons and alcohols. However, for methane fuel processors an incomplete conversion is feasible, which is discussed in Section 2.3.2.

Excess water fed to the reformer decreases efficiency, so this water needs to be removed from the system downstream. In other words, if this excess water is neither consumed by the water–gas shift reaction downstream of the reformer (see Section 3.10.1), nor required to prevent dry-out of the membrane of low temperature PEM fuel cells (see Section 2.3.1), nor to prevent coking in high temperature fuel cells (see Section 2.3.2), the water should be recovered to avoid a negative water-balance of the system. However, the heat of condensation is difficult to recover and commonly lost to the cooling air. Thus, excess water should be minimised in fuel processor/fuel cell systems. Condensers might be integrated upstream or downstream of the fuel cell to ensure water recovery and net positive water balancing of the whole system. Such condensers may also recover water produced by the fuel cell itself.

2.2 Basic Definitions, Calculations and Legislation

Low temperature heat losses are usually mandatory in all fuel cell/fuel processor systems because the efficiency of heat-exchangers is limited and the system needs to work with cooling air, which might even have elevated temperatures in the summer time. The major portion of the heat losses does not originate from the fuel processor but rather from the fuel cell itself for low temperature proton exchange (PEM) fuel cells. Thus the efficiency of the fuel processor is usually high provided heat losses are neglected, because internal heat-exchangers can keep the high temperature heat within the fuel processor.

Heat losses affect the fuel processor efficiency, especially at partial system load. This becomes obvious when taking into consideration that the reactors of the fuel processor require an elevated operating temperature. This temperature will not decrease at partial load. Therefore heat losses remain constant and become dominant at partial load of the system. Smaller system size has similar effects.

To judge fuel processor efficiency, the following operational efficiency factor is thus proposed as a modification to the definition provided by Schmid and Wünning [16] and others:

$$\eta_{\text{Fuel processor, operation}} = 1 - \frac{\dot{Q}_{\text{cond,FP}} + \dot{Q}_{\text{losses}}}{LHV(\text{fuel})n_{\text{Fuel}}} \quad (2.10)$$

$Q_{\text{cond,FP}}$ is the heat generated by condensation of steam, which is not required for fuel cell membrane dry-out. Thus the total fuel processor efficiency is defined as follows:

$$\eta_{\text{Fuel processor, total}} = \eta_{\text{Fuel processor}} \eta_{\text{Fuel processor, operation}} \quad (2.11)$$

The fuel cell efficiency is defined as the ratio of the electrical power output of the fuel cell $P_{\text{Fuel cell}}$ to the lower heating value of the fuel converted by the electrochemical reactions:

$$\eta_{\text{FC}} = \frac{P_{\text{Fuel cell}}}{LHV(H_2)n_{H_2} + LHV(CO)n_{CO}} \quad (2.12)$$

The overall efficiency of a fuel processor/fuel cell system is commonly defined as the ratio of the electrical fuel cell stack power output $P_{\text{Fuel cell}}$ to the LHV of the fuel:

$$\eta_{\text{system}} = \frac{P_{\text{Fuel cell}}}{LHV(\text{Fuel})n_{\text{Fuel}}} \quad (2.13)$$

Some basic thermodynamic definitions and calculations of enthalpy differences when heating and evaporating fluids and for the reaction enthalpy are provided below. The heat capacity c_p is the derivative of enthalpy by temperature:

$$c_p = \frac{dh}{dT} \quad (2.14)$$

The enthalpy difference between the reference temperature T_0 and temperature T is calculated as follows:

$$\int_{T_0}^{T} dh = h(T) - h(T_0) = \int_{T_0}^{T} c_p(T) dT \quad (2.15)$$

and the molar heat capacity is according to:

$$c_{pm}(T) = \frac{1}{T-T_0} \int_{T_0}^{T} c_p(T) dT = \frac{h(T) - h(T_0)}{T - T_0} \qquad (2.16)$$

The enthalpy of the fluid at temperature T is then:

$$h(T) = h(T_0) + c_{pm}(T) \cdot (T - T_0) \qquad (2.17)$$

To heat a substance from temperature T_1 to T_2 requires the enthalpy difference Δh_1:

$$\Delta h_1 = h(T_2) - h(T_1) = c_{pm}(T_2) \cdot (T_2 - T_0) - c_{pm}(T_1) \cdot (T_1 - T_0) \qquad (2.18)$$

The phase transition of a substance from phase A to phase B requires the enthalpy difference Δh_2:

$$\Delta h_2 = h_B(T) - h_A(T) = h_B(T_0) - h_A(T_0) + c_{pm_B}(T) \cdot (T - T_0) - c_{pm_A}(T) \cdot (T - T_0) \qquad (2.19)$$

While the reaction enthalpy Δh_R of the chemical reaction

$$A + B \rightarrow C \qquad (2.20)$$

is calculated as follows:

$$\Delta h = h_C(T) - h_A(T) - h_B(T) = h_C(T_0) - h_A(T_0) - h_B(T_0) + c_{pm_C}(T) \cdot (T - T_0) - c_{pm_A}(T) \cdot (T - T_0) - c_{pm_B}(T) \cdot (T - T_0) \qquad (2.21)$$

Table 2.4 provides the most important physical and chemical properties of substances present in the reformate.

To date, no emission regulations exist for fuel cell systems. European legislation has directives for heating systems based on natural gas or Liquified Petroleum Gas (LPG). They limit nitrous oxides (NO_x) to 200 ppm and carbon monoxide to 100 ppm. However, the legislation in some EU member countries are well below these values, German emission control regulations limit NO_x to 80 ppm and carbon monoxide to 60 ppm.

In instances where homogeneous combustion is applied in fuel processors, the formation of NO_x is inevitable and may lead to emissions exceeding the limitations set by legislation [17]. This is not expected for catalytic combustion.

Based on their experimental results with a methanol fuel processor, Emonts et al. calculated that for a light duty vehicle, carbon monoxide emissions could be reduced to 1%, NO_x emission to 10% and volatile organic compounds (without methane) to 10% using a fuel processor/fuel cell system compared with an internal combustion engine, fulfilling the EU standards of 2005 for the new European driving cycle [18]. The catalytic afterburner was the only source of emissions in the system. It generated 1.8 mg km^{-1} carbon monoxide, 0.3 mg km^{-1} NO_x and 3.2 mg km^{-1} unconverted hydrocarbons. The Super Ultra Low Emissions vehicle regulation allowed much higher values, namely 625 mg km^{-1} carbon monoxide, 12 mg km^{-1} NO_x and 6 mg km^{-1} unconverted hydrocarbons.

Table 2.4 Key chemical properties of gases most relevant for fuel processing (source: IMM, Institut für Mikrotechnik Mainz)

	t °C	C_{pm} J mol^{-1} K^{-1}	h J mol^{-1}	LHV J mol^{-1}	ρ g/L	γ 10^5 Pa s	λ W m^{-1} k^{-1}
$CH_3OH(g)$ 32.043 g mol^{-1}	25	43.71	−201167	675990	1.293	0.959	0.0157
	250	52.64	−189322	–	0.737	1.721	0.0413
	500	61.90	−171766	–	0.498	2.489	0.0853
	750	69.78	−150576	–	0.377	3.191	0.1443
$CH_4(g)$ 16.043 g mol^{-1}	25	35.34	−74873	802284	0.647	1.111	0.0335
	250	41.76	−65478	–	0.369	1.742	0.0699
	500	48.65	−51765	–	0.250	2.287	0.1187
	750	54.94	−35042	–	0.189	2.743	0.1704
$C_2H_5OH(g)$ 46.069 g mol^{-1}	25	65.18	−234810	1277678	1.858	0.891	1.1620
	250	81.83	−216398	–	1.059	1.502	1.1891
	500	98.89	−187836	–	0.717	2.133	1.2195
	750	111.64	−153868	–	0.542	2.734	1.2497
$C_3H_8(g)$ 46.096 g mol^{-1}	25	73.40	−103847	2043972	1.779	0.828	0.0183
	250	96.56	−82121	–	1.014	1.362	0.0473
	500	116.59	−48467	–	0.686	1.849	0.0902
	750	133.12	−7336	–	0.518	2.303	0.1424
$CO(g)$ 28.01 g mol^{-1}	25	29.14	−110541	282964	1.130	1.767	0.0249
	250	29.48	−103908	–	0.644	2.660	0.0399
	500	30.15	−96218	–	0.436	3.447	0.0530
	750	31.03	−88045	–	0.329	4.102	0.0646
$CO_2(g)$ 44.01 g mol^{-1}	25	37.10	−393505	–	1.775	1.493	0.0166
	250	41.40	−384191	–	1.012	2.496	0.0354
	500	45.30	−371986	–	0.685	3.420	0.0547
	750	48.07	−358651	–	0.517	4.193	0.0721
$H_2(g)$ 2.016 g mol^{-1}	25	28.98	0	241826	0.081	0.892	0.1780
	250	28.79	6478	–	0.046	1.306	0.2747
	500	29.02	13784	–	0.031	1.689	0.3686
	750	29.39	21309	–	0.024	2.033	0.4548
$H_2O(g)$ 18.015 g mol^{-1}	25	33.63	−241826	–	0.727	0.908	0.0197
	250	34.40	234087	–	0.414	1.827	0.0383
	500	35.69	−224872	–	0.280	2.854	0.0668
	750	37.16	−214883	–	0.212	3.848	0.1004
$N_2(g)$ 28.013 g mol^{-1}	25	29.13	0	–	1.130	1.785	0.0260
	250	29.31	6595	–	0.644	2.685	0.0396
	500	29.91	14206	–	0.436	3.511	0.0537
	750	30.65	22221	–	0.329	4.217	0.0670
$O_2(g)$ 31.999 g mol^{-1}	25	28.96	0	–	1.291	2.069	0.0260
	250	30.87	6947	–	0.736	3.147	0.0417
	500	31.96	15181	–	0.498	4.139	0.0575
	750	32.72	23720	–	0.376	4.954	0.0717

2.3
The Various Types of Fuel Cells and the Requirements of the Fuel Processor

The principle of the fuel cell was discovered more than 100 years ago by the frequently cited Sir William Grove but also by Christian Friedrich Schoenbein [19, 20]. However, despite its large potential for highly efficient power generation, it still lacks widespread applications due mostly to the economic aspects and some remaining technical problems, such as durability issues.

Only a few aspects of the complex theory of electrical power generation by fuel cells will be discussed briefly below, in order to highlight the consequences of these basic rules on the fuel processor and its design.

2.3.1
PEM Fuel Cells

The most commonly used fuel cell is composed of a membrane that is able to transport protons, the Proton Exchange Membrane (PEM), and of a catalyst, such as platinum, positioned on both sides of the membrane on conducting material that serves as the electrode. This arrangement is termed the Membrane Electrode Assembly (MEA). Nafion® membranes, a fluorocarbon polymer of sulfuric acid developed by DuPont, are the most frequently used membrane materials. Where hydrogen is fed to one side of the MEA and oxygen to the other, hydrogen is oxidised into water in a controlled combustion. In parallel, an electric potential of about 0.9 V is generated, which decreases when current is withdrawn from the arrangement. To achieve a voltage higher than 1 V, several MEAs need to be switched in series, forming a fuel cell stack. Because the hydrogen and oxygen need to be distributed to each MEA and over its entire area, gas distribution layers are also required, which must be manufactured from electrically conductive material, in most instances this is graphite or metal.

The side of the MEA, which catalyses the hydrogen dissociation:

$$H_2 \rightarrow 2H^+ + 2e^- \qquad (2.22)$$

is the anode of the MEA, whereas the opposite side, which converts oxygen into water:

$$O_2 + 4H^+ + 4e^- \rightarrow 2H_2O \qquad (2.23)$$

forms the cathode. It is obvious that it is much more convenient to provide air instead of oxygen to the cathode. However, this implies that not all gas fed to the cathode is converted and that nitrogen and unconverted oxygen need to be removed from the MEA. In practical systems a surplus of oxygen is fed to the cathode to avoid extremely low concentrations at the exit. Frequently a two-fold or higher surplus of the stoichiometric ratio λ:

$$\lambda = \frac{n_{O_2}}{2\,n_{H_2}} = 2 \qquad (2.24)$$

is fed to the cathode.

For the anode, however, it is not typically the stoichiometric ratio but rather the amount of hydrogen (or hydrogen and carbon monoxide, depending on the fuel cell type) converted in the fuel cell as a percentage of the feed that is specified. This amount is termed the hydrogen utilisation. For practical PEM fuel cell systems running on reformate, 80% hydrogen utilisation may be assumed. This value, however, is by no means fixed. When decreasing the electrical power withdrawal from the fuel cell, while keeping the reformate flow constant, hydrogen utilisation may drop to lower values.

Hydrogen utilization of less than 80% might be the preferred option when the energy of unconverted hydrogen is required to supply processes downstream of the anode (see Section 4.2).

The electrical power output of the fuel cell refers to about 50% of its energy generation, the remaining energy is released as heat. The overall efficiency is even lower due to energy losses of the fuel processor and to balance-of-plant components. Such a system might be regarded as having a low overall efficiency. However, a comparison with the overall efficiency of conventional passenger cars driven by internal combustion engines reveals even lower values, 12% for gasoline-powered cars and 15% for diesel-powered vehicles [21].

The unconverted part of the energy is released as heat within the fuel cell stack, which generates a heat removal problem for practical systems. The heat may be removed by water cooling, which in turn makes the system more complex and expensive. Cooling by the anode and cathode gas flows are simpler alternatives. They become more attractive the smaller the fuel cell system is, because cost issues are more stringent in such instances.

Dilution of hydrogen with inert gases such as carbon dioxide and nitrogen, as is the situation for reformate from fuel processors, should not impair the fuel cell power generation by more than 10%, even if fuel utilisation is high and the hydrogen content is as low as 40 vol.% [22].

Conventional low temperature PEM fuel cell anodes are sensitive to carbon monoxide, which poisons the catalyst. In other words, the carbon monoxide is preferentially adsorbed at the catalyst and thus the desired reactions can no longer take place. However, the poisoning is partially reversible [23]. The dilution of the hydrogen in reformate from fuel processors amplifies the poisoning effect unfortunately [24].

To a certain extent this poisoning is suppressed or at least reduced for certain alloys of platinum with other metals, amongst them, most commonly, is ruthenium, but also iron, cobalt, molybdenum and tungsten. This beneficial effect originates from the fact that platinum adsorbs carbon monoxide preferentially, but water is adsorbed at the second metal, which makes the oxidation into carbon dioxide feasible [25]. The long term carbon monoxide tolerance of PEM fuel cells may be increased from a few ppm to values between 50 and 100 ppm at the most by these means. Bimetallic catalysts such as platinum/ruthenium are also more tolerant towards carbon dioxide [26], which is of course essential when reformate is applied as the fuel. Increasing the fuel cell operating temperature decreases the poisoning effect, which also applies to high temperature PEM fuel cells (see Section 2.3.2) [24]. The negative effect of carbon dioxide on fuel cell stability originates from the reaction with the hydrogen adsorbed at the platinum to form carbon monoxide, which means, in other words, that a reverse water–gas shift reaction [see Eq. (3.4), Section 3.1] is taking place [27]. However, this effect is, yet again, less significant over platinum/ruthenium catalysts.

Another measure to reduce the detrimental effect of carbon monoxide on the anode performance is the addition of a small amount of air during normal operation, which is commonly termed "bleed air". It oxidises the carbon monoxide adsorbed on the active sites of a selective oxidation catalyst layer [26] at the anode (see Section 4.1.2). However, similar to the oxygen addition performed for the preferential oxidation of carbon monoxide in a dedicated clean-up reactor (see Section 3.10.2), addition of air to the hydrogen containing reformate generates safety issues.

The poisoning effect of formaldehyde on PEM fuel cells is less dramatic compared with carbon monoxide. The tolerance of PEM fuel cells to small amounts of formic acid is approximately ten times higher [23]. However, Amphlett et al. judge formic acid to be a severe poison for PEM fuel cells [28]. Formic acid causes irreversible performance losses at concentrations as low as 250 ppm.

The poisoning effect of methane is very small for conventional PEM fuel cells [23]. Up to 5 vol.% are known to have no detrimental effect on the performance.

Methanol, which may originate from incomplete conversion in a methanol fuel processor, can be tolerated in concentrations up to 0.5 vol.% according to Amphlett et al. [29]. Kawatsu, from Toyota, suggested that methanol forms formaldehyde in the fuel cell anode and also shows cross-over through the fuel cell membrane [30]. Application of a platinum/ruthenium anode catalyst reduced these detrimental effects [30]. The tolerance of PEM fuel cells towards methyl formate is assumed to be similar to that for methanol according to Amphlett et al. [28].

Ammonia impairs the proton conductivity of the electrode considerably at less than 100 ppm [24], and sulfur containing compounds also affect the catalyst performance [27].

Hydrogen sulfide has a more severe poisoning effect on PEM fuel cells compared with carbon monoxide. It originates from preferential adsorption; 1 ppm leads to significant performance losses [24].

Metallic ions such as copper, iron and sodium, which might be released from a fuel processor or from fuel processing catalysts, impair the fuel cell performance. Hydrogen peroxide might be formed from iron and copper ions, which attacks the membrane [24].

Even if the reformate is purified by catalytic carbon monoxide clean-up to well below 50 ppm carbon monoxide and if other impurities are reduced to the ppb level, performance losses are to be expected when running a fuel cell with reformate. A 7% lower power production was observed by Shi et al. [31] when running a 2 kW PEM fuel cell stack with reformate produced from liquid hydrocarbons.

Direct methanol and direct ethanol fuel cells are alternatives to PEM fuel cells operated with methanol reformers. However, these types of fuel cells are not within the scope of this book, and hence will not be discussed.

2.3.2
High Temperature Fuel Cells

A major drawback to PEM fuel cells, apart from their sensitivity to carbon monoxide poisoning, is the osmotic drag of the Nafion® membrane, which causes water migration from the side of the anode to that of the cathode. This causes performance losses unless the feed is humidified [32]. Polybenzimidazole membrane material doped with phosphoric acid is an alternative, which allows for higher operating temperature of the PEM fuel cell of between 150 and 170 °C and requires no humidification [32]. Additionally, it has significantly higher tolerance towards carbon monoxide in concentrations exceeding 1 vol.%. This membrane material was developed by the former Hoechst Company; the technology is now owned by the BASF Company.

Classical phosphoric acid fuel cells use phosphoric acid as the electrolyte, which is immobilized in a Teflon bonded silicon carbide matrix. Phosphoric acid fuel cells usually work at temperatures around 200 °C and are able to tolerate carbon monoxide levels of up to 2 vol.% [1]. Platinum/ruthenium as the anode catalyst may improve the performance in presence of carbon monoxide, similar to PEM fuel cells [33].

Alkaline fuel cells use an aqueous solution of potassium hydroxide as the electrolyte. When the solution is concentrated (85%), the operating temperature of the alkaline fuel cell can be as high as 250 °C, lower concentrations result in lower operating temperatures, below 120 °C. Both carbon monoxide and carbon dioxide act as poisons, which means that this fuel cell type is not really suited for hydrocarbon reformate, unless the hydrogen is separated by membranes or pressure swing adsorption (see Section 5.2). Hydrogen produced from ammonia is an interesting fuel for this type of fuel cell, because reformate obtained from ammonia cracking contains no carbon oxides [34] (see also Section 3.8).

Solid oxide fuel cells contain solid electrolytes, which are frequently based on zirconia stabilised by yttria. On the cathode side strontium or calcium doped lanthanum manganese oxide ($LaMnO_3$) is most commonly used, while the anode side comprises of yttria stabilized zirconia frequently doped with nickel to achieve electrical conductivity. The oxygen is reduced at the cathode and the oxygen anions diffuse through the electrolyte to the anode, where they oxidise the fuel. Hydrogen or carbon monoxide may serve as the fuel. To achieve sufficient mobility of the oxygen anions, high operating temperatures of between 800 and 1000 °C are usually required [35]. This value may be lowered to 600 °C for thin electrolytes made from certain electrolyte materials, such as lanthanum gallate based perovskites [36]. However, the nickel catalyst is subject to coke formation when carbon monoxide is present in the reformate. Addition of steam is one possible way to reduce coke formation. Through the addition of steam, internal reforming of light hydrocarbons, such as methane, becomes feasible within the solid oxide fuel cell [37]. Internal reforming may either be performed at the catalyst positioned adjacent to the anode or at the anode itself [37]. However, the efficiency of the fuel cell is reduced due to the dilution of anode feed by the steam [35]. On the other hand, steam reforming consumes energy, which helps to cool the fuel cell [38, 39]. Various alternative anode materials, such as copper-based systems, are under investigation, which reduce carbon formation in the solid oxide fuel cell [35]. These issues will not be discussed further in the present book.

Reforming of natural gas for solid oxide fuel cells is achieved either internally as described above or externally by a pre-reformer reactor [40, 41]. Further processing of the fuel is not required, because of the unlimited tolerance of the fuel cell to carbon monoxide. The re-circulation of anode off-gas to the pre-reformer [42] is an interesting option for solid oxide fuel cells. Through these means, addition of water is omitted, which clearly decreases the complexity of the system and reduces cold start problems (see also Section 3.5).

Molten carbonate fuel cells operate at temperatures around 650 °C and are tolerant to unlimited amounts of carbon monoxide. In most instances mixtures of lithium carbonate and potassium carbonate act as the electrolyte. The electrolyte is suspended in an insulating and chemically inert lithium aluminate ceramic. Nickel or nickel–chromium alloys serve as the anode catalysts, while nickel oxide is used as the cathode catalysts.

Autothermal or steam reforming of methane was considered in thermodynamic calculations by Cavallaro and Freni for reformers, which were integrated into a molten carbonate fuel cell [43]. Direct or indirect internal reforming is possible within the molten carbonate fuel cell. The reforming may be performed either by the anode itself or by a dedicated catalyst in the anode compartment in analogy with the solid oxide fuel cell, as has been explained above. Direct reforming of alcohol fuels is also possible in molten carbonate fuel cells [44], whereas processing of liquid hydrocarbons requires a pre-reformer.

Theoretical calculations to evaluate internal reforming of methanol, ethanol and methane for molten carbonate fuel cells were performed by Maggio et al. [45].

3
The Chemistry of Fuel Processing

The most important parameter of reforming as the first step in fuel processing is the feed composition. In this book the feed composition is provided by the following terms:

The steam/carbon ratio S/C, which is the ratio of molar steam flow rate to the molar flow rate of the fuel $C_xH_yO_z$ multiplied by the number of carbon atoms in the fuel, x:

$$\frac{S}{C} = \frac{\dot{n}_{H_2O}}{x\,\dot{n}\,C_xH_yO_z} \quad (3.1)$$

The oxygen/carbon ratio, O/C, which is the ratio of molar oxygen flow rate multiplied by two to the molar flow rate of the fuel multiplied by the number, x, of carbon atoms in the fuel. This definition holds no matter whether there is oxygen contained in the fuel or not:

$$\frac{O}{C} = \frac{2\dot{n}_{O_2}}{x\,\dot{n}\,C_xH_yO_z} \quad (3.2)$$

3.1
Steam Reforming

Steam reforming is the gas phase conversion of energy carriers, such as hydrocarbons and alcohols described by the general formula $C_xH_yO_z$, using steam, into a mixture of carbon monoxide and hydrogen, according to the formula provided below:

$$C_xH_yO_z + (x-z)H_2O \rightarrow xCO + \left(x - z + \frac{y}{2}\right)H_2 \quad (3.3)$$

The product mixture of the reaction is known as the reformate. The reaction is endothermic and thus requires a heat supply. Besides hydrogen and carbon monoxide, the reformate usually contains significant amounts of unconverted steam, and to a lesser extent some unconverted fuel and carbon dioxide, the latter being formed by the consecutive water–gas shift reaction:

$$CO + H_2O \rightarrow CO_2 + H_2 \quad \Delta H^0_{298} = -40.4 \text{ kJ mol}^{-1} \quad (3.4)$$

Fuel Processing for Fuel Cells. Gunther Kolb
Copyright © 2008 WILEY-VCH Verlag GmbH & Co. KGaA, Weinheim
ISBN: 978-3-527-31581-9

The water–gas shift reaction increases the hydrogen concentration of the reformate. This reaction is usually fast enough at the elevated temperatures of hydrocarbon reforming to achieve thermodynamic equilibrium. Owing to its exothermic character, higher reaction temperatures favour the reverse reaction (see Section 3.10).

Methane is frequently formed in significant amounts, up to several percent. Higher reaction temperatures suppresses methane formation, according to the equilibrium of the methanation reaction, which is of course the reverse reaction of methane steam reforming:

$$3H_2 + CO \leftrightarrow H_2O + CH_4 \quad \Delta H^0_{298} = -253.7 \text{ kJ mol}^{-1} \quad (3.5)$$

The equilibrium conversion of methane steam reforming as calculated for different S/C ratios and system pressure is shown in Figure 3.1.

The carbon dioxide reforming reaction:

$$C_xH_yO_z + (x-z)CO_2 \rightarrow (2x-z)CO + \frac{y}{2}H_2 \quad (3.6)$$

is much slower than the steam reforming in the presence of adequate amounts of steam and is thus insignificant compared with water–gas shift [47].

While steam reforming of methane is suppressed at high pressure and is a reversible reaction, this is not so for other fuels. Thus methane steam reforming at pressures of 10–20 bar (1 bar = 10^5 Pa) suffers from low methane conversion due to the thermodynamic equilibrium [48]. Such high pressure is required, for example, if the reforming process is combined with membrane separation using conventional palladium membranes (see Section 5.2.4). Industrial steam reformers work with

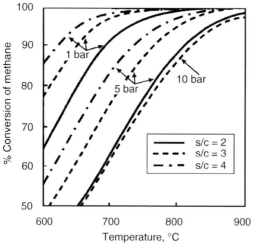

Figure 3.1 Equilibrium conversion of methane steam reforming versus reaction temperature for various S/C ratios and system pressures [46].

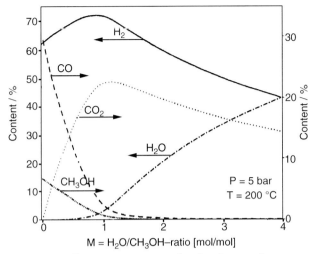

Figure 3.2 Equilibrium gas composition of methanol steam reforming versus S/C ratio of the feed; pressure 5 bar; temperature 200 °C [25].

pressures exceeding 20 bar because of the requirements of the purification processes, such as pressure swing adsorption (see Section 5.2.5) and the processes downstream of the reformer.

Elevated pressure also decreases the equilibrium conversion of methanol steam reforming [49]. At 21 bar pressure, which may be required to run a steam reformer along with a membrane separation system, the equilibrium methanol conversion decreases to 99.2% at a reaction temperature of 280 °C and to 98.1% at 260 °C with a S/C ratio 1.5 [50].

Figure 3.2 shows the equilibrium conversion of methanol steam reforming as a function of the S/C ratio of the feed [25]. It is obvious that the maximum hydrogen concentration in the reformate is gained at S/C 1. However, to minimize the carbon monoxide concentration in a practical system, a surplus of steam is required. Therefore, in practice, systems operate at S/C ratios of between 1.3 and 2.0. Elevated pressure also decreases the selectivity towards carbon monoxide [49].

Further possible by-products of methanol steam reforming are formic acid and methyl formate (CH_3OCHO), which are both harmful, at least for PEM fuel cells (see Section 2.3.1). These by-products are formed because methanol steam reforming probably takes place with both of these species as intermediate products, as proposed by Takahashi et al. [51]:

$$2CH_3OH \rightarrow CH_3OCHO + 2H_2 \tag{3.7}$$

$$CH_3OCHO + H_2O \rightarrow CH_3OH + HCOOH \tag{3.8}$$

$$HCOOH \rightarrow CO_2 + H_2 \tag{3.9}$$

The decomposition reaction of formic acid is usually fast [52]. Another possible by-product of methanol steam reforming is dimethyl ether, which is generated by methanol dehydration (see Section 4.2.1):

$$2CH_3OH \rightarrow CH_3OCH_3 + H_2O \tag{3.10}$$

Dimethyl ether formation is favoured by decreasing the reaction temperature [53].

Steam reforming of all fuels typically under investigation in the research community (except for methanol) is performed at temperatures of at least 400 °C or higher. Under these reaction conditions, light hydrocarbons may be formed, namely methane, ethylene and propylene.

The second important alcohol fuel frequently investigated is ethanol. Ethanol steam reforming:

$$C_2H_5OH + H_2O \rightarrow 2CO + 4H_2 \quad \Delta H^0_{298} = +256 \text{ kJ mol}^{-1} \tag{3.11}$$

usually requires a higher reaction temperature than methanol steam reforming. The ethanol decomposition reaction is one important side reaction, which generates carbon monoxide and methane and is favoured by higher reaction temperatures [54]:

$$C_2H_5OH \rightarrow CO + CH_4 + H_2 \quad \Delta H^0_{298} = +49 \text{ kJ mol}^{-1} \tag{3.12}$$

Ethanol steam reforming has the disadvantage of numerous side reactions, which are summarised below.

Formation of ethylene is critical because ethylene is known to be a coke precursor, favouring deactivation of the catalyst:

$$C_2H_5OH \rightarrow C_2H_4 + H_2O \quad \Delta H^0_{298} = +45 \text{ kJ mol}^{-1} \tag{3.13}$$

Acetaldehyde may also be formed depending on the catalyst selectivity:

$$C_2H_5OH \rightarrow CH_3CHO + H_2 \quad \Delta H^0_{298} = +71 \text{ kJ mol}^{-1} \tag{3.14}$$

Acetaldehyde may then further decompose to carbon monoxide and methane:

$$CH_3CHO \rightarrow CH_4 + CO \quad \Delta H^0_{298} = -21.9 \text{ kJ mol}^{-1} \tag{3.15}$$

For a low S/C ratio, ethanol decomposition may occur according to [54]:

$$2\,C_2H_5OH \rightarrow 3\,CH_4 + CO_2 \quad \Delta H^0_{298} = -74 \text{ kJ mol}^{-1} \tag{3.16}$$

Hydrocarbon fuels require a higher reaction temperature compared with ethanol. As illustrated in Figure 3.3, the equilibrium dry gas composition of reformate gained

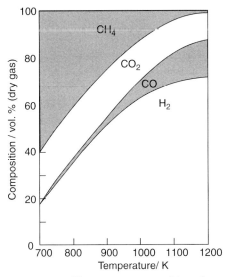

Figure 3.3 Equilibrium composition of dry reformate from n-heptane steam reforming versus temperature; pressure 30 bar; S/C ratio 4 [55].

from steam reforming of n-heptane shows an increasing carbon monoxide content and a decreasing methane content with increasing temperature, while the carbon dioxide concentration remains almost constant.

The methane formation through the reforming process is not only dictated by thermodynamics but also by the properties of the catalyst that is applied.

Reactions other than methanation may well be responsible for the formation of methane from hydrocarbon fuels [56]:

$$C_x H_y \rightarrow CH_4 + C_{x-1} H_{y-4} \tag{3.17}$$

$$C_x H_y + H_2 \rightarrow CH_4 + C_{x-1} H_{y-2} \tag{3.18}$$

However, methane can be tolerated by most fuel cell systems up to a concentration of 5 vol.% without damage.

Unconverted fuel and by-products are undesired because in most instances they cannot be converted by fuel cells and so they reduce the efficiency of the system. In addition, they usually also poison the fuel cell catalyst and the catalyst of subsequent carbon monoxide clean-up systems (see Sections 2.3 and 4.5).

In general, steam reforming of higher hydrocarbons is usually performed at S/C ratios exceeding the stoichiometry (S/C = 2) in order to suppress coke formation. An S/C ratio of 3 may be required for higher hydrocarbons in the absence of oxygen in the feed as for autothermal reforming (see Section 3.3 below). However, excess steam reduces the overall efficiency of the system (see Sections 2.2 and 5.4.3).

3.2
Partial Oxidation

Partial oxidation is the conversion of fuels under an oxygen deficient atmosphere:

$$C_xH_yO_z + \frac{(x-z)}{2}(O_2 + 3.76N_2) \rightarrow xCO + \frac{y}{2}H_2 + 3.76\frac{(x-z)}{2}N_2 \qquad (3.19)$$

The reaction is significantly faster than steam reforming and is usually performed in the diffusion limited regime [57]. The catalytic reaction of hydrocarbons such as octanes over rhodium coated foams starts at relatively low temperatures of around 250 °C [58].

An obvious advantage of partial oxidation is that only an air feed is required, apart from the fuel. This makes the system simpler because evaporation processes, as required for steam reforming, are avoided. On the other hand, the amount of carbon monoxide formed is considerably higher compared with steam reforming. This puts an additional load onto the subsequent clean-up equipment, but only where CO-sensitive fuel cells are connected to the fuel processor. When fuels are converted by partial oxidation, some total oxidation usually takes place as an undesired side reaction [46]. In practical applications, an excess of air is fed to the system and consequently even more fuel is subject to total oxidation. The water formed by the combustion process in turn gives rise to some water–gas shift. Another typical by-product of partial oxidation is methane, which is formed according to reaction (3.5). Coke formation is a critical issue (see Sections 4.1.1 and 4.2.11). Coke may be formed by reaction of carbon monoxide with hydrogen:

$$H_2 + CO \rightarrow H_2O + C \quad \Delta H^0_{298} = -131 \text{ kJ mol}^{-1} \qquad (3.20)$$

This reaction is suppressed in the presence of substantial amounts of steam due to the thermodynamic equilibrium.

Carbon monoxide alone may form carbon according to the Boudouart reaction:

$$2CO \leftrightarrow CO_2 + C \quad \Delta H^0_{298} = -172 \text{ kJ mol}^{-1} \qquad (3.21)$$

Pennemann et al. performed partial oxidation of propane and identified the Boudouart reaction to be responsible for significant coke formation, which was observed in particular downstream of the catalyst on the steel surface of their microchannel test reactors [59]. The nickel contained in the stainless steel clearly served as the active species. Coating the steel with α-alumina suppressed the coke formation.

All hydrocarbons can form coke by cracking reactions, as exemplified by methane:

$$CH_4 \rightarrow 2H_2 + C \quad \Delta H^0_{298} = -75 \text{ kJ mol}^{-1} \qquad (3.22)$$

Two reaction mechanisms do exist in the literature for partial oxidation. One of these proposes that the reaction begins with catalytic combustion followed by reactions of lower rate, namely steam reforming, CO_2 reforming and water–gas shift [60]. The other mechanism proposes direct partial oxidation at very short residence times [61].

Ioannides and Verykios coated only the inner surface of one tubular reactor with rhodium catalyst, while another reactor was coated on both sides. Partial oxidation of methane was performed in such a manner that the reactants were fed to the tube and then passed the tube outer surface counter-currently, in an annular gap between the tube and the reactor housing. The reactor that was coated on both sides showed an even temperature profile, while the reactor where only inside of the tube was coated showed hot spot formation. The different performances were attributed to endothermic reactions in the second annular flow path. The endothermic reactions consumed the energy that had been generated in the first flow path inside the tube by fast initial exothermic reactions [62].

Lyubovsky *et al.* performed partial oxidation of methane at an O/C ratio of 1.2 over Microlith® metallic screens coated with ceramic catalyst carrier, which were impregnated with rhodium [63]. They took gas samples over the short length of this catalyst bed. As shown in Figure 3.4, the oxygen was completely consumed after the first 4 mm of the bed, while only a small amount of methane had been converted. The primary products were steam, carbon monoxide and hydrogen, carbon dioxide was formed to a lesser extent. The steam was then consumed downstream of the reactor, converting residual methane. As expected from the thermodynamic equilibrium, conversion decreased with increasing pressure. Figure 3.5 shows the decreasing selectivity of the partial oxidation reaction towards steam with increasing conversion. The dominant selectivity towards carbon monoxide remained almost unchanged when the conversion increased. The hot spot in the reactor reached 1000 °C.

Specchia *et al.* performed partial oxidation of methane over rhodium/α-alumina fixed catalyst beds for short contact times over a time range of between 10 and 40 ms [64]. With increasing catalyst particle sizes, conversion decreased, which was attributed to transport limitations. Higher reactor temperature was observed for larger particles and thus more exothermic reactions took place. When increasing the particle size and the weight hourly space velocity (see Section 4.1), the water content in the product increased, while less carbon monoxide was found and carbon dioxide remained at an unchanged low concentration. Similar to the results of Lyubovski discussed above, steam seemed to be a primary product of the reaction.

Partial oxidation is highly exothermic, which makes heat removal a critical issue in order to prevent damage to the catalyst structure (see Section 4.2).

Some important aspects of the partial oxidation reaction were highlighted by Panuccio *et al.*, from the group working with L.D. Schmidt [65]. They investigated the conversion of octane isomers over rhodium coated ceramic foams made from α-alumina. It could be demonstrated, experimentally, that branched hydrocarbons such as isooctane (2,2,4 trimethylpentane) show slightly lower selectivity towards the desired products carbon monoxide and hydrogen, but generate higher amounts of the total oxidation products independently of the feed composition (O/C ratio, shown as its inverse C/O in Figure 3.6). However, at O/C ratios exceeding unity (C/Os ratios less than one) the differences were more pronounced. Under these conditions, full conversion of the fuel was always achieved but significant amounts of light hydrocarbons such as ethylene, propylene and butylenes were formed. Isooctane feed

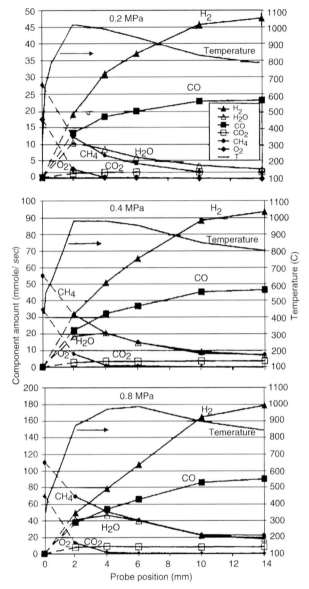

Figure 3.4 Species concentration and temperature profile over a metal screen catalyst bed coated with rhodium catalyst for partial oxidation of methane; O/C ratio 1.2; pressure 2 bar (top), 4 bar (centre) and 8 bar (bottom) [63].

generated significantly less ethylene and more propylene and isobutene compared with n-octane, especially at O/C ratios of less than one.

The reaction temperature ranged between 900 (O/C ratio 1) and 1150 °C (O/C ratio 1.25) in the adiabatic reactors [65].

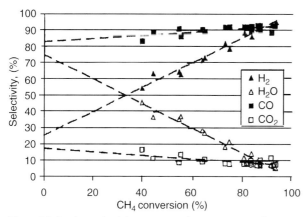

Figure 3.5 Product selectivity versus methane conversion for partial oxidation of methane over rhodium catalyst; O/C ratio 1.2 [63].

Furthermore, these workers investigated the effect of the different pore sizes of the foams on selectivity. As shown in Figure 3.7, the larger pore size [45 ppi (pores per inch) or 470 μm pore size] generated lower selectivity towards hydrogen and carbon monoxide compared with the smaller pores (45 ppi or 250 μm pore size). Not shown here is the selectivity towards light hydrocarbons, which was approximately double in the large pores. These effects were attributed to the higher contribution of

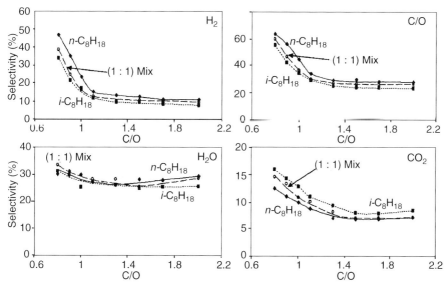

Figure 3.6 Selectivities as determined for n-octane, isooctane and a 1:1 mixture thereof over a rhodium coated ceramic foam [65].

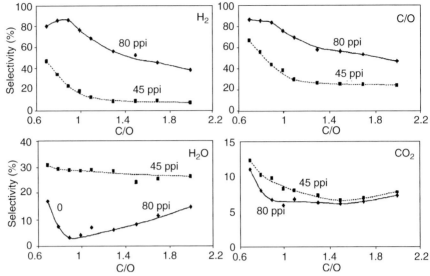

Figure 3.7 Selectivities as determined for n-octane and isooctane over a rhodium coated ceramic foam of different pore sizes, described as ppi (pores per inch) [65].

homogeneous reactions in the larger pores favouring the generation of light hydrocarbons rather than syngas (carbon monoxide and hydrogen). However, numerical simulations revealed that homogeneous gas phase chemistry alone could not explain the experimental results. The catalyst also played a significant role in product distribution.

Some workers add steam to the feed of partial oxidation reactors (running at an O/C ratio higher than 1) and term this autothermal reforming. This is, however, misleading, because oxygen will be consumed initially in the reactor. Thus to a large extent the fuel molecules are already converted and there is little feed left for steam reforming. Such operating conditions decrease the system efficiency significantly, as shown in Section 3.3. This has the consequence that it is mainly water–gas shift that will take place in the reactor downstream of the inlet section, owing to the presence of steam. Because both partial oxidation and water–gas shift are exothermic, this mode of operation does not help to reduce hot spot formation in the reactor substantially. However, coke formation is certainly suppressed in the presence of steam.

Running a fuel processor at an O/C ratio of 1.0 or higher in the presence of steam feed should be termed steam supported partial oxidation.

Figure 3.8 shows methane conversion, hydrogen yield and adiabatic temperature rise of a methane partial oxidation reactor as a function of the O/C ratio (here expressed as the air ratio, λ) [66]. Owing to coke formation below O/C = 1.2 ($\lambda = 0.3$) and it being more pronounced below O/C = 1 ($\lambda = 0.25$), the reactor temperature rises only moderately until O/C reaches 1.2. Beyond this value the adiabatic

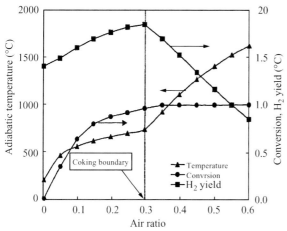

Figure 3.8 Effect of the O/C ratio [expressed as an air ratio, $\lambda = (O/C)/4$] on the equilibrium values of conversion and hydrogen yield, and on the adiabatic temperature rise of the reactor for methane partial oxidation; pre-heating temperature 200 °C; pressure 1 bar [66].

temperature increases rapidly because catalytic combustion becomes dominant, which also decreases the hydrogen yield considerably of course. The adiabatic temperature of 743 °C calculated at O/C = 1.2 is relatively low for a partial oxidation reactor, which originates from the low pre-heating temperature of 200 °C assumed for the feed. Seo *et al.* also demonstrated, by their calculations, that increasing the feed temperature does not affect the O/C ratio required to prevent coke formation [66]. However, the feed temperature increases the adiabatic temperature of the reactor. At a feed temperature of 500 °C and an O/C ratio of 1.2, the adiabatic reactor temperature readily exceeded 900 °C.

Similar thermodynamic equilibrium calculations were performed by Docter and Lamm for gasoline feed (C_7H_{12}) at various O/C ratios (again expressed as the λ value for total combustion), which are shown for S/C ratios of 0 and 0.7 (partial oxidation) in Figure 3.9 (steam supported partial oxidation) [7]. This makes it clear that coke is only stable up to a certain O/C ratio and coke formation decreases with an increased concentration of steam. However, coke formation is not only dictated by the thermodynamic equilibrium but also affected by other factors such as catalyst properties (see Section 4.2). The decreasing content of methane in the reformate with increasing reactor temperature becomes obvious in Figure 3.9, which originates from the thermodynamic equilibrium. In a practical system the extent of methane formation also depends on the performance of the catalyst. The maximum hydrogen concentration was calculated for air ratios between 0.3 and 0.35, which corresponds to an O/C ratio of 1. This is equivalent to the stoichiometric composition of partial oxidation. It is important to realize that a much higher hydrogen content in the reformate is achieved in the presence of steam, owing to the water–gas shift reaction. In the presence of steam, steam reforming may occur below $\lambda = 0.3$,

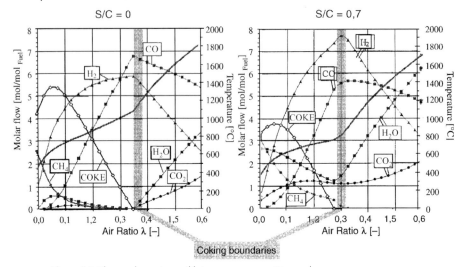

Figure 3.9 Thermodynamic equilibrium gas composition and reformer adiabatic temperature versus air ratio (λ) for gasoline reforming [7]; feed temperatures were 400 °C for air, 200 °C for steam and 20 °C for the fuel; left, S/C = 0; right, S/C = 0.7.

because the oxygen is then insufficient to convert the fuel completely through partial oxidation.

The suppression of methane formation with increasing reaction temperature and the increasing formation of carbon monoxide according to thermodynamics was also determined experimentally by Moon et al. [67] as shown in Figure 3.10.

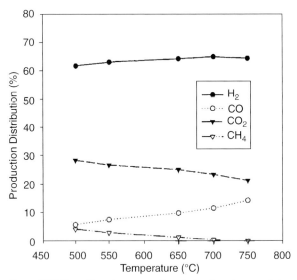

Figure 3.10 Dry reformate composition during isooctane partial oxidation in the presence of steam; S/C = 3.0; O/C = 1.0; GHSV = 8776 h^{-1} [67].

3.3
Oxidative Steam Reforming or Autothermal Reforming

Oxidative steam reforming is the general term for the operation of a steam reformer, to which a certain amount of additional air is fed:

$$C_xH_yO_z + n(O_2 + 3.76N_2) + (x - 2n - z)H_2O$$
$$\rightarrow xCO + \left(x - 2n - z + \frac{y}{2}\right)H_2 + 3.76nN_2 \quad (3.23)$$

Air addition should be performed in very limited amounts in order to prevent coke formation at the catalyst, similar to the "air bleed" which is added to the anode feed of fuel cells running on reformate (see Section 2.3.1). When the amount of oxygen is increased to a level where the energy generation by partial oxidation reaction balances the energy consumption of steam reforming, the overall reaction is theoretically self-sustaining or autothermal.

However, this is not the case in a practical system because heat losses need to be compensated for. Usually an optimum atomic oxygen/carbon (O/C) ratio exists for each fuel under thermally neutral conditions to achieve optimum efficiency. This value amounts to O/C = 0.88 (or 0.44 for the O_2/C ratio here named x) as shown for methane in Figure 3.11 [68].

The maximum efficiency at this optimum ratio amounts to 93.9% for methane, 6.1% of the efficiency being lost mainly for the evaporation of water. The general formula for the optimum efficiency is:

$$\eta = \frac{x \dfrac{\Delta H_{f,\,CO_2}}{\Delta H_{f,\,H_2O(l)}} + \dfrac{y}{2} + \dfrac{\Delta H_{f,\,fuel}}{\Delta H_{f,\,H_2O(l)}}}{x \dfrac{\Delta H_{f,\,CO_2}}{\Delta H_{f,\,H_2O(g)}} + \dfrac{y}{2} + \dfrac{\Delta H_{f,\,fuel}}{\Delta H_{f,\,H_2O(g)}}} \quad (3.24)$$

where $\Delta H_{f,i}$ is the heat of formation of species i, $\Delta H_{f,\,H_2O(l)}$ and $\Delta H_{f,\,H_2O(g)}$ are the heat of formation of liquid and gaseous water, respectively. Table 3.1 shows the optimum O/C ratio and maximum efficiency as calculated by Ahmed and Krumpelt

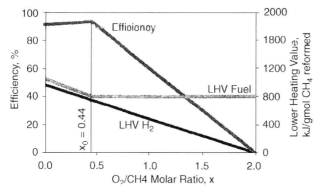

Figure 3.11 Effect of O/C ratio on the lower heating value of hydrogen and the efficiency of the methane reforming process as calculated by Ahmed and Krumpelt [68].

Table 3.1 Optimum O/C ratio and maximum efficiency under thermo neutral conditions for various fuels according to Ahmed et al. [68].

Fuel	Optimum O/C ratio	Maximum efficiency [%]
Methanol (l)	0.46	96.3
Methane (g)	0.89	93.9
Ethanol (l)	0.61	93.7
Iso-Octane (l)	0.74	91.2
Gasoline $C_7H_{15}O_{0.1}$	0.72	90.8

for various fuels [68]. The highest efficiencies can be achieved for alcohol fuels, which require less oxygen to reach thermo neutral conditions. In fact, the formula above and the values provided in Table 3.1 are even valid for steam reforming and partial oxidation. For steam reforming the amount of additional fuel that needs to be combusted (separately) to supply the endothermic process is then included in the calculations, whereas for partial oxidation the amount of water required to run a subsequent water–gas shift needs to be included.

The maximum efficiency that is obtainable from autothermal reforming of various hydrocarbon fuels was calculated by Hagh [13] at fixed a S/C ratio and pressure. It revealed the highest maximum efficiency for methane and slightly decreased maximum efficiency with increased feed temperature (see Figure 3.12). However,

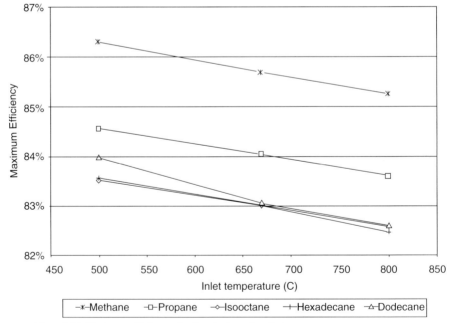

Figure 3.12 Maximum efficiency of various hydrocarbon fuels as a function of the feed temperature at S/C = 3; pressure 1.36 bar [13].

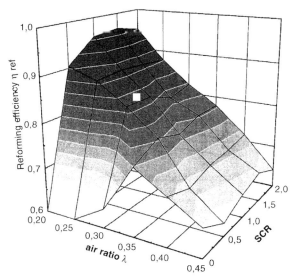

Figure 3.13 Efficiency of diesel reforming depending on the air ratio, λ, and the S/C ratio (SCR) at 800 °C, as determined by Hartmann et al. [69].

practical reasons such as increased heat losses at higher temperature affect the efficiency even more.

Hartmann et al. performed a comparison of the reformer efficiency for diesel fuel [69] under conditions of steam reforming and partial oxidation with and without steam addition at a reaction temperature of 800 °C. The efficiency shown in Figure 3.13 was calculated according to the definition provided in Eq. (2.2), Section 2.2. For partial oxidation in the absence of steam in the feed, the optimum air ratio was calculated to be 0.35 [69], which corresponds to an O/C ratio of 1.07, when pentadecane was assumed to be the feedstock. The efficiency was calculated to be 75% under these conditions. In the presence of steam, the optimum conditions were determined to an O/C ratio of 1.07 and an S/C ratio 0.4, which corresponded to an efficiency of 82%. A much higher efficiency of greater than 95% was found when the S/C ratio was increased beyond 1.0 and the O/C ratio decreased below 0.6 (air ratio 0.2), which proves once more that excessive air addition at an O/C ratio of around 1.0 impairs the efficiency of the reforming process.

Ahmed amd Krumpelt also performed a comparison of the maximum efficiency achievable for autothermal reforming, steam reforming and partial oxidation [68]. Their conclusion was that steam reforming has the lowest efficiency, mainly due to the fact that steam was lost from the system on the burner side of the reformer. This conclusion, however, is no longer valid when residual hydrogen from the anode off-gas is utilised for combustion. It will be shown in Section 5.4.4 that steam reforming has the highest overall system efficiency.

However, amongst many others, zur Megede [70] from XCellsis (a former joint-venture of Daimler-Chrysler and Ballard) drew exactly the opposite conclusion to that

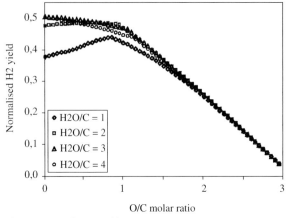

Figure 3.14 Hydrogen yield versus O/C ratio at various S/C ratios for isooctane reforming [71].

of Ahmed et al.: steam reforming is most efficient followed by autothermal reforming and partial oxidation.

Thermodynamic calculations by Villegas et al. revealed that the optimum hydrogen yield was achieved close to an O/C ratio of 1 for S/C ratios below 2 [71]. These calculations, however, do not agree with the experimental results by the same workers, which revealed a lower optimum O/C ratio (see Section 4.2.7). For higher values of the S/C ratio, the hydrogen yield decreased continuously when starting from pure steam reforming (O/C = 0) with increasing oxygen addition (see Figure 3.14).

Today it can be regarded as certain that steam reforming generates the highest system efficiency, followed by autothermal reforming and partial oxidation. However, steam reformers are usually more complex systems, but this will be discussed in Chapter 5.

For autothermal reforming of higher hydrocarbons, the main hydrocarbon by-product usually observed is methane. The formation of light alkenes is favoured over alkanes in the case of incomplete conversion towards carbon oxides and methane [10, 72, 73].

Flytzani-Stephanopoulos and Voecks found that for autothermal reforming of n-tetradecane and benzene, the aromatic feed produced only methane, whereas aliphatic feed also produced light olefins and paraffins as shown in Figure 3.15 [74]. Ethylene was only an intermediate product in benzene reforming.

Figure 3.16 shows how the average temperature of a porous ceramic block increases for the autothermal reforming of isooctane with increasing O/C ratio. However, conversion of isooctane increased both with the S/C ratio and the O/C ratio, as shown in Figure 3.17 as determined by Ellis et al.

Methane formation was decreased by increasing the S/C and the O/C ratios as shown in Figure 3.18, whereas the carbon monoxide content usually equals the thermodynamic equilibrium of the water–gas shift and consequently decreases with increasing S/C ratio. It is almost independent of the O/C ratio provided the O/C is smaller than 1 [75].

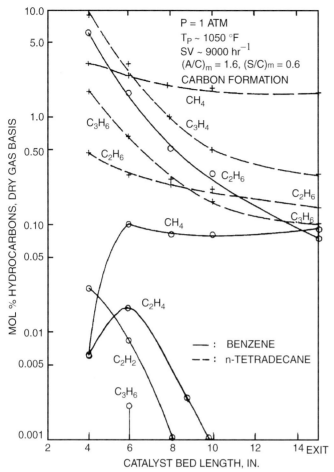

Figure 3.15 Formation and consumption of light hydrocarbons as determined during autothermal reforming of n-tetradecane and benzene [feed pre-heating temperature = 565 °C, p = 1 bar, S/C = 0.6; O/C = 0.67; GHSV (Gas-Hourly Space Velocity) = 9000 h^{-1}] [74].

Seo et al. investigated thoroughly the interplay of the O/C ratio and S/C ratio on carbon formation and hydrogen yield for methane oxidative steam reforming by thermodynamic calculations [66]. A minimum S/C ratio of 1.4 had been identified previously as being necessary to prevent carbon formation in the absence of oxygen [66] (see Section 3.1). When air was fed to the system on top of steam and methane, this minimum value decreased, as shown in Figure 3.19.

Chan et al. calculated that an O/C ratio of 0.67 was required to prevent carbon formation regardless of the S/C ratio [76].

However, from many practical applications coke formation is known to occur at even higher values of the O/C ratio. The calculations of Chan et al. also showed very

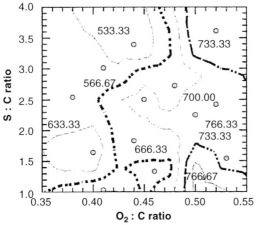

Figure 3.16 Average temperature of a ceramic block coated with proprietary catalyst for various O/C ratios (expressed as O_2/C ratio) and feed temperature at an S/C ratio of 2 [75]; feed temperature 330 °C.

similar results to those of Seo et al. discussed above, that the S/C ratio of 1.5 or higher prevented carbon formation in the absence of oxygen with natural gas. Under stoichiometric conditions of partial oxidation, which correspond to an O/C ratio of 1.0 (and an air ratio of 0.25), an S/C ratio between 0.2 and 0.4 was required. When switching to autothermal conditions, which correspond to an O/C ratio of 0.88 (see above), the S/C ratio required to prevent coke formation was 0.8.

A question frequently discussed concerning autothermal reforming deals with the sequence of reactions taking place. The overall reaction could be either total oxidation

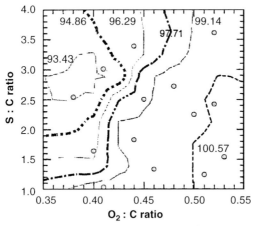

Figure 3.17 Isooctane conversion as achieved over a ceramic block coated with proprietary catalyst for various O/C ratios (expressed as O_2/C ratio) versus the S/C ratio [75]; feed temperature 330 °C.

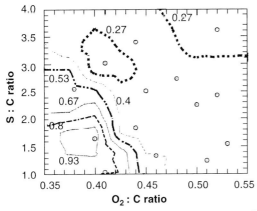

Figure 3.18 Methane dry content as achieved over a ceramic block coated with proprietary catalyst for various O/C ratios (expressed as O_2/C ratio) and S/C ratios; feed temperature 330 °C [75].

of the fuel followed by steam reforming and reverse water–gas shift or partial oxidation followed by steam reforming. Springmann *et al.* [56] performed measurements under conditions of autothermal reforming of 1-hexene and took samples of the reaction mixture over the reactor length (see Figure 3.20). These measurements revealed two remarkable facts. Firstly, the concentration of steam increased in the first part of the reactor, indicating that substantial total oxidation occurred but there

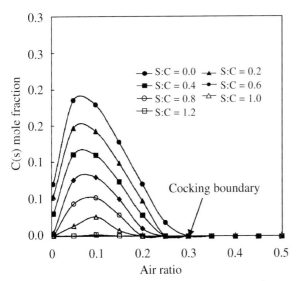

Figure 3.19 Effect of S/C ratio and O/C ratio [here expressed as the air ratio $\lambda = (O/C)/4$] on carbon formation for methane steam reforming; pressure, 1 bar; feed pre-heating temperature 400 °C [66].

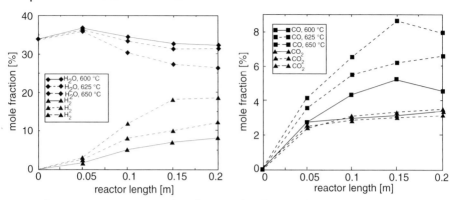

Figure 3.20 Gas composition along the reactor length axis as determined experimentally for autothermal reforming of 1-hexene: S/C = 2.2; O/C = 0.9; T = 600–650 °C [56].

was only little water consumption by steam reforming. Secondly, the carbon dioxide was formed almost exclusively in the first section of the reactor and this was almost independent of the reaction temperature. Thus the conclusion could be that in autothermal reformers partial oxidation and total oxidation are taking place in parallel and steam reforming takes place consecutively. Numerical calculations were performed by Pacheco *et al.* for autothermal reforming of isooctane [77]. The results of the calculations were validated with experimental data from the Argonne National Laboratory, however only reformate compositions determined at the reactor outlet were used to judge the model quality. Concentration profiles were calculated over the reactor length axis revealing a peak of the steam content close to the reactor inlet [77], similar to the results of Springmann discussed above. The carbon monoxide concentration also increased gradually and carbon dioxide quickly reached a plateau, in excellent agreement with the experimental results of Springmann (see Figure 3.21).

Because partial oxidation reactions take place faster compared with steam reforming, the heat balance of autothermal reforming is only neutral on an overall basis. This leads to local overheating, similar to partial oxidation (see Section 4.2).

Increasing pressure has a negative effect on the autothermal reforming reaction. Figure 3.22 shows the maximum hydrogen yield that is achievable, in theory, in an adiabatic autothermal reformer at various feed pre-heating temperatures and pressures. Owing to the volume expansion of the reforming reaction, increasing pressure reduces hydrogen formation.

An important issue when handling hydrocarbon/steam/air mixtures are the flammability limits. Auto ignition temperatures were determined for reformulated gasoline by Ellis *et al.* [75]. The auto ignition temperature of the air/fuel mixture was determined to be 360 °C. Addition of steam increased this value to at least 500 °C at O/C 1.0 and S/C 1.5.

Autothermal reformers may be started when pre-heated to temperatures of 350 °C, even in the case of gasoline [75].

3.3 Oxidative Steam Reforming or Autothermal Reforming

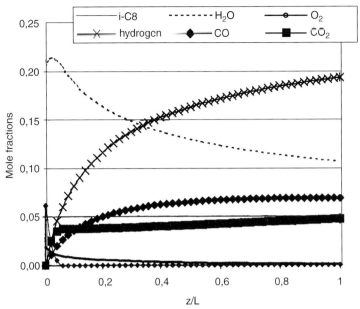

Figure 3.21 Gas composition for autothermal reforming of isooctane as calculated by Pacheco et al.: S/C 1.42; O/C 0.84; $T = 800\,°C$ [77].

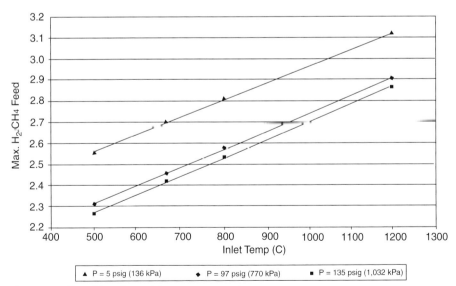

Figure 3.22 Effect of the reformer inlet temperature and pressure on the theoretically achievable hydrogen yield [78]: S/C ratio 3.0.

3.4
Catalytic Cracking of Hydrocarbons

The endothermic cracking of hydrocarbons [see also Eq. (3.22) in Section 3.2], shown here for propane, consumes only about 5% of the lower heating value of this fuel:

$$C_3H_8 \rightarrow 4H_2 + 3C \quad \Delta H^0_{298} = -103.8 \text{ kJ mol}^{-1} \quad (3.25)$$

However, 55% of the lower heating value of the fuel is stored in the carbon product, which considerably reduces the efficiency of hydrogen generation by cracking [79]. The catalytic cracking process produces mainly coke, hydrogen, methane, carbon monoxide and small amounts of carbon dioxide. The carbon oxides stem from the reduction of the oxidised regenerated catalyst (see Section 4.2.9). Figure 3.23 shows the equilibrium concentration for propane cracking at various operating temperatures and pressures as calculated by Ledjeff-Hey et al. [79]. Favourable conditions are achieved at ambient pressure and temperatures exceeding 800 °C. Catalytic cracking is not used on an industrial scale, because the hydrogen yield is lower compared with steam reforming [80]. However, theoretically, the hydrogen produced via the cracking process itself has a potential purity of up to 95%, the remainder being methane for the situation when the catalyst is not regenerated by air treatment [6].

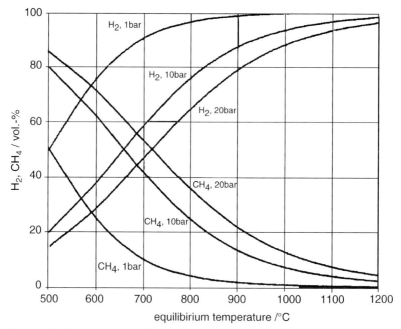

Figure 3.23 Thermodynamic equilibrium gas composition versus reaction temperature and pressure for propane cracking [79].

3.5
Pre-Reforming of Higher Hydrocarbons

Pre-reforming is the conversion of higher hydrocarbons into a mixture of light hydrocarbons. It is performed by different techniques such as the cool flame technology, pyrolysis, catalytic cracking and steam cracking.

Catalytic pre-reforming is usually operated at much lower temperatures compared with the conventional reforming processes, specifically between 350 and 550 °C to prevent carbon formation [46].

Adiabatic pre-reforming is widely used in the chemical and petrochemical industry for hydrogen and synthesis gas production [81]. The industrial process is normally performed by addition of steam to the hydrocarbon mixture according to reactions (3.3) to (3.5). When hydrogen is the target product, the S/C ratio of the feed varies from 2.5 to 5.0 depending on the feedstock [81]. For synthesis gas production, much lower S/C ratios of 0.3 for natural gas and between 1.0 and 2.0 for naphtha and liquefied petroleum gas are applied. If hydrocarbons are used as the feedstock, the overall reaction is endothermic, however for heavier feedstock blends, such as naphtha, it may even be exothermic. Similar to steam reforming, nickel based catalyst are in use on the industrial scale because of the lower cost, despite the lower coke formation tendency of precious metal catalysts [81]. Figure 3.24 shows the operating window of commercial nickel reforming catalyst for natural gas reforming at low S/C ratio [81]. The carbon formation does take place above a certain temperature, which means that whisker carbon formation, typical for nickel catalysts, is the critical issue in this particular case (see Section 4.1.2).

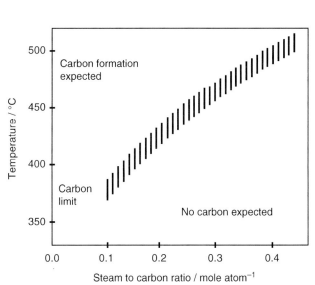

Figure 3.24 Limit for carbon formation in natural gas pre-reforming as a function of S/C ratio and operating temperature [81].

On the other hand, a minimum operating temperature of 450 °C was determined for naphtha feed due to coke formation by polymerisation ("gum" formation, see Section 4.1.2) [81]. Thus the process needs to be operated between the coke formation boundaries of polymeric and high temperature coke.

However, nickel-based catalysts tend to form nickel sulfides when in contact with hydrogen sulfide present in the feed at the low operating temperature of pre-reforming:

$$H_2S + Ni \rightarrow NiS + H_2 \qquad (3.26)$$

Thus hydrodesulfurisation is required upstream of the pre-reformer in the industrial process.

The cool flame technology for diesel fuels has been developed by the German Oel und Wärmeinstitut (OWI) for the pre-reforming of fuels dedicated to fuel cell applications. It enables the generation of a homogeneous mixture of fuel, air and steam [69]. This is a critical issue in order to avoid soot formation. The principle of the cool flame is based on homogeneous auto-ignition of the fuel when atomised into a pre-heated air stream. Partial oxidation of the fuel occurs leading to chain shortening of the hydrocarbons [82]. A large variety of products are formed, amongst them olefins, alcohols, acids, peroxides, aldehydes and carbon monoxide [83]. The pale blue luminescence of the cool flame is a result of the chemiluminescence of electronically excited formaldehyde [83].

The regime of the cool flame is determined by two governing equations. Firstly, the heat generated $\dot{R}$ through the fuel oxidation reaction, which increases exponentially with temperature, and secondly, the heat losses $\dot{L}$ to the environment by heat conduction through the gas itself and ultimately through the housing, which hosts the mixture. Certain fuels including diesel and industrial gas oil have the property of decreasing heat generation within an intermediate temperature range. Figure 3.25

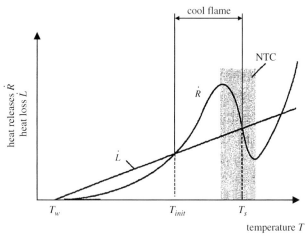

Figure 3.25 Heat generation and heat losses for a mixture of diesel, air and steam in a closed vessel according to Hartmann et al. [69].

shows the relationship between heat generation and heat losses versus the wall temperature of the vessel. Below the initiation temperature T_{init}, which is usually around 300 °C for diesel fuel, the heat losses dominate and no significant chemical reactions occur. When this temperature is exceeded, a temperature rise of up to 180 K is observed because of the surplus of heat generation [69], which is almost independent of the type of fuel [83]. Increasing the system pressure increases the cool flame temperature by about 10 K per 100 mbar [82]. Decreasing the air ratio increases the cool flame temperature, because nitrogen and unconverted oxygen act as diluent for the reaction mixture [82]. A stable state is then established, which returns to stable behaviour, even if the vessel temperature changes slightly. This stable low temperature homogeneous reaction is termed cool flame. It originates from the chemical equilibrium of peroxide-radical formation, which is one of the initial steps of the oxidation reactions [82, 83]:

$$RH + O_2 \leftrightarrow R^{\cdot} + HO_2^{\cdot} \tag{3.27}$$

$$R^{\cdot} + O_2 \leftrightarrow RO_2^{\cdot} \tag{3.28}$$

The equilibrium of the reaction is shifted back to the alkane radical on the left side at elevated temperatures. However, this is a very simplified explanation for the decreasing reaction rate of cool flames observed with increasing reaction temperature. The complex radical reactions taking place during cool flame oxidation will not be discussed in the present book.

The stability range of cool flames from n-hexane is shown in Figure 3.26. The succeeding reactions observed by da Silva *et al.* correspond to auto ignition of the

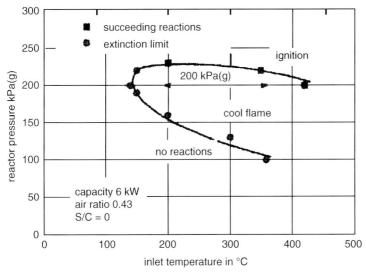

Figure 3.26 Stability range of cool flames depending on feed temperature and system pressure: air ratio $\lambda = 0.43$; fuel, n-heptane [82].

system. It is certainly also limited on the right side of the diagram, which is not shown in Figure 3.26.

Diesel and industrial gas oils may be evaporated in the cool flame unit under conditions of partial oxidation or autothermal reforming [69]. The evaporated feed may then serve as feed for homogeneous or catalytic reforming downstream of the cool flame unit. The cool flame area requires a flame arrester for separation from the reformer. The temperature of the cool flame of course increases with increasing air/fuel ratio. Addition of inert gas decreases the temperature of the cool flame; r is the mass ratio of inert gas flow to the air flow:

$$r = \frac{\dot{m}_{\text{inert}}}{\dot{m}_{\text{air}}} \tag{3.29}$$

Auto-ignition (explosion) of the fuel occurs for stoichiometric air/fuel (λ) values below 0.68 without addition of inert gas, because the limiting upper temperature for ignition of the fuel is exceeded. This limit is around 500 °C. At temperatures exceeding 550 °C the auto-ignition is only delayed by the typical time [82]. The requirement to stay below this delay time while enabling complete evaporation of the fuel limits the modulation of conventional fuel evaporation and air mixing systems [82]. Cool flames enable a modulation range of 1 : 30 when suitable injection systems are used (Figure 3.27) [82].

The OWI cool flame technology has been adopted for diesel reforming as the fuel supply of a solid oxide fuel cell by Nordic Power systems.

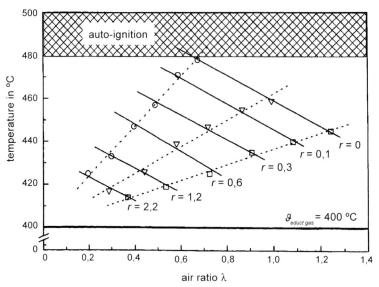

Figure 3.27 Effect of air ratio and mass ratio, r, on the temperature of the cool flame for a feed temperature of 400 °C and an evaporator diameter of 100 mm [69].

3.6
Homogeneous Plasma Reforming of Higher Hydrocarbons

Non-thermal plasma reforming is performed at very high reaction temperatures above 2000 °C without a catalyst. Temperatures greater than 2500 K tend to generate higher hydrocarbons, even from methane. In contrast, thermal plasmas work with even higher energy input and at temperatures of around 5000–10 000 °C, but these will not be considered here. A minimum temperature of 2000 °C is required to avoid soot formation in a non-thermal plasma reformer. A major issue of plasma reforming is the formation of NO_x, well known from internal combustion engines. Another issue for homogeneous partial oxidation, which is different from catalytic partial oxidation, could be ammonia formation [84].

The high temperature makes the residence time and consequently the space requirements of the reactor low, but also generates heat loss issues for small scale systems. About 10% of the electrical power input is lost in plasmatron reformers according to Cohn et al. [85].

The reformate generated by plasma reformers contains only carbon monoxide, hydrogen and residual steam or nitrogen, as far as the thermodynamic equilibrium is concerned [86]. This makes them well suited for carbon monoxide tolerant high temperature fuel cells, such as molten carbonate and solid oxide systems. Plasma reformers are powered by electricity, and thus their energy demand needs to be minimised for fuel cell applications [86]. Plasma reformers have been realized with an efficiency exceeding 90% [86], sometimes close to 100% [87]. The electrical energy input into certain plasma reformers, as published in the literature, was in the range of 0.1–0.2 kWh m^{-3} hydrogen produced – when a heat-exchanger was switched to the reactor for feed pre-heating/product cooling [87]. Approximately 1 kW of electrical energy may be generated from a hydrogen flow of 1 m^3 h^{-1} in a fuel cell. This means that about 20% of the electrical energy production would be required to run such a plasma reformer, which of course lowers the overall efficiency of the process. According to Bromberg et al., the system efficiency of a plasmatron/fuel cell is highest when partial oxidation is performed in the plasmatron. Steam reforming results in lower efficiency [86], because additional energy is required for the endothermic reaction, which increases the electrical power demand of the plasma reformer.

Benilov and Naidis performed numerical calculations for partial oxidation of octane in a plasma reformer [88]. They revealed that the reformer should be operated at a high O/C ratio of 1.3–1.5 to maximise the hydrogen yield. In this instance the electrical discharge served more as an ignition source for the homogeneous partial oxidation reaction. With 0.045 kWh m^{-3} hydrogen, the electrical power input was rather low.

The electric energy input to plasma reformers is performed by a corona discharge, microwaves [89], dielectric barrier discharge [90] or using gliding arc technology [86, 91]. A corona discharge and a dielectric barrier discharge are static, while gliding arc and microwave technology are dynamic processes [92].

Spray-pulse techniques improve the vaporisation of liquid hydrocarbons when fed to the reformer reactor [93].

The practical development of plasmatron reformers is discussed in Section 7.1.6.

3.7
Aqueous Reforming of Bio-Fuels

Because bio-fuels such as bio-ethanol are formed in dilute aqueous solution, their gasification requires extensive amounts of energy. An alternative is the reformation of these fuels in the liquid phase or their treatment with supercritical water.

Another possibility is aqueous phase reforming of methanol, glycol, glycerol, sorbitol or glucose over platinum-based catalysts at 215 °C and 21 bar, which is being developed by Virent Energy Systems [94]. According to Cortright, water–gas shift takes place in parallel in the reactor and catalyst stability is not an issue [94]. Aqueous phase reforming of sorbitol ($C_6H_{14}O_6$) generates a gaseous product, which contains 62 vol.% hydrogen, 37 vol.% carbon dioxide, 0.4 vol.% hydrocarbons and less than 100 ppm carbon monoxide after separation from the liquid phase. Methanol has the highest hydrogen selectivity of all fuels considered. About 5 L hydrogen can be produced per gram of catalyst an hour [94].

Hydrogen production is possible from glucose solutions using micro-organisms such as *Escherichia coli* [95]. One example of such a "bio-fuel processor" will be shown in Section 9.8.

3.8
Processing of Alternative Fuels

3.8.1
Dimethyl Ether

Dimethyl ether (CH_3OCH_3) is an interesting alternative fuel. Its properties are similar to liquefied petroleum gas (LPG), it burns with a visible blue flame and does not require odorants, because it has a sweet ether-like odour [96]. It is not toxic or carcinogenic, has less global warming potential compared with carbon dioxide and is in the same price range as gasoline when produced from natural gas [96]. In addition, it can also be produced from other sources such as biomass, and stored in conventional LPG storage equipment with only minor modifications [96]. However, its volume based lower heating value of 18.9 kJ cm^{-3} is about 50% of the lower heating value of gasoline, ranging between methanol and ethanol. When dimethyl ether is used as fuel for a fuel cell system, its clear advantage is the lower reforming temperature compared with liquid hydrocarbons, although this is still higher than for methanol.

Dimethyl ether may be converted into hydrogen by steam reforming, which actually has methanol as an intermediate product:

$$CH_3OCH_3 + 3H_2O \rightarrow 6H_2 + 2CO_2 \qquad \Delta H^0_{298} = +135 \text{ kJ mol}^{-1} \qquad (3.30)$$

The dimethyl ether conversion into methanol could be performed over an acid catalyst such as alumina, while the second step is then methanol steam reforming (see Sections 3.1 and 4.2.1). However, suitable catalysts for dimethyl ether reforming are still an issue. More details about catalyst technology for dimethyl ether reforming will be provided in Section 4.3.

3.8.2
Methylcyclohexane

Another interesting alternative hydrogen carrier for fuel cell systems is methylcyclohexane, which was proposed amongst others by Heurtaux et al. [3]. It can act as a hydrogen carrier when converted into toluene, which takes place at a reaction temperature of approximately 380 °C, preferably over a platinum catalyst [3]:

$$CH_3C_6H_{10} \rightarrow CH_3C_6H_5 + \frac{5}{2}H_2 \quad \Delta H^0_{298} = +205 \text{ kJ mol}^{-1} \quad (3.31)$$

This process had previously been studied in the 1980s for a hydrogen source for internal combustion engines [97]. Since 2000, eight research teams worldwide have started work on the methylcyclohexane dehydrogenation reaction to provide a hydrogen source, mostly for fuel cells. The reaction is endothermic and this energy could be provided by combustion of a small portion of the toluene product. The remaining toluene could be stored in a separate tank and fed back to the petrochemical industry for re-hydrogenation. The toluene requires separation from the hydrogen, which can be performed by pressure swing adsorption, according to Heurtaux et al. [3].

3.8.3
Sodium Borohydride

Borohydrides such as $NaBH_4$ in an aqueous solution of caustic soda have the highest gravimetric and volumetric storage densities compared with all other hydrogen storage options, including liquid hydrocarbons (see Section 2.1). The reaction with water is exothermic:

$$NaBH_4 + 2H_2O \rightarrow 4H_2 + NaBO_2 \quad \Delta H^0_{298} = -300 \text{ kJ mol}^{-1} \quad (3.32)$$

The alkaline borohydride solution is stable to air exposure and reacts rapidly in the presence of a catalyst, preferably ruthenium or platinum/lithium/cobalt oxide [98] at temperatures as low as 0 °C [99], while the neutral or acidic solution undergoes self-hydrolysis [100]. The product of the reaction in the alkaline solution is sodium tetrahydroxyborate $NaB(OH)_4$, which is environmentally innocuous [99]. In one litre of a 35 wt.% sodium borohydride solution, approximately 74 g hydrogen may be stored, which corresponds to a power output of approximately 800 Wh. According to Kojima et al. from Toyota [100], the concentration of the sodium borohydride should be kept below 16 wt.%, to keep the sodium borate product in solution. Crystallisation of the discharged product to $NaBO_2 \cdot 2H_2O$ also occurred at higher feed concentration during bench-scale tests by Zhang et al. (see Section 9.8). They thus concluded that 10–15 wt.% is the maximum possible feed concentration to prevent the liquid product from crystallisation in the discharge tank. This limits the hydrogen weight content in the fuel to 3 wt.% [101]. However, crystallisation problems were not expected for the decomposition reactor, because the temperatures are much higher there.

The price of sodium borohydride is currently too high by far for a practical application. Compared with hydrogen production from natural gas its price is 130 times higher. However, the idea of applying sodium borohydride as a fuel has the prerequisite of recycling the sodium borate product, and this could lower the price on the basis of mass production [98]. An alternative to hydrogen generation from sodium borohydride is the direct borohydride fuel cell, which is not within the scope of this book, so will not be discussed.

3.8.4
Ammonia

The catalytic decomposition of ammonia, preferably performed at around 600 °C, is another interesting alternative to fuel processing based upon fossil fuels [34]:

$$2NH_3 \rightarrow N_2 + 3H_2 \qquad \Delta H^0_{298} = +46 \text{ kJ mol}^{-1} \qquad (3.33)$$

Ammonia may be stored at 8 bar as a liquid. Because proton exchange fuel cells are very sensitive to ammonia poisoning, hydrogen produced from ammonia decomposition is mostly applied in connection with alkaline fuel cells (see Section 2.3.2).

Another path to hydrogen generation from ammonia works with magnesium hydride over homogeneous platinum and palladium chloride catalysts [103]:

$$2 NH_3 + 3MgH_2 \rightarrow 6H_2 + Mg_3N_2 \qquad \Delta H^0_{298} = -33.5 \text{ kJ mol}^{-1} \qquad (3.34)$$

Alternatively, lithium aluminium hydride could be used [104]:

$$4NH_3 + 3LiAlH_4 \rightarrow 12H_2 + 3AlN + Li_3N \qquad (3.35)$$

Higher pressure is unfavourable for the thermodynamic equilibrium of the reaction, however, an elevated system pressure up to 10 bar is beneficial for the kinetics of the reaction [12]. At 1 bar and 400 °C only negligible amounts of ammonia still remain in the reaction mixture according to thermodynamics.

The energy demand of the process is low with 12% of the higher heating value of the hydrogen product. Therefore, fuel cell anode off-gas would deliver sufficient energy to operate the process as thermoneutral after start-up.

3.9
Desulfurisation

Because fossil fuels and bio-fuels also contain significant amounts of sulfur, their desulfurisation is necessary.

3.9 Desulfurisation

Odorised natural gas may contain 4–6 ppm sulfur, while liquefied petroleum gas typically contains 30 ppm sulfur [105]. Short term peaks of up to 150 ppm are permitted by legislation. Wang et al. reported that the sulfur content of the gas released from their liquefied petroleum gas tanks was different for fresh and re-used tanks [106]. Initially the fresh tank released 1.5 ppm ethyl mercaptane, the re-used one 6.9 ppm. This value increased to 14 ppm for the half-empty tank and to as high as 140 ppm for the empty tank.

Odorants are blends of butyl and propyl mercaptanes sometimes containing dimethylsulfide and methylethylsulfide in the US [107]. In Europe the preferred odorants are dimethylsulfide and tetrahydrothiophene [107]. European natural gas contains up to 20 ppm carbon oxide sulfide, owing to the higher hydrogen sulfide concentration of the gas, which then reacts with carbon dioxide.

Sulfur-free odorants have been developed and commercialized, but are not yet commonly used [108]. They contain a mixture of 37 wt.% methylacrylate, 60 wt.% ethylacrylate and 2.5 wt.% methylethylpyrazine. It has been proven that these components do not affect the performance of a rhodium catalyst during methane steam reforming.

In addition, gasoline contains benzothiophenes and alkylated benzothiophenes, while dibenzothiophene and alkylated dibenzothiophenes can be detected in jet fuels and diesel. Polycyclic sulfur compounds with three or more rings are found in heavy fuel oils [109]. A typical sulfur compound of diesel fuel is 4,6-dimethyldibenzothiophene [109].

Desulfurisation is performed preferably in the gas phase. For gaseous fuels such as natural gas or LPG, the desulfurisation is rather carried out upstream of the reformer, whereas in case of liquid fuels it is better switched downstream.

Most sulfur compounds present in hydrocarbons are converted into hydrogen sulfide in a non-oxidising atmosphere at temperatures between 300 and 400 °C [110]. In an oxidising atmosphere, they are converted into sulfur dioxide.

The hydrogen sulfide formed in the reformer may then be removed by a zinc oxide catalyst/adsorbent between 350 and 400 °C:

$$H_2S + ZnO \rightarrow ZnS + H_2O \qquad \Delta H^0_{298} = -75 \text{ kJ mol}^{-1} \qquad (3.36)$$

The equilibrium constant only depends on the partial pressure of the steam and hydrogen sulfide:

$$K_{eq} = \frac{p_{H_2O}}{p_{H_2S}} \qquad (3.37)$$

Its value is about 6×10^{-6} at 300 °C [111]. Thus, theoretically, hydrogen sulfide may be removed to values below 1 ppm, regardless of the steam partial pressure at this temperature. The equilibrium hydrogen sulfide content in the presence of 20 vol.% steam for zinc oxide at 400 °C is about 1 ppm, for copper oxides lower values are achieved but copper is usually not present as the oxide in reformate at this temperature [112].

Catalysts and adsorbents for sulfur removal will be discussed in Section 4.4.

3.10
Carbon Monoxide Clean-Up

Where low temperature PEM fuel cells are the consumers of the reformate, its carbon monoxide content has to be minimised. The situation is different when high temperature fuel cells, such as solid oxide fuel cells, are applied, because they are also capable of converting carbon monoxide (see Section 2.3.2).

3.10.1
Water–Gas Shift

The moderately exothermic water–gas shift reaction [WGS, Eq. (3.4) in Section 3.1] is usually limited by its thermodynamic equilibrium, which may be calculated according to the formula provided below [113]:

$$K_{eq} = \exp\left(\frac{4577.8}{T} - 4.33\right) \tag{3.38}$$

Figure 3.28 shows the adiabatic temperature rise, which is the increase in gas temperature in a perfectly insulated reactor, for a typical reformate for increasing conversion of carbon monoxide by a water–gas shift [57]. Owing to thermodynamic limitations and its exothermicity, the reaction is divided into two consecutive steps, known as high temperature and low temperature water–gas shifts on the industrial scale. However, two water–gas shift stages are only mandatory for fixed bed or monolithic reactors, which will be discussed in Section 5.2.1. High temperature

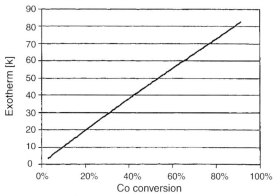

Figure 3.28 Adiabatic temperature rise (termed "exotherm" here) of reformate containing 53.1 vol.% hydrogen, 7.7 vol.% carbon monoxide, 7.5 vol.% carbon dioxide, 31.4 vol.% steam and 0.3 vol.% methane versus carbon monoxide conversion by water–gas shift [57].

water–gas shift is usually performed at temperatures between 375 and 450 °C and low temperature water–gas shift operates between 200 and 300 °C. Depending on the feed composition, the carbon monoxide concentration of the high temperature water–gas shift product ranges between 2 and 3 vol.%, whereas the low temperature water–gas shift product contains between 0.05 vol.% for larger scale systems [46] and 0.3–1 vol.% for small scale applications. As the reaction takes place without any change in volume, the pressure dependence of the equilibrium is negligible.

Side reactions are limited with the water–gas shift. Depending on the catalyst technology, methanation [see Eq. (3.5) in Section 3.1] may occur in addition to deactivation reactions of the catalyst (see Section 4.5.1).

3.10.2
Preferential Oxidation of Carbon Monoxide

The removal of small amounts of carbon monoxide present in pre-cleaned reformate of hydrocarbon and ethanol reformers is commonly performed by the preferential oxidation with air (PrOx):

$$CO + 0.5O_2 \rightarrow CO_2 \qquad \Delta H^0_{298} = -283 \text{ kJ mol}^{-1} \qquad (3.39)$$

The reaction is accompanied by the undesired side reaction by some of the hydrogen present in the reaction mixture:

$$H_2 + 0.5O_2 \rightarrow H_2O \qquad \Delta H^0_{298} = -242 \text{ kJ mol}^{-1} \qquad (3.40)$$

Preferential oxidation usually requires a minimum of excess air corresponding to an O/CO ratio or λ value of between 1.5 and 2.0. Under these conditions and full conversion of carbon monoxide, between 0.5 and 1.0 mole of hydrogen is lost for each mole of carbon monoxide converted.

Another problem occurring in preferential oxidation reactors at low concentration of carbon monoxide is the re-formation of carbon monoxide over the catalyst by the Reverse Water–Gas Shift Reaction (RWGS) in an oxygen deficient atmosphere:

$$H_2 + CO_2 \rightarrow H_2O + CO \qquad \Delta H^0_{298} = +41 \text{ kJ mol}^{-1} \qquad (3.41)$$

Most catalysts applied for the preferential oxidation of carbon monoxide have some activity for water–gas shift and also its reverse reaction. This means that over-sizing the reactor presents the potential of impaired conversion, because the absence of oxygen favours the reverse water–gas shift. Unfortunately the same applies for partial loading of the reactor. Because the acceptable concentration of carbon monoxide is usually in the region of 100 ppm or less, even low catalytic activity for the reverse shift becomes an issue.

Owing to the negative reaction order of carbon monoxide for the preferential oxidation over noble metal based catalysts such as platinum [114, 115] and platinum/rhodium [116], the amount of catalyst required to achieve 90 or 99.9% conversion does not increase significantly [114].

3.11
Catalytic Combustion

Combustion reactions are performed frequently in fuel processor systems. The combustion is carried out either homogeneously or heterogeneously over catalysts. In the former, reaction temperatures easily exceed 1000 °C, which may generate NO_x emissions, a problem well known from internal combustion engines.

The catalytic combustion of hydrogen over a platinum catalyst starts even at room temperature and full conversion of diluted hydrogen (as present in anode off-gas) is easily achieved at temperatures exceeding 100 °C. The same applies to alcohol fuels, for example, methanol. Hydrocarbon fuels such as LPG require higher reaction temperatures between 250 and 300 °C to achieve full conversion, over specially designed catalysts (see Section 4.6), whereas complete methane combustion takes place at even higher temperatures exceeding 500 °C. The temperatures provided above refer to reasonably high space velocity over the catalysts. Complete combustion is possible at lower temperature of course but at the expense of excessive amounts of catalyst.

Homogeneous and heterogeneous burners are usually operated with an excess of air to ensure complete combustion of the fuel.

The lower heating values for hydrogen and selected alcohol and hydrocarbon fuels are provided in Table 2.1, Section 2.1.

Over catalytic surfaces, catalytic combustion is possible with milliseconds contact time, which has been demonstrated by numerous research groups. The large temperature and concentration gradients that develop under such conditions were demonstrated impressively by the numerical calculations of Redenius et al. [118]. They developed a laboratory radiant burner, which is shown in Figure 3.31. It was composed of a single strip of Fecralloy foil, which was 50 μm thick and 80 mm wide. This foil was folded around an inlet manifold tube. In this way the feed gases were preheated from ambient temperature in a central duct of the burner before they entered the combustion zone, which was wash-coated with a γ-alumina/chromium/platinum catalyst doped with yttrium. The ducts had a width of 2 mm. Temperature gradients of about 200 K between the gas phase and the catalyst were calculated (see Figure 3.31). The catalyst surface temperature was determined experimentally up to 930 °C. Large concentration gradients were calculated, indicating the massive mass transport limitation at the 2 mm scale (see Figure 3.32). Redenius et al. reported that their platinum catalyst had low dispersion, particles of 1 μm size were observed, which had no negative effect on its performance, because the reaction took place only at the outer surface of the wash-coat.

3.12
Coke Formation on Metal Surfaces

Steel and other reactor materials such as nickel based alloys have a certain tendency to form coke deposits, especially when exposed to higher hydrocarbons. The coke

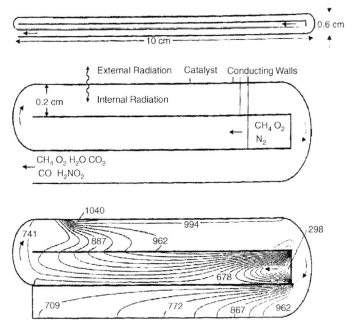

Figure 3.31 Laboratory-scale radiant burner as developed by Redenius et al. [118]; top, drawing to scale; centre, sketch showing three ducts separated by only the 50 μm Fecralloy metal foil; bottom, temperature distribution as determined by numerical calculations, temperatures in K.

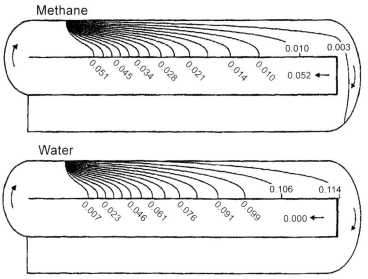

Figure 3.32 Mass fraction of methane and water calculated for their radiant burner by Redenius et al. [118].

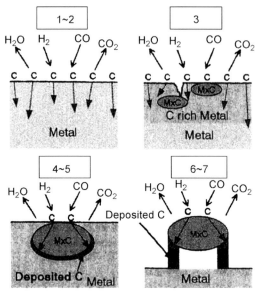

Figure 3.33 Generalised carbon deposition mechanism as proposed by Sone et al. [120].

formation is a two-phase procedure, where initially the metal acts as a catalyst forming a coke deposit, which is followed by the heterogeneous non-catalytic growth of coke molecules on their own surface by radical reactions [119].

Sone et al. proposed a generalised mechanism for carbon formation on metal surfaces from reformate containing hydrogen and carbon oxides [120], which is shown in Figure 3.33. It starts with adsorption of carbon oxides on the metal surface, the carbon atoms are separated from oxygen (step 1–2 in Figure 3.33). They diffuse into the metal surface and form carbides (step 3). The gases then adsorb onto the surface of these carbides, carbon diffuses into them (step 4–5) and precipitates again causing filamentous growth of carbon fibres (step 6–7). These workers investigated the behaviour of 19 wt.% Cr, 9 wt.% Ni stainless steel in a mixture of 75 vol.% hydrogen, 15 vol.% carbon monoxide and 10 vol.% carbon dioxide at 550 °C reaction temperature and 1 bar pressure. In this reaction mixture, it was mainly the water–gas shift reaction and the Boudouart reactions that were expected to take place. The water–gas shift reaction was expected to be much faster. Thus the C–H–O phase diagram was calculated for the reaction conditions provided above, which is shown in Figure 3.34. Carbon filament growth was observed at the stainless steel surface. The filaments had diameters between 7 and 100 nm. In-depth profiles of the stainless steel in the fresh state and after reaction showed an increase in the chromium content on the metal surface, while the iron content decreased. Thus iron was assumed to form the metal carbides preferably, hence its concentration on the surface decreased. Nickel was not found at the steel surface regardless of the reaction time.

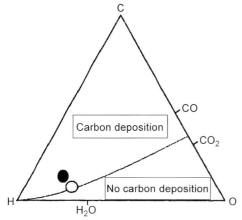

Figure 3.34 C–H–O phase diagram for a mixture of 75 vol.% hydrogen, 15 vol.% carbon monoxide, the balance being carbon dioxide; the boundary for carbon deposition is indicated by the straight line; (o) composition of natural gas reformate; (•) composition of water-free natural gas reformate [120].

Springmann *et al.* [56] observed significant isooctane conversion on the steel surface of their testing reactor at temperatures exceeding 800 °C under steam reforming conditions (S/C ratio 2.7). The main products were lower hydrocarbons.

4
Catalyst Technology for Distributed Fuel Processing Applications

4.1
A Brief Introduction to Catalyst Technology and Evaluation

Catalysts for heterogeneous reactions usually consist of a carrier material and one or several active species. In most instances the fine dispersion of the latter is a critical issue in catalyst technology. The finely dispersed active species form the so-called active sites of the catalyst. The dispersion is achieved in many cases by providing a high surface area of the carrier material. The most frequently used carrier material for catalysts is alumina, which can easily reach specific surface areas of up to $300 \, m^2 \, g^{-1}$. However, even higher values, exceeding $750 \, m^2 \, g^{-1}$, may be achieved by other carriers such as silica or zeolites (silica aluminates). The active species are all types of metals and metal oxides. Thus one possible method of bringing the active species onto the carrier material is to generate a solution of the metal as a soluble metal salt and impregnate the carrier material with this solution. Usually a temperature treatment termed calcination is then performed with the impregnated catalyst, the conditions of which are crucial in order to achieve optimum and stable dispersion of the active species. Characterisation methods do exist to measure the surface area of the catalyst and the dispersion of the active species, which is frequently distributed in small metal clusters of nanometre size.

The most important parameters used to evaluate catalyst performance are conversion, yield and selectivity. In a specific chemical reactor, a difference between the concentration $c_{a,0}$ of species A fed to the reactor, and its concentration at the reactor exit $c_{a,1}$ is achieved through the chemical reaction. The conversion is the ratio of this difference to the concentration of the species in the feed, usually provided in percent, sometimes also as degree of conversion, in the latter as a number between 0 and 1:

$$X[\%] = \frac{c_{a,0} - c_{a,1}}{c_{a,0}} \cdot 100 \qquad (4.1)$$

By chemical reaction, different products (here termed as examples B and C) are formed according to the rules of stoichiometry and appear with a certain concentrations c_j at the reactor exit:

$$n_a A \rightarrow n_b B + n_c C \qquad (4.2)$$

Fuel Processing for Fuel Cells. Gunther Kolb
Copyright © 2008 WILEY-VCH Verlag GmbH & Co. KGaA, Weinheim
ISBN: 978-3-527-31581-9

Basically, it makes sense only to provide yields when the different products are formed competitively from the feed molecule. The yield Y, normally expressed in percent, is the ratio of the concentration of a product species determined at the reactor exit to the concentration of the feed species at the reactor inlet normalised by the stoichiometric factors:

$$Y[\%] = \frac{n_a}{n_b} \frac{c_{b,1} - c_{b,0}}{c_{a,0}} \cdot 100 \tag{4.3}$$

Finally, the selectivity S is the ratio of yield to conversion. It illustrates the distribution of the converted species to the different products when taking stoichiometric factors and degree of conversion into consideration:

$$S[\%] = \frac{Y}{X} = \frac{n_a}{n_b} \frac{c_{b,1} - c_{b,0}}{c_{a,0} - c_{a,1}} \cdot 100 \tag{4.4}$$

4.1.1
Catalyst Activity

Specific surface area and active species dispersion of a catalyst are not the only features by which its performance are judged. On the contrary, it might be misleading to assume that high surface area and dispersion alone guarantee the best performance. The most important feature is the overall activity.

Basically, catalyst activity is defined as a ratio of reaction rates. To quantify the deactivation behaviour of a catalyst, the rate of the desired reaction at a certain time on stream $r(t)$ is related to the rate of reaction as determined for the fresh catalyst. This leads to the definition of activity:

$$a(t) = \frac{r(t, \underline{c})}{r(t = 0, \underline{c})} \tag{4.5}$$

As indicated in Eq. (4.7), the concentration vectors, which include all species concentrations in the reaction system, need to be identical to allow the correct determination of activity. This is hard to achieve even in a laboratory environment, because the deactivation procedure obviously changes the concentrations. Thus much simpler means are required to judge catalyst activity in practical use, which will be discussed briefly in Section 4.1.2 below.

One possible method of comparing the activity of different catalysts is to provide the flow rate under which complete and stable conversion is achieved over the catalysts. When this flow rate is related to the catalyst volume or mass, it is termed space velocity. Numerous definitions may be applied to define space velocity:

- The Gas-Hourly Space Velocity (GHSV) is the simplest definition, and is frequently provided for commercial systems, because catalysts are usually used as fixed beds.

It is defined as the ratio of the volume flow of feed under normalised conditions to the catalyst volume:

$$GHSV\ [\mathrm{h}^{-1}] = \frac{\dot{V}_{feed}}{V_{catalyst}} \quad (4.6)$$

It is easy to use this definition for large scale fixed beds, because the porosity of the catalyst and of the catalyst bed are easy to measure and in most instances the reactor housing contributes to the overall volume only to a minor extent. However, the definition becomes doubtful when the catalyst is coated onto a monolith or foam. Here the question arises as to which volume is then being referred to: the volume of the catalyst coating itself, of the monolith void fraction (channels) or of the entire monolith? All these questions need to be clarified before a fair comparison of catalytic activity is feasible when gas hourly space velocity is applied for the calculations.

The inverse value of gas hourly space velocity is known as the residence time τ.

- The best definition is the Weight Hourly Space Velocity (WHSV), which is defined as the ratio of the volume flow of feed calculated back to normalised conditions to the weight of the catalyst:

$$WHSV\ [\mathrm{L(h\ g_{catalyst})}^{-1}] = \frac{\dot{V}_{feed}}{M_{catalyst}} \quad (4.7)$$

This value is independent of the way in which the catalyst is introduced into the system, whether it is as a coating or a pellet. The inverse value of the weight hourly space velocity is termed the modified residence time.

- Amongst other definitions, the Turn-Over Frequency (TOF) is frequently provided. It is the ratio of the number of feed molecules converted per hour to the number of atoms of the active species:

$$TOF\ [\mathrm{h}^{-1}] = \frac{\dot{n}_{feed}}{n_{active\ species}} \quad (4.8)$$

This value is less convenient for practical applications, because calculations back to technical conditions are required.

For all of the definitions above, care has to be taken as far as the feed volume flow is concerned, because it is sometimes related not to the total feed flow rate but to the flow rate of the key reactant under consideration. For example, the fuel alone, but not the air required to combust it, may be used for the calculation. However, in most instances the total feed flow rate is used.

The same applies to the mass of the catalyst, which may sometimes not be related to the overall mass of the catalyst but to the mass of the active species, such as the precious metal deposited onto the carrier. For ceramic monoliths it is actually the convention to provide the mass of the entire monolith as "catalyst mass", which makes, however, comparison of performance data almost impossible.

In this book the weight hourly space velocity will be used as the measure for comparing the performance of different catalysts. The total feed flow rate will be used for its calculation. Literature data have been converted into weight hourly space velocity as far as possible. The catalyst mass is defined as the mass of carrier and active species.

4.1.2
Catalyst Stability

The durability of the catalyst is another issue that needs to be addressed carefully. It is affected by various factors.

Poisoning of the active sites originates from either permanent or reversible adsorption of species present in the feed at the active sites of the catalyst. One typical poison in the field of fuel processing is carbon monoxide, which blocks precious metal sites such as platinum.

Various types of catalyst sintering exist. There has to be careful distinction between sintering of the carrier itself and sintering of the dispersed active species. The former is a procedure that usually takes place at elevated temperatures through phase transition of the carrier material and with loss of surface area as a typical result. Hence, it is mostly limited to high temperature reaction systems performed at temperatures exceeding at least 600 °C. For example, alumina is present as the so-called γ-alumina phase at temperatures between 750 and 800 °C.

This phase typically has a surface area of around $80 \, m^2 \, g^{-1}$ or higher. When it is exposed to temperatures greater than 800 °C, a transition to δ-alumina takes place, around 1000 °C θ-alumina is formed and above 1200 °C α-alumina is the stable species, which has a surface area of around $10\text{--}20 \, m^2 \, g^{-1}$ or less, obviously resulting in a loss of the original high surface area of the catalyst.

γ-alumina and silica are the most stable carriers under an oxidising atmosphere [121].

However, sintering of dispersed metal may take place at much lower temperatures, examples are the sintering of platinum under the conditions of the water–gas shift (see Section 4.5.1) and sintering of gold when applied as a catalyst for the preferential oxidation of carbon monoxide at temperatures well below 150 °C (see Section 4.5.2)

Coke is a general name for carbonaceous species, which are formed by reaction of a hydrocarbon or alcohol fuels during the reforming process itself. Further details of coke formation of reforming catalysts will be discussed in Section 4.2.11.

Sintering procedures usually lead to a linear loss in activity and a decline in the conversion, whereas coking procedures are normally described by exponential or power-law equations. Even though the activity may decline only to a certain degree when coke formation occurs, the continuing growth of coke on the catalyst surface presents the danger of complete blockage of the catalyst pores or even of the flow paths of the reactor.

There are various ways of judging the stability of a catalyst. The most obvious and convincing method is the performance of long-term tests under realistic feed conditions. However, the effect of feed impurities and temperature cycling owing to start-up/shut-down procedures during real operation need to be taken into consideration for these tests, especially for smaller scale systems such as fuel processors. The duration of these tests, of course, strongly depends on the application. The typical duration of a car engine does not exceed 2 000–4 000 h with a lifetime, normally, of 10 years. Higher values are required for trucks of course. For residential heating systems, the durability required is usually in the range of at least 20 000 h and the reliability requested by the customer is even higher for obvious reasons.

A more convenient method, especially for catalyst development, is accelerated aging under harsh conditions. For example, treatment with steam at temperatures exceeding the typical operating conditions by several hundred Kelvin for one day may simulate a much longer exposure to realistic operating conditions. However, sound experience and data for comparison are required to judge the results of such accelerated tests because they obviously deviate significantly from reality.

4.1.3
Catalyst Coating Techniques

The most prominent coating technique for catalysts is wash-coating, which allows both the coating of a catalyst carrier and the coating of a ready-made catalyst onto a substrate. It may be applied for ceramic and metallic monoliths (see Section 6.2) as well as for plates of a stainless steel or aluminium plate heat-exchangers.

Materials that are routinely coated with catalyst wash-coats are ceramics such as cordierite, which is the construction material of ceramic monoliths, metals such as Fecralloy, the construction material of metallic monoliths (see Section 6.2) and stainless steel [57]. The amount of catalyst material that can be coated onto a monolith ranges between 20 and 40 $g\,m^{-2}$, while plate heat-exchangers may even take up more catalyst when coated prior to the sealing procedure, because the access to the channels is better.

When metallic surfaces are coated with catalyst, a pre-treatment to improve the adherence is usually applied [122]. Besides mechanical roughening, chemical and thermal pre-treatments are methods that are frequently used. Fecralloys, which are the construction material for metallic monoliths, are usually pre-treated at about 900–1000 °C, because they form an alumina layer of about 1-µm thickness on their surface. This is an ideal basis for catalyst coatings. However, metal oxide layers are formed on stainless steel and may also serve as an adhesion layer.

Aluminium substrates are frequently pre-treated by anodic oxidation to generate a porous surface, which may serve as the catalyst support itself or as an adhesion layer for a catalyst support [122]. The surface area of the obtained alumina layer may reach 25 $m^2\,g^{-1}$, with a thickness of 70 µm or higher. Assembled alumina microchannel reactors can also be oxidised. The layer generated by anodic oxidation of aluminium and aluminium alloys is amorphous hydrated alumina. It is believed to contain boehmite, pseudoboehmite and physically adsorbed water [123]. It has the morphology of a packed array of hexagonal cells, each containing a pore in the centre. Thus the layer has a highly ordered porous structure and uniform thickness, when an equally sized anode is positioned close to the oxidised surface. However, in order to apply anodic oxidation as a cheap method suitable for the mass production of ready-made microreactors, which are mounted and sealed, e.g., by diffusion bonding, the oxidation procedure requires some modification. Wunsch et al. demonstrated anodic oxidation of aluminium or aluminium alloys (AlMg$_3$) in 1.5% oxalic acid for a complete stack of microstructured foils [124]. Aluminium wires were positioned at the inlet and outlet of the stack, which served as the cathodes. After 3–4 h a satisfactory even film distribution in the range of 10–12 μm was achieved up to 14 mm deep into the axial direction of the stack of channels. The film thickness grew almost linearly with time. Deeper inside the channel system, a decrease in the layer thickness from 7 to 3 μm at 15 and 40 mm, respectively, was determined. However, owing to the inaccessibility of the catalyst generated *in situ*, surface area data of the layers were available as surface magnification factors (400 m^2 m^{-2}) only [124]. Ganley et al. demonstrated a method to increase the surface area and thickness of the alumina layers that had been generated by anodic oxidation [123]. They found that the thickness of the oxide layer depended not on the concentration of oxalic acid but only on the anodisation potential and electrolyte temperature, whereas the pore size depended on the concentration of oxalic acid and on anodisation potential; 30 V, 0.6 M oxalic acid concentration and 18 °C electrolyte temperature were identified as the optimum conditions. By a two-step anodisation and hydrothermal treatment of the alumina layers, up to 60-μm thick coatings were achieved. A hydrothermal treatment was then performed by immersion in water followed by drying and dehydration in air at 550 °C for 16 h. This increased the surface area of the layer by a factor of 10. In this way, the boehmite alumina layer formed after oxidation was converted into δ- and γ-alumina [123].

Gorges et al. introduced anodic spark deposition as a modification of anodic oxidation for the formation of polycrystalline ceramic oxide layers on passivating metals [125]. The method was therefore limited to titanium, aluminium and zirconium. The voltages exceeded 100 V. By local melting of the oxide layer into the substrate, strongly adhering coatings of up to 40 μm thickness could be generated. A titania coating was generated, which was composed of rutile and anatase and had a surface area of 51 m^2 g^{-1}. A ceramic green body was coated with titanium by physical vapour deposition. Anodic spark deposition was then performed on the titanium.

Presting et al. from Daimler Chrysler reported the anodic etching of silicon as a method of introducing porous catalyst carriers into microreactors on the smallest scale [126]. A pore depth of 250 μm was achieved by these means, the pore diameter was as low as 1.1 μm and the pore wall thickness much less than 100 nm (see Figure 4.1).

Once the surface is pre-treated, the coating slurry needs to be prepared. The most frequently used method is to prepare a dispersion of the finished material, which sometimes includes gelification steps [122]. Ceramic monoliths are usually washcoated by these means. The catalyst carrier or the catalyst itself [127] is mixed with a binder, such as poly(vinyl alcohol) or methylhydroxyethyl cellulose [128], acid and solvent, usually water. Lower particle sizes improve the adhesion [129,130]. For ceramic monoliths, the particle size should be in the same range as the

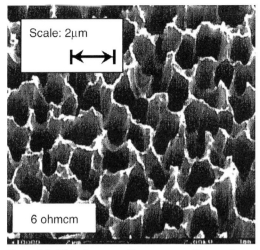

Figure 4.1 SEM of porous silicon generated by anodic etching [126].

macro-pores of the ceramic support [130]. A carrier material such as alumina reacts with the acid, which increases the pH value of the slurry over the course of time. This aging procedure may last more than 24 h [131]. Alumina powder only partially reacts with the acid and creates a gel. Cristiani *et al.* observed that up to almost 20 wt.% of their γ-alumina powder reacted to give gel within 24 h and the viscosity of the slurry increased with time, then decreased again after 27 h to reach a stable value after 2 days [131].

As an example, the wash-coating method for microstructured steel substrates as used by Zapf *et al.* will be described. Microstructured stainless steel plates (X2CrNi-Mo17-12-2 and MoTi17-12-2) structured by photochemical etching were used. The slurry was composed of 20 g γ-alumina (average particle size 3 μm), 75 g deionised water, 5 g poly(vinyl alcohol) binder and 1 g acetic acid. The suspension was stirred at 60 °C for 2 h and left overnight in a vessel for the air bubbles to be released. Alumina and acetic acid were added and stirred overnight. The microstructured substrates were cleaned with isopropanol in an ultrasonic bath. Areas of the substrate that should not be coated with the catalyst were protected by a mask. The channels were filled completely and surplus slurry was removed. After drying of the slurry, the layer thickness in the channels had shrunk and an open cross-section not filled with catalyst had developed. The final coating thickness could be adjusted by the alumina content of the aqueous suspension. The wash-coats were calcined at 600 °C and a porous coating with a specific surface area of approximately 70 m^2 g^{-1} was achieved. A falling test was developed and applied for the coatings. At an impingement velocity of 3 m s^{-1} losses of the wash-coat did not exceed 1%. The deviation of the coating thickness over a group of parallel microchannels at the same axial position was less than 5%. To achieve coating profiles of equal thickness in the radial direction, an optimisation of the slurry viscosity was required. Wash-coatings of various aluminas were prepared using this procedure.

Meille et al. demonstrated that the slurry viscosity determines the thickness of the coating [132]. Similar results were found by other workers [133, 134]. The viscosity itself is determined by the concentration of the particles, pH value and on the surfactant added [132].

Additives to suppress particle agglomeration may be added to the suspension. This is crucial for low particle sizes [129]. Pfeifer et al. described a technique for wash-coating copper/zinc oxide catalysts onto aluminium microchannels [135]. Copper oxide nanoparticles of 41 nm average particle size were mixed with zinc oxide nanoparticles of 77 nm average particle size either by wet mixing with aqueous hydroxy ethyl cellulose or hydroxy propyl cellulose in isopropyl alcohol. Alternatively, the particles were typically milled and then dispersed in aqueous hydroxy ethyl cellulose. The dispersion then filled in the microchannels, resulting in a catalyst layer of 20 µm thickness, which was then calcined in air at 450 °C. The surface area of the samples was around $20 \, m^2 \, g^{-1}$.

Sol–gel methods do include a gelation procedure, also known as peptisation of the sol. The time demands for this procedure may vary considerably from h to weeks. A sol is prepared by polycondensation of alkoxides [136]:

$$M(OR)_n + nH_2O \rightarrow M(OH)_n + nROH; \quad M = Al, Si, Ti, Zr \tag{4.9}$$

$$M - OH + HO - M' \rightarrow -M - O - M' - + H_2O \tag{4.10}$$

Alumina sol may be prepared from aluminium alkoxide or pseudo-boehmite [AlO(OH)xH$_2$O]. Addition of additives such as urea provides the porosity of the catalyst layer through thermal composition during calcination [136]:

$$NH_2CONH_2 + H_2O \rightarrow 2NH_3 + CO_2 \tag{4.11}$$

The sol then serves as a binder for the particles, which form the coating. Active metals can be readily incorporated into the sol. Usually sol–gel methods produce coatings of lower thickness in the range of a few µm. Therefore, hybrid methods between sol–gel and wash-coating are sometimes applied, which make higher coating thicknesses possible. Hwang et al. used an alumina layer prepared by the sol–gel method as the adhesion layer for their catalyst, which was coated onto silicon microchannels [134].

Haas-Santo et al. optimised the operating conditions for sol–gel coating onto microstructured metal foils [137]. Alkoxides such as aluminium-sec-butylate were used to generate titania, alumina and silica coatings. Acetyl acetone was used as the stabiliser and nitric acid as the catalyst, whereas an alcohol served as the solvent instead of water. Fecralloy foils (German code 1.4767) with an aluminium content of 5 wt.% were treated for 15 h at 1000 °C before coating. In this way an alumina layer of about 1 µm thickness was created on the steel surface, as is known from Fecralloys. Dip-coating was used for the single metal foils, whereas stacks of foils were coated by pipetting the sol into the channel system placed into a vertical position. To remove the carbonaceous species from the coatings, a temperature treatment at 500 °C was required. Higher temperatures decreased the surface area of the coatings, which was attributed to alumina sintering and phase transition to corundum, especially at temperatures exceeding 800 °C. Sols made from aluminium-sec-butylate mixed with acetyl acetone in ethanol were stable for months and formed coatings of similar surface area and pore volume distribution. The viscosity of silica sols was very sensitive to pH and to the aging time. The surface area of silica sol coatings increased with increasing pH from $250\,m^2\,g^{-1}$ at pH 6 to $800\,m^2\,g^{-1}$ at pH 8. However, above pH 8 gelation took place immediately. Thus, a trade-off was required between time demand for coating and coating quality, because gelation took place earlier for higher pH and surface areas. Alumina coatings were deposited into a ready-mounted microstructured reactor. The coatings had a thickness of 2–3 µm.

The sol–gel coating technique may be well suitable for the coating of highly porous substrates such as metal foams, because it generates uniform coatings, which are difficult to achieve by wash-coating using foams.

Bravo *et al.* described a catalyst coating technique, which was applied for coating of commercial copper/zinc oxide catalyst onto quartz and fused silica capillaries [138]. The catalyst was milled with boehmite alumina and deionised water in a mass ratio of 44:11:100. The thickness of the coating, which was only 1 µm for this gel formulation, could be increased to some 25 µm by addition of hydrochloric acid. The catalyst was still active after the coating procedure. The development work was moving towards coating of ready-made microreactors in the future [138].

Once the wash-coat slurry or sol is prepared, it needs to be deposited onto the substrate. This is frequently achieved by dipping the substrate into the slurry or sol. This is the usual procedure for ceramic automotive exhaust treatment monoliths. The slurry excess is then removed by blowing off with air. An alternative is spray-coating, which requires a reduction of the viscosity of the slurry or sol [122]. Spray-coating has been used to improve the coating distribution on Fecralloy fibres [132]. Flame spray deposition was used by Thybo *et al.* for microchannels [139]. The organometallic precursor compounds are guided through a flame and then driven toward the surface of the cooled substrate by thermophoresis. Spin coating was also used for wafer substrates.

Electrophoretic deposition is a colloidal process used to coat either aluminium or stainless steel [122]. Either adhesion layers or complete coatings can be achieved. To-date mainly alumina suspensions have been used for electrophoretic deposition. Wunsch *et al.* coated microchannels with alumina nanoparticles dissolved in oxalic acid and mixed with alumina gel or glycerol [124]; 2–4 µm thick adhesion layers were formed. Födisch *et al.* applied electrophoretic precipitation of industrial catalyst powders at 100 V (dc) [140]. After 2 min a uniform deposition of the catalyst powder on the surface was achieved.

Pfeifer *et al.* performed electrophoretic deposition of alumina nanoparticles onto a stack of microstructured metal foils [141]. The alumina nanoparticle dispersion was pumped through the stack for 20 h while the voltage was set to 50 V at $0.5\,A\,cm^{-2}$. The composition of the dispersion best suited to achieving well-adhering, homogeneous coatings was water and glycerol. A surface enhancement factor of $102\,m^2\,m^{-2}$ and 3-µm coating thickness were achieved.

Electrochemical deposition, also known as electroplating, generates metallic or metal oxide coatings from their metal salts [122]. It can be used when metallic surfaces are required as the catalyst but also to generate metal oxide layers such as zirconia and lanthana. Electroless plating uses redox reactions to deposit metals. It is also used to generate metallic membranes for hydrogen separation (see Section 7.4).

After the deposition, drying and calcinations steps usually follow, the latter is a temperature treatment in air or other gases for a defined duration. Calcination may or may not include temperature ramps. However, normally the dried samples are not immediately put into a hot furnace, but, rather, are heated up gradually. The final temperature of calcination needs to ensure that organic materials such as binders are removed completely. Many binders decompose entirely by 350 °C, however, others such as poly(vinyl alcohol) may require significantly higher temperatures [128].

Then, for a catalyst carrier that has been coated onto the surface, and not for a ready-made catalyst, the active species of the catalyst has to be impregnated onto the carrier. For porous monoliths and other high surface area materials such as glass fibres [122], a catalyst carrier is not required and thus the active species may be impregnated immediately onto the uncoated substrate. Other methods of incorporating active species are crystallisation and precipitation, to name but a few.

A second drying and calcination procedure finalises the catalyst preparation where an impregnation has been performed. In a thorough study, Villegas et al. described that the drying procedure is crucial for the homogeneity of the active metals on the catalyst carrier [142]. Microwaves and room temperature drying resulted in a more homogeneous distribution of nickel on cordierite monoliths compared with oven drying at 100 °C. Surface enrichment was observed for all drying procedures. A higher content of large nickel particles was evident for the samples dried by microwaves and in an oven.

Apart from catalyst coatings, incorporated catalysts can be introduced into high surface area ceramic monoliths. Incorporated catalysts are prepared by either incorporating the catalyst into the material before extrusion (see Section 6.2) or by impregnation of a high surface area monolith body with the active species of the catalyst. In this instance, the monolith itself thus forms the catalyst carrier. However, this type of ceramic monolith has lower thermal resistance compared with low surface area monoliths.

Kim and Kwon reported an interesting coating strategy for microreactors [143], which is shown in Figure 4.2. The catalyst was coated onto a glass substrate by introducing a layer of poly(vinyl alcohol) onto the glass (step 4), which had previously been covered with photoresist (PR, step 2) and structured by lithography (step 3). After lift-off of the photoresist with acetone (step 5), the poly(vinyl alcohol) remained only where no photoresist had been present before. The milled catalyst was mixed with bentonite binder in a ratio of 2.5:1 and a 7 wt.% slurry was prepared by sonication and stirring. The catalyst was then precipitated onto the poly(vinyl alcohol) layer (step 6) followed by subsequent drying. After removal of the poly(vinyl alcohol) by calcination at 550 °C (step 7), only the remaining part of the substrate was covered with the catalyst. The amount of catalyst deposited into the microchannels was in the range of from 50 to 80 g m^{-2}, which is similar to other coating procedures described above. The thickness of the catalyst layer was determined to be 30 μm. However, Kim et al.

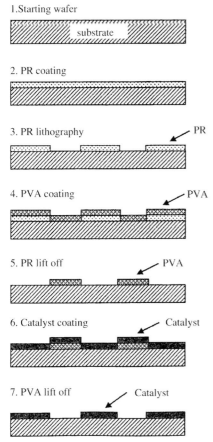

Figure 4.2 Catalyst coating technique as developed by Kim and Kwon [144].

report that they could increase the amount of catalyst in the channels up to $150\,\mathrm{g\,m^{-2}}$ by repeating the precipitation procedure. The adhesion of the catalyst coating could be improved by controlling the parameters of precipitation. According to Kim *et al.*, the quality of the coatings achieved by this precipitation method was superior to coatings gained by spin-coating or dip-coating.

Thin layers of catalyst can be deposited onto the surface of silicon microchannels by physical vapour deposition. Silicon is the preferred material, because the equipment for physical vapour deposition is available at microelectronics fabrication sites, which can also produce silicon microreactors. Physical vapour deposition such as cathodic sputtering, electron beam evaporation and pulsed laser deposition but also chemical vapour deposition create uniform metal surfaces with thicknesses in the nm range. Such coatings are rarely suitable as catalysts. However, a few exceptions such as hydrogen oxidation [145] and reactions in the very high temperature range do exist.

4.1.4
Specific Features Required for Fuel Processing Catalysts in Smaller Scale Applications

In the open literature, frequently catalyst formulations are applied for fuel processing that has been derived from technology which had been developed for large scale industrial processes. In these processes, only a minor part of the catalyst pellet is utilised in many instances. The reaction is limited by intraparticle (or intrapellet) diffusion. The effectiveness of the utilisation of nickel steam reforming catalysts for industrial processes was calculated to about 20% by Christensen [81]. This low catalyst utilisation is usually counterbalanced by a surplus of catalyst and the low cost of the catalyst itself. Non-precious metals such as nickel for steam reforming (see Section 4.2) or copper/zinc oxide for a water–gas shift (see Section 4.5.1) make low prices possible. The situation is different when the catalysts are dedicated to fuel cell applications on a smaller scale, because the mass of the catalyst and the size of the reactor are critical issues. Thus, formulations of higher activity, such as precious metals, may well be required despite the higher cost, and this has to be counterbalanced by better utilisation of the catalyst. Automotive exhaust systems, which rely on precious metal catalyst technology, are a widespread application with similar requirements and technical solutions.

Better utilisation of the catalyst can be achieved by coating a thin layer onto inert non-porous carrier bodies, as described in Section 4.1.3. The resulting catalyst coating has reduced dimensions for the molecular diffusion of the reactants. Up to two orders of magnitude lower catalyst mass is required according to theoretical calculations, owing to the full utilisation of the catalyst [146].

Transport limitations also limit the catalyst effectiveness for reactions in the lower temperature range. Water–gas shift was already mentioned above and methanol steam reforming is another example. Karim *et al.* performed methanol steam reforming over commercial copper/zinc oxide catalysts in fixed beds of different diameter [147]. When decreasing the diameter of the fixed bed from 4 to 1 mm, increasing conversion was observed, indicating reduced transport limitations. In another paper, Karim *et al.* calculated temperature gradients in a small fixed bed of only 1 mm diameter, which were in the region of 20 K, as shown in Figure 4.3 [148]. When small channels in the size range from 500 µm to 4 mm were coated with catalyst, no mass and heat transport limitations could be observed for coating thicknesses in the range between 5 µm and 25 µm. Calculations revealed that even for a coating thickness of 95 µm no transport limitations should be expected [147]. However, mechanical stability issues of catalyst coatings limits their thickness to about 50 µm in most instances (see Section 4.1.3). Purnama *et al.* reported reduced methanol conversion by mass transport limitations for methanol steam reforming in catalyst particles, with diameters of 700 µm to 1000 µm [149].

Because conditions are less well defined in small energy generation devices during shut-down and start-up, the catalysts need to be more robust towards exposure to air and moisture compared with a catalyst dedicated to large scale production. Reactors of industrial scale are ideally running under constant conditions for several years. For example, during start-up of a small system, reduction of the catalyst with hydrogen,

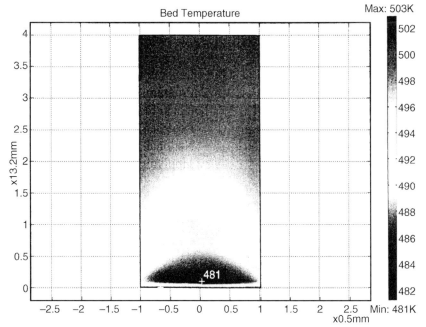

Figure 4.3 Two-dimensional temperature profile as calculated by Karim et al. for a small fixed bed catalyst during methanol steam reforming: flow velocity, 0.0184 m s^{-1}; catalyst particle size, 150 μm; weight hourly space velocity, 5.3 L (h g$_{cat}$)$^{-1}$.

a pre-treatment frequently applied in industrial systems, is not possible at all, because no hydrogen is available. On the contrary, it might be necessary to heat up the reactor by hot air or combustion gases, before the reforming process is started and the catalyst activity must not suffer from this oxidising treatment in a practical system. All these aspects require consideration when selecting a catalyst formulation for a fuel processor smaller than one on an industrial scale.

4.2
Reforming Catalysts

Catalyst formulations suitable for various fuels are discussed below. When fixed beds or monolithic reactors are applied, a hot spot is formed at the reactor inlet for partial oxidation, because the reaction is highly exothermic. This hot spot is less pronounced for oxidative steam reforming, due to the presence of steam.

Very early work of Flytzani-Stephanopoulos and Voecks [74], which had been performed in the early 1980s, has already dealt with the hot spot formation over various nickel based catalysts under conditions of autothermal reforming of hexane, tetradecane and benzene. The hot spot was determined to be 1030 °C for benzene and to 930 °C for tetradecane at an O/C ratio of 0.67 and a low S/C ratio 0.6. The gas hourly space velocity was set to 9000 h^{-1}, the pressure was close to ambient and the feed was pre-heated to 560 °C. The lower temperature observed for paraffinic feed was explained by cracking reactions, which produced less energy compared with partial oxidation [74].

Cheekatamarla and Lane [8] observed a temperature rise from 400 °C to less than 800 °C in an adiabatic testing reactor for autothermal reforming of diesel fuel.

Roychoudhury et al. [150] determined a hot spot up to 950 °C when operating their diesel autothermal reactor at an S/C ratio of 1.44 and an O/C ratio of 0.96, which is close to full partial oxidation conditions. During start-up and without steam addition the hot spot even approached 1100 °C.

Hot spot formation was observed by Rampe et al. [151] up to 900 °C for their monolithic propane reformer when operated at a 2.7 kW power equivalent.

Aicher et al. [72] measured the temperature profile in their autothermal diesel reformer reactor (see Figure 4.4). Temperatures up to 900 °C were detected upstream of the catalyst honeycomb, while the reformate temperature never exceeded 700 °C at the reactor exit.

Simulation work of Springmann et al. [152] revealed a gas phase temperature up to 900 °C for autothermal reforming of gasoline in a metallic monolith. However, the temperature of the catalyst itself was much higher and reached 1000 °C.

The hot spot formation and the high operating temperature of partial oxidation led to the development of high temperature resistant catalyst formulations based on perovskite structures. $CaTiO_3$ is natural perovskite. Perovskites are formulations of the type ABO_3, where the site A is a metal with a larger ionic radius, typically of

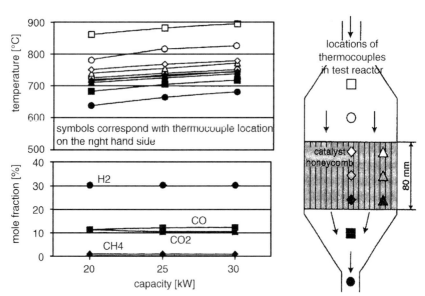

Figure 4.4 Temperature profile as determined by Aicher et al. in a steam supported partial oxidation reactor: S/C = 1.4; O/C = 0.98 [72].

the rare earth type, and the cation B, typically a transition metal, is substituted partially, which creates lattice defects. These defects in turn create the catalytic activity.

Common catalyst carriers such as alumina may also be stabilised against extremely high temperatures by addition of so-called promoter metals, which will be discussed below.

Another important issue of reforming catalysts is their tendency towards carbon formation. It is important to know that carbon formation is frequently observed when the thermodynamic equilibrium still predicts that carbon is an unstable species [74,153].

4.2.1
Catalysts for Methanol Reforming

The equilibrium composition for methanol steam reforming is shown in Figure 4.5. It becomes obvious that the process should be operated at an S/C ratio of 1.5 at least, to minimise the carbon monoxide content of the reformate.

Copper/zinc oxide is the catalyst technology most frequently used for methanol steam reforming. Numerous publications have dealt with this type of catalyst and only a few are cited here. The maximum operating temperature of copper/zinc oxide catalysts for methanol steam reforming is limited to 300 °C.

Wiese et al. reported full conversion of methanol over a commercial copper/zinc oxide/alumina catalyst, with no further specifications given, at 280 °C and a maximum weight hourly space velocity of 5 LH_2 (h g_{cat})$^{-1}$ in a tubular fixed bed [154]. This is within the usual range of space velocities for this type of catalyst. The power density of the fixed catalyst bed was calculated to be 12.5 kW L^{-1}. The increase in carbon monoxide concentration at partial load, which is typical over copper/zinc oxide catalysts, is illustrated in Figure 4.6. Naturally, the lower the space velocity, the higher the conversion becomes. As soon as the space velocity is considerably below the value required to achieve full conversion, selectivity of the catalyst towards water–gas shift

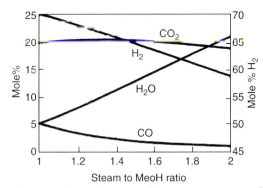

Figure 4.5 Thermodynamic equilibrium gas composition for methanol steam reforming versus S/C ratio: reaction temperature, 280°C; pressure, 5 bar [46].

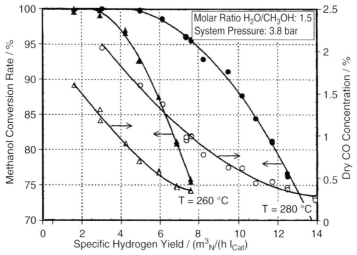

Figure 4.6 Hydrogen yield versus methanol conversion and carbon monoxide content in the dry reformate for a commercial copper/zinc oxide catalyst [154].

becomes significant, which in turn increases the carbon monoxide content. This issue was also discussed by Agrell et al. [155]. It makes high turndown ratios difficult. Long-term experiments performed with the catalyst revealed a slow deactivation, which was described as a gradual loss of activity beginning at the inlet of the catalyst bed. At 3.8 bar pressure, 280 °C reaction temperature and a S/C ratio of 1.5, a 5.5 mg h^{-1} loss of active catalyst was calculated at 3.8 mol h^{-1} feed flow rate.

Mass transport limitations also increase the selectivity of copper/zinc oxide/alumina catalysts towards carbon monoxide in larger particles [149]. Purnama et al. therefore proposed catalyst coatings as a means of reducing carbon monoxide formation.

Wiese et al. investigated the dynamic behaviour of the commercial copper/zinc oxide/alumina catalyst. Hydrogen production changed quickly, within 25 s, during load changes, while the carbon monoxide concentration followed slowly within 5 min. Thus, different reaction pathways were assumed by these workers for the formation of both species [154]. Wiese et al. observed up to 5 vol.% carbon monoxide in the reformate during start-up of their reformer, which was regarded as an inherent feature of the copper/zinc oxide/alumina catalyst. Wiese et al. assumed that methanol was converted by decomposition, which lead to the extensive carbon monoxide concentration observed. Dimethyl ether was also formed during start-up [154].

The main advantage of commercial copper/zinc oxide catalysts is their relatively high activity at operating temperatures below 300 °C. This advantage actually becomes a drawback, because the catalyst is not stable at temperatures exceeding 300 °C. Start-up procedures by hot combustion gases easily create temporary temperature levels above 300 °C, even if the catalyst is not directly exposed to oxygen containing gas mixtures. In other words, it is difficult to heat a reactor to an operating temperature close to 300 °C within a few minutes without exceeding this temperature at least locally. Furthermore, copper/zinc oxide catalysts are pyrophoric, which means they show temperature fluctuations when exposed to air in the active state. This originates from the oxidation of copper. On top of that, the copper/zinc oxide catalysts require pre-treatment in hydrogen to achieve full activity right after start-up, when they have been previously exposed to air. Although the full activity of the catalyst may be gained during the reforming process itself, the time demand for this process is considerable. Copper/zinc oxide catalysts are sensitive to sulfur and chlorine poisoning [156]. Lindström and Petterson reported severe deactivation of a commercial copper/zinc oxide/alumina catalyst during methanol steam reforming when it was exposed to 2 ppm sulfur in the methanol feed [156]. The zinc oxide is converted into zinc sulfide, similar to the process utilised for desulfurisation (see Section 4.4). Further problems with the practical operation of copper/zinc oxide catalysts, which are more related to their operation as catalysts for water–gas shift, will be discussed in Section 4.5.1.

However, copper/zinc oxide catalysts are also active for autothermal reforming of methanol [157], the copper is then present as the oxide. Some examples will be described below.

Men and Gnaser and coworkers investigated methanol steam reforming over copper/ceria/alumina catalysts [158–161] in a ten-fold screening reactor for microstructured plates developed by Kolb *et al.* [162]. The alumina carrier was wash-coated onto the plates followed by subsequent impregnation steps of ceria and copper salt solutions. The atomic ratio of copper/ceria was varied from zero to 0.9, while the total loading of copper and ceria was kept constant at 13 wt.%. Activity measurements were performed at a reaction temperature 250 °C, a S/C ratio of 1.1 and a rather high weight hourly space velocity of 320 L $(h\, g_{cat})^{-1}$. The lowest conversion was observed for pure ceria with a sharp maximum of about 50% conversion for an atomic copper/ceria ratio of 0.1. The carbon monoxide selectivity was well below 2% for all samples. Substantial formation of dimethyl ether was observed over all samples, the highest selectivity of 23% was found for pure ceria on alumina. The dimethyl ether formation was attributed to separate dehydration of methanol on the alumina surface. However, the steam reforming activity of the catalyst was attributed to the copper/ceria support interface. The improved dispersion of copper in the presence of ceria was supported by XPS (X-ray photon spectroscopy) measurements and both XPS and SEM-EDX (scanning electron microscopy–electron dispersive X-ray analysis) measurements suggested the enrichment of copper and ceria on the alumina surface. Excess copper, however, is known to form bulk particles, which do not contribute substantially to the overall catalyst activity. For low copper loadings an enhanced reducibility of copper was regarded as the possible origin of the higher catalyst activity. However, Cu^+ was believed to be the most favourable oxidation state to achieve methanol adsorption for this specific type of catalyst.

Dimethyl ether formation was also observed by Men *et al.* over copper/zinc oxide/alumina catalysts at lower values of the weight hourly space velocity [163]. A low weight hourly space velocity of 10.9 L $(h\, g_{cat})^{-1}$ was required at a S/C ratio of 2 in order to gain full methanol conversion without formation of any by-products such as dimethyl ether. Under these conditions, around 1.5 vol.% carbon monoxide was formed.

In general, copper-based catalysts are extremely sensitive to oxidising conditions, which makes their application for autothermal reforming of methanol difficult in practical systems [164]. The issue is that copper is present as copper oxide at the reactor inlet leading to methanol combustion, temperature fluctuations and severe sintering of the copper oxide [165]. The elevated temperature consequently deactivates the catalyst [166]. Thus, early publications dealing with autothermal reforming of methanol relied on dual catalyst beds such as platinum/alumina for the initial methanol oxidation followed by copper/zinc oxide/alumina catalyst for steam reforming [167].

Lindström [168] doped their self-developed copper/zinc oxide catalyst containing 40 wt.% copper and 60 wt.% metallic zinc with 2 wt.% zirconia. Then Lindström and Petterson [169] prepared copper/zinc oxide catalysts with different loading of both metals with counterparts, which were doped with 10 wt.% zirconia on top of copper and zinc oxide. A sample with a copper/zinc oxide ratio of 60:40 showed the highest activity, while the addition of zirconia decreased the activity, but decreased the carbon monoxide selectivity. Later, Lindström et al. investigated the performance of copper/zinc oxide, copper/zirconia and copper/chrome oxide catalysts, which had been coated onto ceramic cordierite monoliths for steam reforming and autothermal reforming of methanol [170]. The monoliths were coated with γ-alumina and then the various active species were impregnated onto the catalyst carrier. The catalysts were calcined at 350 °C [170]. The catalysts contained 10 wt.% of each active species on the carrier and the weight distribution between copper and the second species was 60:40 [170]. The activity of all catalysts increased with increasing oxygen addition to the feed. However, in most instances carbon monoxide selectivity was lowest under conditions of steam reforming. The copper/zinc oxide catalyst showed highest activity under conditions of steam reforming, while the copper oxide/chromium oxide catalyst had highest activity under conditions of autothermal reforming. The copper/zirconia catalyst showed the lowest carbon monoxide selectivity under all operating conditions. Jeong et al. also found lower carbon monoxide selectivity over their copper/zinc oxide/zirconia catalysts, compared with the zirconia-free counterparts for methanol steam reforming in microchannels, which was attributed to higher copper dispersion over the zirconia containing samples, which also had a significantly higher surface area [171].

Murcia-Mascaros et al. performed partial oxidation and autothermal reforming over self-developed and commercial copper/zinc oxide catalysts [172]. The temperature was limited to 300 °C to prevent degradation of the catalyst. At this temperature about 80% conversion was achieved at a low weight hourly space velocity of 36 L $(h\,g_{cat})^{-1}$ over the commercial catalyst. This was the case both for partial oxidation conditions (O/C ratios of 0.6 and 0.8) and autothermal conditions (O/C ratio 0.6 and S/C ratio 1.1). The lowest selectivity towards carbon monoxide was achieved under autothermal conditions.

Velu et al. reported the performance of copper/zinc oxide/alumina and copper/zinc oxide/zirconia catalysts for the autothermal reforming of methanol at S/C 1.3 and high O/C between 0.55 and 1.1 [173]. Zirconia turned out to be the better support showing the highest activity, and also the highest selectivity towards carbon monoxide. However, the higher selectivity towards carbon monoxide, which contradicts several findings described above, was not observed provided alumina was

 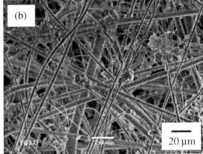

Figure 4.7 Optical and SEM image of copper/zinc oxide catalyst paper prepared by Koga et al. [175].

present in the catalyst. Lattner *et al.* also used a copper/zinc oxide/zirconia catalyst on an alumina support for the autothermal reforming of methanol [174].

Koga *et al.* [175] prepared copper/zinc oxide catalysts by a paper-making technique from catalyst powder and ceramic fibres in an alumina sol added to a pulp suspension. The resulting catalyst paper, which is depicted in Figure. 4.7, was compared with powder samples of the original catalyst. It showed an equivalent methanol conversion and hydrogen selectivity under conditions of autothermal reforming (S/C ratio 1.5, O/C ratio 0.25), but significantly lower carbon monoxide selectivity, which was attributed to improved transport mechanisms in the paper-like catalyst.

Palladium is the precious metal most frequently applied for methanol steam reforming [176–178]. Despite its higher price compared with the copper-based systems, it is an attractive alternative owing to the potential for higher activity and greater robustness, which are key features for small scale reformers. The combination of palladium and zinc showed superior performance and soon the formation of a palladium–zinc alloy was identified as a critical issue for optimum catalyst performance [179]. Besides palladium/zinc oxide, palladium/ceria/zinc oxide may well be another favourable catalyst formulation [177]. However, precious metal based catalysts have a tendency to show higher carbon monoxide selectivity than copper–zinc oxide catalysts, because it is a primary product of the reforming reaction over precious metals.

Pfeifer *et al.* investigated several copper/zinc oxide catalysts prepared from nanoparticles and copper/zinc oxide/titania catalysts, which were either prepared from sintered nanoparticles or from titania nanoparticles impregnated with copper/zinc nitrate [180]. These workers compared the performance of these catalysts with a palladium/zinc oxide sample, which consisted of zinc oxide nanoparticles impregnated with palladium acetate. All samples were coated onto steel foils carrying microchannels. The noble metal catalyst showed higher activity, but all samples suffered from deactivation. Because deactivation could be stopped by addition of air to the feed in case of the palladium catalyst, the activity loss was attributed to coke formation. Another drawback of the palladium catalyst was its higher carbon monoxide selectivity. The authors then focused on palladium/palladium–zinc/palladium/zinc oxide systems [181]. The formation of the palladium–zinc alloy at higher reduction temperatures was identified as the crucial feature in gaining lower carbon monoxide selectivity. In a later study, Pfeifer et al. [182] prepared palladium/zinc oxide catalysts both by pre- and post-impregnation of wash-coated zinc oxide particles with palladium. For a standard sample containing 10 wt.% palladium, the stability of the catalyst towards an oxidative atmosphere up to 200 °C was verified by X-ray diffraction. Pre-impregnation increased the carbon dioxide yields compared with post impregnation, which was observed for samples containing 10 wt.% palladium. The origin of the high carbon monoxide selectivity was finally attributed to an interaction of palladium with the metal foils during the post-impregnation procedure. This could be proven by the preparation of a powder catalyst (no coating). These samples showed carbon monoxide concentrations below the equilibrium of water–gas shift. Isolated Pd(0) was assumed to be responsible for the increased formation of carbon monoxide. For both preparation routes, the highest activity was determined for the samples containing 10 wt.% palladium, which were also the most stable against deactivation. The weight hourly space velocity amounted to 18 L $(h\, g_{cat})^{-1}$ for the activity tests. Dimethyl ether was observed only in trace amounts of a few hundred ppm by Pfeifer et al. over their palladium/zinc oxide catalyst [183].

Chin and coworkers [184, 185] studied palladium/zinc oxide catalysts for methanol steam reforming, which contained 4.8, 9.0 and 16.7 wt% Pd deposited onto zinc oxide powder by impregnation. Similar to the results of Pfeifer et al. discussed above, a palladium–zinc alloy was identified, which dispersed on the zinc oxide matrix under the conditions of methanol steam reforming and was once more recognised as the origin of the low carbon monoxide selectivity observed. In agreement with Pfeifer et al., the presence of metallic palladium was thought to be the origin of the high carbon monoxide selectivity. Interestingly, the palladium–zinc alloy was not only formed during the initial reduction step but also in situ in the hydrogen rich reaction mixture of methanol steam reforming, according to Chin et al. [186].

Palo et al. investigated methanol steam reforming at a 375 °C reaction temperature, applying proprietary catalyst formulations [187], which minimised the carbon monoxide concentration to 1.2 vol.%. This value was significantly lower compared with the 3 vol.% found for a copper/zinc oxide catalyst [188]. It could well be assumed that the proprietary catalyst was also a palladium/zinc oxide formulation.

Liu et al. observed significant activity gains up to a palladium content of 10 wt.% for their palladium/zinc oxide catalysts [189].

Over palladium/zinc oxide based catalysts, autothermal reforming of methanol is also feasible. Air addition forms heat and this generates an additional degree of freedom for the operation of the fuel processor. However, the advantage of low carbon monoxide formation, which was described for steam reforming over palladium/zinc oxide catalyst above, is lost under these conditions. This is believed to originate from methanol decomposition at the elevated reaction temperature of autothermal reforming [190]. The behaviour of the palladium catalyst is similar to rhodium and platinum catalysts under autothermal conditions [190]. Iwasa et al. tested catalysts containing 10 wt.% of palladium, platinum, iron, cobalt, nickel, iridium and ruthenium on zinc oxide under conditions of autothermal reforming [166]. The palladium containing sample showed highest activity at a 220 °C reaction temperature. At 300 °C, both platinum and palladium containing samples converted the methanol completely. A comparison of palladium catalysts supported by different carrier materials, namely zinc oxide, silica and ceria, revealed that only the palladium/zinc oxide sample showed significant activity and selectivity towards hydrogen, while the other samples produced mostly water and carbon oxides [166]. When increasing the palladium loading over zinc oxide supported samples from 1 to 10 wt.%, the activity of the catalysts increased with increasing palladium content, while the carbon monoxide selectivity decreased both under conditions of steam reforming and autothermal reforming. Addition of small amounts of copper, iron or chromium might decrease the carbon monoxide selectivity of palladium/zinc oxide catalysts according to Liu et al. [191].

Chen et al. prepared a hybrid copper/zinc oxide/alumina/palladium/zinc oxide catalyst by wash-coating a copper/zinc oxide catalyst supported by alumina into microchannels [192]. Palladium/zinc oxide powder was then coated onto this catalyst. The activity tests revealed complete methanol conversion at an S/C ratio of 1.2, a rather high O/C ratio of 0.6 and a gas hourly space velocity up to 15 000 h^{-1}. Despite the very high reaction temperature of between 450 and 600 °C and low residence time of 7.2 ms, the carbon monoxide content in the reformate was well below the thermodynamic equilibrium, namely 2 vol.% at 450 °C.

Borup et al. [193] demonstrated that it is possible to heat up an autothermal methanol reformer equipped with a precious metal based catalyst from room temperature, when the O/C ratio of the feed exceeds 1.45. The S/C ratio was set to 1.0 for these investigations. The exothermic oxidation reactions clearly started even at ambient temperature and caused light-off of the reformer.

Partial oxidation of methanol is less frequently reported in the open literature. Cubeiro et al. investigated the performance of palladium/zinc oxide, palladium/zirconia and copper/zinc oxide catalysts for partial oxidation of methanol in the temperature range between 230 and 270 °C [194]. Increasing selectivity towards hydrogen and carbon dioxide was achieved with increasing conversion, while selectivity towards steam and carbon monoxide decreased. The palladium/zinc oxide catalyst showed lower selectivity towards carbon monoxide compared with the palladium/zirconia catalyst. However, the lowest carbon monoxide selectivity was determined for the copper/zinc oxide catalyst.

Wanat et al. investigated methanol partial oxidation over various rhodium containing catalysts on ceramic monoliths, namely rhodium/alumina, rhodium/ceria, rhodium/ruthenium and rhodium/cobalt catalysts [195]. The rhodium/ceria sample performed best. Full methanol conversion was achieved at reaction temperatures exceeding 550 °C and with O/C ratios of from 0.66 to 1.0. Owing to the high reaction temperature, carbon monoxide selectivity was high, exceeding 70%. No by-products were observed except for methane.

4.2.2
Catalysts for Ethanol Reforming

Nickel-based catalysts on various carriers such as alumina, lanthana, magnesia and zinc oxide have been studied intensively for ethanol steam reforming [196].

Bimetallic nickel containing catalysts are frequently combinations of nickel and copper [196].

Klouz investigated ethanol steam reforming and autothermal reforming over a nickel/copper catalyst on a silica carrier in the temperature range between 300 and 600 °C [197]. While the catalyst suffered from coke formation under steam reforming conditions, the addition of oxygen to the feed reduced both coke formation and carbon monoxide selectivity. By-products such as ethylene and acetaldehyde were not reported by these workers.

Similar results were reported by Frusteri *et al.* for their nickel/magnesia and nickel/ceria catalysts [198]. At a reaction temperature of 650 °C, coke formation was significant under conditions of steam reforming. In the nickel/magnesia catalyst, which contained 15 wt.% nickel, less than 0.1 wt.% carbon was formed within a 20-h test duration when operated under autothermal conditions. Consequently, no deactivation of the catalyst was observed. However, the O/C ratio that was required to achieve this stable performance was rather high at 1.2, and the S/C ratio of 4.2 was also very high. Besides methane, small amounts of acetaldehyde were formed as a by-product.

For nickel-based catalysts, Fatsikostas *et al.* reported that the alumina carrier promoted carbon formation, which was suppressed by the addition of lanthana [199,200]. Nickel, as the active species, promoted reforming to carbon dioxide but the also water–gas shift and methanation [199].

Batista *et al.* performed ethanol steam reforming over cobalt/alumina and cobalt/silica catalysts containing 8 and 18 wt.% cobalt [201]. Even with a reaction temperature of 400 °C, 70% conversion could be achieved. Methane was the main by-product, ethylene was only formed over samples containing 8 wt.% cobalt. Then a bed of an iron oxide/chromium oxide water–gas shift catalyst was switched behind the cobalt/silica catalyst. The carbon monoxide was converted as expected, but also less methane was found in the product [202]. Even less carbon monoxide was formed when both catalysts were mixed. Sahoo *et al.* varied the cobalt content of the cobalt/alumina catalyst from 10 to 20 wt.%. The highest activity was determined for the sample containing 15 wt.% cobalt [203].

Cobalt/zinc oxide catalysts show low temperature activity for ethanol steam reforming even below 400 °C, along with low or no selectivity towards carbon monoxide, which has been reported by Llorca and coworkers [204,205]. However, residence time over the catalysts needs to be sufficiently low and the S/C ratio rather high to achieve this performance.

Homs *et al.* reported that nickel supported on zinc oxide is not a favourable catalyst formulation for ethanol steam reforming, but the addition of nickel to a cobalt/zinc oxide catalyst promoted with sodium increased the catalytic activity [206]. At S/C 6.5 and only 300 °C reaction temperature, full ethanol conversion could be achieved without by-product formation, apart from methane.

Noble metal catalysts are alternatives to the nickel and cobalt formulations discussed above.

Liguras *et al.* investigated platinum, palladium, rhodium and ruthenium catalysts containing 1 wt.% of the noble metal on an alumina carrier at S/C 1.5 in the temperature range between 600 and 850 °C [207]. The rhodium catalyst was most active and showed the lowest selectivity towards acetaldehyde and ethylene. By-product formation could be further suppressed when the rhodium content was increased to 2 wt.%. However, because rhodium is the most expensive noble metal, the ruthenium catalyst was further optimised by Liguras *et al.* Increasing the ruthenium content to 3 wt.% improved the activity and selectivity, while a further increase to 5 wt.% no longer had a beneficial effect. As a next step, ruthenium catalysts made from alumina, magnesia and titania carrier material were compared, with each containing 5 wt.% ruthenium. The alumina carrier showed the highest activity and least selectivity towards ethylene and acetaldehyde.

Toth *et al.* investigated rhodium catalysts on different carrier materials such as alumina, ceria, magnesia, silica, titania and zirconia for ethanol steam reforming [208]. At reaction temperature of 450 °C, only the rhodium/alumina sample, which was also most active, showed short-term stability for 2 h. However, Wanat *et al.* compared the performance of rhodium and rhodium/ceria catalysts for ethanol steam reforming and found a lower selectivity towards methane at reaction temperatures in the range of 700–800 °C for the ceria containing sample. Less than 1 vol.% by-products were observed provided the S/C ratio exceeded 1.0 [209]. Deluga *et al.* investigated autothermal reforming of ethanol over a rhodium/ceria catalyst. A reaction temperature exceeding 700 °C was required to suppress by-product formation [210]. However, the residence time was below 10 ms.

Men *et al.* investigated ethanol steam reforming over cobalt, nickel, rhodium and ruthenium catalysts on different carrier materials such as alumina, silica, magnesia and zinc oxide [211] at S/C 1.5 and a weight hourly space velocity of 90 L (h g_{cat})$^{-1}$ in the temperature range between 400 and 600 °C. All these monometallic catalysts showed high selectivity mainly towards acetaldehyde and to a lesser extent towards ethylene. Only the rhodium sample required a 600 °C reaction temperature to achieve 80% hydrogen selectivity. A rhodium/nickel/ceria catalyst containing 5 wt.% rhodium, 10 wt.% nickel and 15 wt.% ceria on alumina showed full conversion at 500 °C and was selective only towards methane and carbon oxides. This catalyst showed full ethanol conversion at 650 °C for more than 100 h test duration without apparent deactivation [211].

Liguras *et al.* investigated autothermal reforming of ethanol over ruthenium and nickel catalysts on structured supports such as ceramic foams and monoliths [212,213]. Conditions chosen were an O/C ratio of 0.61 and an S/C ratio of 1.5. The reaction was performed at a very high pre-heating temperature of the monoliths and consequently substantial conversion occurred even upstream of the reactor, which created a hot spot of up to 950 °C in the monoliths. A ceramic monolith coated with 5 wt.% ruthenium formed in addition to carbon oxides methane as the main by-product, but there were also small amounts of acetaldehyde, ethylene and ethane [212]. When the S/C ratio was increased to 2.0, the by-products could be suppressed. Increasing the O/C ratio had a similar effect and also suppressed the methane formation. The ruthenium catalyst showed stable conversion for a 75-h test duration. Nickel/lanthana catalysts containing 13 wt.% nickel on a lanthana carrier showed similar performances with respect to activity, selectivity and stability [213].

Wanat et al. performed ethanol partial oxidation over different rhodium containing catalysts on ceramic monoliths [195]. Similar to partial oxidation of methanol (see Section 4.2.1), the rhodium/ceria sample performed best, 95% ethanol conversion was achieved at reaction temperatures exceeding 700 °C and O/C ratios between 0.66 and 1.0. Similar to methanol, carbon monoxide selectivity was high, exceeding 85%, while methane and ethylene were observed as by-products.

4.2.3
Overview of Catalysts for Hydrocarbon Reforming

Catalysts that are suitable for natural gas reforming are, in principle, also suitable for reforming of higher hydrocarbons. A common overview of reforming catalysts for methane and higher hydrocarbons will be provided below, while selected catalyst development work dedicated to specific hydrocarbons will be presented in the following sections.

Catalysts for industrial scale methane steam reforming are commonly based on nickel/nickel oxide or cobalt on alumina or magnesia alumina spinel [164]. Nickel catalysts in their active form are pyrophoric, which means that they generate excessive amounts of heat when exposed to air [107]. This is less of an issue in a large-scale industrial process but causes problems in the practical operation of small fuel processors. Additionally, the heat generated by oxidation of the metallic nickel may cause degradation of the catalyst through sintering [107]. Nickel catalysts require reduction to achieve full activity, which is not the case for precious metal catalysts, which will be discussed below.

Higher in activity, but also more costly, are catalysts that contain precious metals such as rhodium, ruthenium, platinum, palladium and rhenium or mixtures thereof [107], while alumina or magnesia [214] and rare earth oxides such as ceria and zirconia or mixtures thereof serve as the carrier material. Rare earth metals have an oxygen storage capability, they interact with the precious metal and generate active sites for hydrocarbon activation [164].

Turn-over numbers have been reported for methane steam reforming over alumina-supported catalyst systems [214]:

$$Rh\ (13) > Ru\ (9.6) > Pd\ (1.0) > Ni\ (1.0) > Pt\ (0.9)$$

Rhodium is certainly the best catalyst for hydrocarbon reforming, but also the most expensive. Thus, the rhodium content of the catalysts must be minimised [215]. Promoters such as alkali or alkali earth metals are added to accelerate carbon removal, increase activity and suppress methane by-product formation (the latter in case of higher hydrocarbons) [164]. Alkali promoters potentially become volatile in high temperature steam environments and may cause pore blockage owing to this migration [164].

Giroux et al. and Farrauto et al., from Engelhard, presented a proprietary catalyst formulation for autothermal reforming, which was actually composed of a layer of platinum/rhodium steam reforming catalyst covered with a platinum/palladium partial oxidation catalyst [57,107]. The catalyst was capable of reforming natural gas, liquefied petroleum gas, gasoline and diesel, the last even in the presence of 2000 ppm sulfur.

The temperature required for ignition of the hydrocarbons over a catalyst, which is termed the light-off temperature, is an important feature when operating an auto-thermal reformer or partial oxidation reactor, because it affects the start-up strategy and time demand. The auto-ignition or light-off temperature under conditions of partial oxidation decreases with increasing carbon number.

While homogeneous auto-ignition occurs at 600 °C for methane, a temperature of only 200 °C is required for alkanes larger than undecane [58]. Surface ignition at an O/C ratio of 1.0 occurs over platinum at 430 °C with methane, while 175 °C is required for butane. Williams et al. investigated the light-off temperature of their rhodium coated ceramic foams at an O/C ratio 1.0. A pre-heating temperature of 240–250 °C was sufficient for isooctane, n-octane and n-decane, while only 220 °C was required for hexadecane; 300 °C was the light-off temperature of methane [58].

Borup et al. investigated the light-off behaviour of various gasoline components over precious metal catalysts in the presence of steam [193]. At a S/C ratio of 1.0 and an O/C ratio of 0.9, the light-off temperature of n-alkanes (C_6–C_{10}) was in the range between 150 and 180 °C. However, isooctane showed a significantly higher light-off temperature of 210 °C. This value decreased when n-alkanes were mixed in with the isooctane. The addition of alcohols also decreased the light-off temperature significantly, a temperature of 140 °C was determined for a mixture of isooctane and 20 vol.% methanol. Aromatic compounds increased the light-off temperature. Kerosene had a light-off temperature of 220 °C, and for diesel fuel 270 °C was measured. However, significant carbon formation was observed during the light-off experiments. Thus, higher O/C ratios may well be required in the initial pre-heating period.

4.2.4
Catalysts for Natural Gas/Methane Reforming

Methane or natural gas steam reforming performed on an industrial scale over nickel catalysts is described above. Nickel catalysts are also used in large scale productions for the partial oxidation and autothermal reforming of natural gas [216]. They contain between 7 and 80 wt.% nickel on various carriers such as α-alumina, magnesia, zirconia and spinels. Calcium aluminate, 10–13 wt.%, frequently serves as a binder and a combination of up to 7 wt.% potassium and up to 16 wt.% silica is added to suppress coke formation, which is a major issue for nickel catalysts under conditions of partial oxidation [216]. Novel formulations contain 10 wt.% nickel and 5 wt.% sulfur on an alumina carrier [217]. The reaction is usually performed at temperatures exceeding 700 °C. Perovskite catalysts based upon nickel and lanthanide allow high nickel dispersion, which reduces coke formation. In addition, the perovskite structure is temperature resistant.

Find et al. developed a nickel-based catalyst for methane steam reforming in microchannels [218]. AluchromY® steel, which is a Fecralloy (see Section 10.2.1), was used as the construction material for the microstructured plates. The catalyst was based upon a nickel spinel support ($NiAl_2O_4$) for stabilisation. The active nickel was added to the catalyst slurry as a surplus to the amount required for the spinel stoichiometry. The sol–gel technique was used to coat the plates with the catalyst slurry. Good catalyst adhesion was shown through mechanical stress tests and thermal shock tests. The catalyst was tested in a fixed bed at a S/C ratio 3 and reaction temperatures of between 527 and 750 °C. The feed was composed of 12.5 vol.% methane, 37.5 vol.% steam and a balance of argon. At a reaction temperature of 700 °C and 32 h^{-1} space velocity, conversion close to the thermo-dynamic equilibrium could be achieved. During 96 h of operation, the catalyst showed no detectable deactivation. The catalyst was then incorporated into a combined steam reformer/catalytic combustor composed of the microchannel plates. This work is discussed in Section 7.1.3.

Nurunnabi et al. observed improved performance of their $Ni_{0.2}Mg_{0.8}O$ catalyst for oxidative steam reforming of methane when doping it with 0.1 wt.% palladium [219].

Noble metal catalysts are more active for partial oxidation of methane [216]. Rhodium is the best performing catalyst, ruthenium and iridium being alternatives of lower cost.

Berman et al. demonstrated short-term stability of their ruthenium catalysts, which contained α-alumina as the carrier [220]. The catalysts were composed of 2 wt.% ruthenium stabilised by 2.8 wt.% manganese oxide or 4 wt.% magnesia. The stabilised catalysts showed no apparent deactivation during 17-h operation at a very high reaction temperature of 1150 °C, which was not the case for their un-stabilised counterparts.

Wang et al. prepared rhodium catalysts for methane steam reforming, which were stabilised with 6 wt.% magnesia [221]. Samples were prepared containing 1, 5 and 10 wt.% rhodium. The sample containing 5 wt.% rhodium showed the highest activity. Johnson et al reported on the stable performance of their rhodium catalysts, which were prepared from α-alumina shims by a tape-cast method and then impregnated with rhodium [222]. A sample containing 3.7 wt.% rhodium showed stable performance for a 100-h test duration at 900 °C.

Short contact time partial oxidation of methane has been the subject of numerous publications, with the group of Schmidt being one of the pioneers in this field. Only noble metal catalysts are applied for short contact times.

The effect of particle size on methane conversion was investigated by Hohn and Schmidt [223] for the partial oxidation of methane for short contact times. Conversion losses were observed for catalyst particle sizes exceeding 1 mm (see Figure 4.8). This observation was later supported by numerical calculations [224]. Hence, in most instances short contact time partial oxidation is obviously limited by heat and mass transfer [224].

It is evident that catalyst particle diameters as low as 1 mm lead to extensive pressure drops in a fixed catalyst bed and thus most of the research work on short contact time partial oxidation is focussed on catalyst coatings in ceramic and metallic monoliths or microchannels.

Hickman and Schmidt were amongst the first to propose partial oxidation of methane over noble metal catalysts with short contact times of few milliseconds [225]. They used platinum and rhodium coated onto porous alumina foam monoliths. Rhodium showed superior performance, namely higher activity and hydrogen

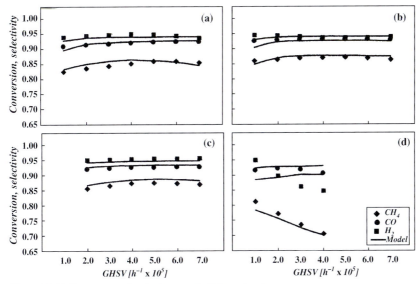

Figure 4.8 Methane conversion, hydrogen selectivity and carbon monoxide selectivity versus gas hourly space velocity as determined by Hohn and Schmidt [223] (symbols) and calculated by Bizzi et al. [224] (straight lines) for short contact time partial oxidation of methane in fixed beds of different particle diameters: (a) 0.4 mm; (b) 0.8 mm; (c) 1.2 mm; (d) 3.2 mm.

selectivity compared with platinum. Over the rhodium catalyst, reaction temperatures considerably exceeded 900 °C and an O/C ratio higher than 1.2 was required to achieve full conversion at 10-ms contact time. Huff et al. investigated the performance of platinum, nickel, iridium and rhodium catalysts deposited onto porous ceramic foams for partial oxidation of methane [226]. Rhodium and nickel showed the highest activity and hydrogen selectivity.

Bodke et al. studied the effect of ceramic foam supports on activity and selectivity of rhodium catalysts for partial oxidation of methane and butane [227]. α-Alumina, which was wash-coated onto the alumina foams and then impregnated with rhodium, revealed higher conversion and hydrogen selectivity for methane partial oxidation compared with porous α-alumina foams, which were impregnated with rhodium immediately. However, butane conversion was lower than methane conversion. The hydrogen selectivity was still higher for butane. When the pore size of the foams was reduced from 8 to 31 pores cm^{-1}, methane conversion and hydrogen selectivity increased [227]. This was probably due to reduced mass transport limitations. However, the opposite was the case for the partial oxidation of butane, suggesting the absence of mass transfer limitations with butane. Various foam materials were tested and the highest activity was obtained with zirconia/tetra alumina, which was composed of 20 wt.% zirconia and 80 wt.% alumina. Rhodium catalysts supported by neat zirconia were less active, followed by a sample supported by α-alumina, which contained 8 wt.% silica.

Bizzi et al. performed numerical calculations of the short contact time partial oxidations of methane over fixed-bed catalysts. Very high hot spot temperatures of 1100 °C at an O/C ratio of 2 were calculated for their test reactor, which was also verified by experiments [224].

4.2.5
Catalysts for Reforming of LPG

Less work on catalyst development for reforming of liquefied petroleum gas has been published compared with natural gas, but some examples will be discussed below.

Nickel catalysts are rarely applied for the partial oxidation of higher hydrocarbons. One example is the work of Li et al. [228], which converted propane completely at 700 °C at an O/C ratio of 1.2 at a moderate weight hourly space velocity of 120 L $(h\,g_{cat})^{-1}$. The catalyst was a hydrotalcite ($Mg_{2.5}Ni_{0.5}AlO$) doped with different amounts of ruthenium. The catalyst containing 0.1 wt.% ruthenium turned out to be the most active formulation.

Beretta et al. investigated the partial oxidation of light paraffins to synthesis gas [229]. A catalyst containing 5 wt.% platinum on γ-alumina was mostly selective towards the total oxidation of ethane under conditions of partial oxidation (O/C ratio 1.0) and also when steam was added to the feed (O/C ratio 1.0, S/C ratio 0.5). A sample containing 0.5 wt.% rhodium on α-alumina showed superior performance in the partial oxidation of propane even at a substoichiometric feed composition for partial oxidation (O/C ratio 0.66) and under quasi autothermal conditions (O/C ratio 0.66, S/C ration 0.33). Rhodium showed higher selectivity towards hydrogen than towards steam at temperatures exceeding 600 °C in the presence and in the absence of steam. Surprisingly, no coke formation issues were reported by these workers for the rhodium sample, despite the low oxygen and steam contents.

Kolb et al. used small sandwich-type microreactors to screen catalyst coatings for propane steam reforming [230]. Amongst the platinum, palladium and rhodium catalysts, the last showed superior activity, selectivity and stability. The rhodium sample was fully selective towards propane steam reforming by a reaction temperature of 550 °C. Then ceria containing samples were prepared, which showed a lower tendency towards coke formation, most pronounced for a rhodium/platinum/ceria catalyst. The stability of this catalyst was then tested at an S/C ratio of 3.2 and weight hourly space velocity of 300 L $(h\,g_{cat})^{-1}$, which corresponded to a residence time of 7 ms in the microchannels. For of 6-h duration, full conversion was achieved and no light hydrocarbons such as methane or ethylene were observed: 62 vol.% hydrogen, 7 vol.% carbon dioxide and 12 vol.% carbon monoxide were formed; 9.6 L h^{-1} hydrogen was produced over 13.2 mg of catalyst (including the alumina carrier). Thus, the turnover frequency amounted to around 700 L h^{-1} or 63 g h^{-1} hydrogen per gram of catalyst an hour or 1575 g hydrogen per g rhodium an hour, which is a fairly high number.

Subsequently, 1000-h durability for butane steam reforming could be achieved at a S/C 3.2 over an improved catalyst formulation, as shown in Figure 4.9.

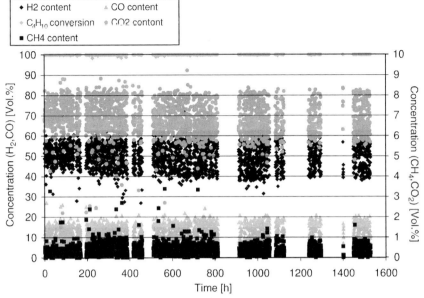

Figure 4.9 1000-h durability of a catalyst coating for butane steam reforming at an S/C 3.2; weight hourly space velocity 300 L $(h\,g_{cat})^{-1}$ (source: IMM).

Pennemann et al. developed and optimised noble metal catalysts for the partial oxidation of propane in wash-coated microchannels [59]. Rhodium turned out to be by far the most suitable active metal compared with platinum and palladium. Over a catalyst containing 1 wt.% rhodium supported by alumina, the O/C ratio was increased from 1.0 to 1.3. Hydrogen selectivity increased with increasing oxygen addition, because methane formation could be suppressed. Increasing the reaction temperature also decreased the selectivity towards methane according to the thermodynamic equilibrium of methanation (see Section 3.2). A bimetallic catalyst containing 1 wt.% rhodium and 1 wt.% platinum showed higher stability compared with pure rhodium. Full conversion was achieved at an extremely high weight hourly space velocity of 1700 L $(h\,g_{cat})^{-1}$ and 700 °C reaction temperature. The addition of only 0.1 wt.% rhodium to a catalyst containing 1.9 wt.% platinum lead to a significant increase in activity compared with pure platinum. The selectivity towards hydrogen was also improved, while the methane selectivity was still moderate. Thus an optimum catalyst formulation was identified as 1–2 wt.% platinum and 0.1–0.2 wt.% rhodium on alumina. However, all catalytic tests suffered from carbon formation, which lead in most instances to blockage of the small test reactors by carbon deposits. The coke formation, which was mostly observed at the reactor outlet, was finally attributed to carbon monoxide reduction in the presence of hydrogen at the reactor metal surface (see also Section 3.2).

Silberova et al. performed partial oxidation and oxidative steam reforming of propane over a rhodium catalyst containing only 0.01 wt.% rhodium on alumina, which was coated onto a ceramic foam [231]. Under conditions of partial oxidation (O/C ratio 1.3) full conversion of the feed could be achieved for 9-h duration and reactor furnace temperatures exceeding 700 °C, which corresponded to a hot spot around 900 °C in the foam. Under conditions of autothermal reforming, the catalyst required a higher temperature to achieve full conversion and suffered from deactivation in the short term, which was attributed to the presence of steam in the feed causing sintering and migration of the rhodium particles. However, the residence time in the foam was rather high, at 100 ms. Side products observed were, apart from methane, ethylene, propylene and, surprisingly, small amounts of acetylene [232].

Pino et al. reported significant activity and favourable selectivity patterns for their platinum/ceria catalyst. However, the experiments were performed at a very high S/C ratio of 3.6 and a high O/C ratio of 1.3, which shifts the conditions into the field of steam supported partial oxidation [233].

As a cheap alternative to noble metal and even nickel catalysts, Laosiripojana et al. proposed high surface area ceria [234]. Complete conversion of liquefied petroleum gas (prepared as a sulfur-free 60 wt.% propane/40 wt.% butane mixture) could be achieved above a reaction temperature of 800 °C. The S/C ratio was low at 1.45 and ethylene as a by-product could be suppressed at O/C ratios exceeding 0.6. The weight hourly space velocity was relatively low at 120 L $(h\, g_{cat})^{-1}$, however, catalyst deactivation was moderate within 70-h test duration, which is a promising result.

4.2.6
Catalysts for Pre-Reforming of Hydrocarbons

Catalysts for pre-reforming of hydrocarbons are basically similar to reforming catalyst formulations. Nickel oxide catalysts, other base metal catalysts and noble metal catalysts have been reported in the literature. Some examples of the performance of catalysts for pre-reforming of hydrocarbon feedstock are provided below, which does not claim to be a complete list.

Sperle et al. investigated the minimum S/C ratio required to guarantee carbon-free operation for a pre-reformer catalyst containing 18 wt.% nickel supported on a magnesia/alumina spinel [235]. The investigations revealed, as shown in Figure 4.10, that the required S/C ratio increased significantly when C_2 and even more when C_3 hydrocarbons were added despite their low content in the feed. Unsaturated hydrocarbons had a much higher coking tendency than saturated hydrocarbons, as expected (see Section 4.2.11). While a S/C 0.7 was sufficient to suppress coke formation with methane feed, a S/C around 3.0 was required for a mixture of methane, propane and propylene [235]. The reaction temperature where minimum coke formation was observed proved to be 480 °C for methane, while a sharp minimum occurred at 500 °C for a mixture of methane, propane and propylene.

Chen et al. prepared a catalyst for pre-reforming of gasoline [236]. It was composed of 50 wt.% nickel oxide, 18 wt.% lanthana and the balance alumina. The catalyst was tested at a weight hourly space velocity of between 40 and 168 L $(h\, g_{cat})^{-1}$ at 5-bar pressure and 480 °C reaction temperature. The gasoline surrogate was composed of 74 wt.% isooctane, 20 wt.% xylene, 5 wt.% cyclo-hexane and 1 wt.% 1-octene (average

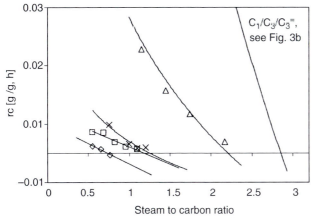

Figure 4.10 Rate of carbon formation for various hydrocarbon mixtures under conditions of hydrocarbon pre-reforming: temperature, 500 °C; pressure, 20 bar; ◊, pure methane; □, methane/ethane molar ratio 1/0.8; ×, methane/propane molar ratio 1/0.035; △, methane/ethane/ethylene molar ratio 1/0.08/0.08; straight line, methane/propane/propylene molar ratio 1/0.08/0.08; hydrocarbon partial pressure, 3.9 bar [235].

formula $C_{7.86}H_{15.58}$). At a S/C ratio 2.7, 95.6% conversion was achieved over the full range of space velocities tested. The main products were methane, carbon dioxide and hydrogen (see Figure 4.11). The catalyst showed stable performance for 100 h at a weight hourly space velocity of 40 L $(h\, g_{cat})^{-1}$.

Beta (SiO_2/Al_2O_3 ratio 50; 680 m^2 g^{-1} surface area) and ZSM-5 zeolites (SiO_2/Al_2O_3 ratio 25; 425 m^2 g^{-1} surface area) and manganese/alumina catalysts (surface area 151 m^2 g^{-1}) were evaluated by Campell et al. for catalytic cracking of JP-8 fuel [9]. The catalysts were tested as fixed beds and, alternatively, coated onto the walls of

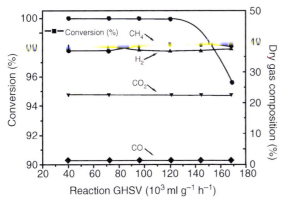

Figure 4.11 Dry gas composition and gasoline conversion obtained for gasoline pre-reforming (T = 480 °C, P = 5 bar, S/C = 2.7) [236].

ceramic tubes. Up to 80% conversion was achieved over the catalysts, methane, ethylene and propene being the main products. The manganese catalyst produced heavier products compared with the zeolites. Sulfur compounds were converted into hydrogen sulfide and methanethiol, however, thiophenes remained unconverted.

Adiabatic pre-reforming of desulfurised diesel and jet fuels was demonstrated for commercial nickel catalysts at 480 °C and S/C ratios around 2.5 for 1000- and 500-h duration, respectively [81]. The feed was not completely sulfur-free, but still contained about 1 ppm sulfur and thus catalyst deactivation originated from sulfur poisoning and rather than from coke formation.

4.2.7
Catalysts for Gasoline Reforming

Nickel and perovskite catalysts have been reported in the literature for gasoline reforming. The alternatives are noble metal catalysts, amongst them mostly platinum, rhodium and ruthenium as shown by some examples below.

Schwank et al. developed nickel catalysts supported by ceria/zirconia, $Ce_{0.75}Zr_{0.25}O_2$. The catalysts were coated onto metal foams and tested for the autothermal reforming of isooctane and a gasoline surrogate [237]. They contained 5–15 wt.% nickel, had 15 $m^2 g^{-1}$ surface area and showed a nickel dispersion of 1% [238]. The metal foams were 500-µm thick layers of Fecralloy with a pore density of 24 pores cm^{-1}. The foams were pre-treated at 800 °C in air for 24 h to generate an oxide layer for better adhesion of the coating (see also Section 10.2.1). The catalysts were tested both in fixed beds and as coated foams at a high gas hourly space velocity of up to 770 000 h^{-1}. The S/C ratio was 2.0, the O/C ratio of around 1.0 was in the range of partial oxidation. The catalyst required some *in situ* activation because it was exposed to air during the pre-heating phase at the beginning of the experiments. Nickel oxide had to be reduced to nickel by the reformate hydrogen [238]. Coke formation was not observed during 8 h of operation even with gasoline surrogate (43% isooctane, 35% toluene, 22% n-heptane) [237]. A hot spot temperature of 800 °C was measured in the fixed bed. The hot spot was lower in the metal foam, which led to some coke formation. When the metal foams were insulated from the reactor housing, full conversion could be achieved at a reaction temperature of 600 °C and 560 000 h^{-1} gas hourly space velocity. The reaction light-off was at 400 °C [237] (see also below). The catalyst formulation was also later used for autothermal reforming of diesel surrogates (see Section 4.2.8).

Chen et al. prepared a catalyst for autothermal reforming of gasoline [239]. It was composed of 8 wt.% nickel oxide, 2 wt.% lanthana on an α-alumina carrier stabilised by 23 wt.% magnesia. The air feed was introduced into the inlet of the autothermal test reactor. Both gasoline surrogate and a mixture of pre-reformed gasoline, as described in Section 4.2.6, were used as feedstock. Under conditions that were well steam supported partial oxidation (O/C = 1, S/C = 2.7), 770 °C reaction temperature, 5 bar pressure and 28 L $(h\, g_{cat})^{-1}$ weight hourly space velocity, full conversion over the catalyst was observed. The conversion deteriorated within 6 h of test duration to 99% and various hydrocarbons were detected in the product. The origin of the deactivation was the formation of filamentous coke, which is frequently observed over nickel-based catalysts. When pre-reformed feedstock was used, which contained only methane as a stable hydrocarbon, full conversion was observed for 100 h under these conditions [240].

Kalia performed autothermal reforming of n-heptane, n-dodecane, toluene and methylcyclohexane over nickel oxide/alumina catalysts containing 15 wt.% nickel oxide [241]. The S/C ratio of the experiments was 3.0, while O/C amounted to 0.68. They observed that toluene was most difficult to convert. Experiments were then performed with a gasoline surrogate composed of n-heptane, methylcyclohexane and toluene. Over the nickel catalyst, and a catalyst containing 0.5 wt.% platinum on zirconia, even at 900 °C the toluene conversion was only 85%, while the other hydrocarbons were completely converted. However, complete conversion of toluene could be achieved over fresh rhodium catalyst containing 0.5 wt.% rhodium on zirconia. Besides carbon oxides and methane, ethylene and alkenes and alkanes not further specified were formed as by-products. Both the nickel catalyst and the rhodium/zirconia catalyst suffered from deactivation during autothermal reforming of the gasoline surrogate.

Qi *et al.* developed perovskite catalysts coated onto cordierite monoliths for autothermal reforming of gasoline [242]. They started with the substitution of the B-site metal (see Section 4.2). Among $LaNiO_3$, $LaCoO_3$ and $LaMnO_3$, the nickel containing sample showed the highest activity. When modifying the A-site metal, the desired perovskite structures could only be synthesised from lanthanum, which was not the case for ceria, samarium and gadolinium. The composition of the catalyst was optimised to $La_{0.8}Ce_{0.2}NiO_3$. This catalyst showed a stable performance for n-octane reforming for more than 200 h at S/C 2.0 and O/C 0.76. However, it deactivated within a time of 50 h on stream when 5 ppm sulfur were added to the n-octane feed.

Murata *et al.* performed extensive catalyst screening for isooctane steam reforming; 13 samples were tested for their activity and selectivity (Figure 4.12) [243]. The

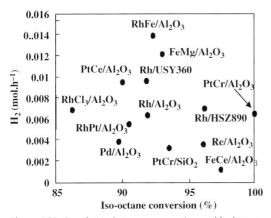

Figure 4.12 Correlation between conversion and hydrogen production for steam reforming of isooctane at 550 °C reaction temperature: S/C 1.4; 1 bar pressure [243].

experiments were performed at 550 °C, which is a rather low temperature for steam reforming of higher hydrocarbons, a low S/C of 1.4, ambient pressure and a very low weight hourly space velocity of 7.8 L $(h\,g_{cat})^{-1}$. Surprisingly, a platinum/chromium catalyst supported by alumina showed the highest conversion and superior selectivity pattern compared with the other catalysts. No light hydrocarbons were found in the product for a test duration of 5 h.

Villegas et al. performed autothermal reforming of isooctane over a 2 wt.% platinum catalyst on a ceria/zirconia mixed support ($Ce_{0.67}Zr_{0.33}O_2$) [71]. The tests were performed at a gas hourly space velocity of 150 000 h^{-1}. Full conversion of isooctane was achieved at 710 °C under conditions of steam supported partial oxidation (O/C = 1; S/C = 2). The optimum hydrogen content in the reformate was achieved at a lower O/C ratio of 0.75 and S/C ratio of 1.5. The carbon dioxide content in the reformate was always higher than the carbon monoxide content, which contradicted the thermodynamic calculations performed by the same workers [71], see Section 3.3. This supports the assumption that platinum preferably catalyses catalytic combustion.

Carbon formation could be stopped in the presence of oxygen, while the catalyst deactivated under conditions of steam reforming (S/C = 1.5), which was attributed to sintering of the carrier material [71].

It is widely accepted that steam reforming of higher hydrocarbons requires S/C ratios higher than 2.5 to prevent coke formation. Thus, coke formation might well have been the origin of catalyst deactivation observed by Villegas et al. [71].

The catalyst was also deactivated by coke formation under conditions of partial oxidation (O/C = 1), which was proven by oxidation experiments.

Trimm et al. applied a two-zone catalyst bed for autothermal reforming of gasoline surrogate [244], a mixture of 5% cyclohexane, 35% isooctane, 20% n-octane, 35% toluene and 5% hexane. The catalysts contained 2–3 wt.% platinum on ceria in the first zone to catalyse catalytic combustion. A commercial 34 wt.% nickel (as nickel oxide) catalyst was applied for steam reforming downstream. The operating conditions were an O/C ratio of 1.2, which is once more a significantly steam supported partial oxidation and S/C ratios of between 2.1 and 3.4. However, reaction temperatures were very low and never exceeded 600 °C for all experiments. Consequently, full conversion could not be achieved. Strong coke formation was observed when commercial gasoline was tested as the feedstock [244].

Springmann et al. [56] performed mechanistic investigations for autothermal reforming of gasoline compounds over a reforming catalyst from the OMG Company (subsequently Umicore). The catalyst formulation was not disclosed in detail, but it contained 0.1–2 wt.% rhodium on an alumina carrier along with small amounts of platinum. A high tendency towards coke formation and activity losses were observed for steam reforming of toluene even at a high S/C ratio of 4.0, which was not the case for isooctane at a much lower S/C ratio of 1.8. No coke formation was observed under autothermal conditions. Then a mixture of 40 vol.% isooctane, 20 vol.% 1-hexene and 40 vol.% toluene was used as the feedstock. Under conditions of autothermal reforming, namely a S/C ratio 2.2, O/C ratio 0.87 and temperatures between 625 and 675 °C, the hexene was easily converted, while isooctane showed much lower conversion and toluene was mostly converted into benzene. The benzene, which had been formed, showed very low conversion into lower hydrocarbons.

Kolb *et al.* performed catalyst development in microchannel reactors for autothermal reforming of isooctane. Rhodium, nickel, ruthenium and palladium catalysts supported by zirconia and alumina were tested [73]. Rhodium on alumina turned out to be the most active catalyst, which also showed lowest selectivity towards methane. The rhodium content was then varied from 0.1 to 2 wt.%; 1 wt.% rhodium on alumina was considered to be the optimum catalyst formulation with respect to both performance, stability and cost. A minimum S/C ratio of 3.3 was required to prevent coke formation. The catalyst was incorporated into an autothermal reforming reactor of kilowatt size (see Section 7.1.2).

Ferrandon and Krause investigated the effect of the catalyst support on the performance of rhodium catalysts wash-coated onto cordierite monoliths for autothermal reforming of gasoline [245]. The samples contained 2 wt.% rhodium on gadolinium/ceria, and lanthanum-stabilised alumina [245]. The latter sample showed higher activity and superior selectivity. Only 30 ppm of light hydrocarbons (C > 1) were detected. However, the sample also had higher surface area and rhodium dispersion. Both samples showed stable performance for more than 50-h test duration.

Krumpelt *et al.* provided an overview of Argonne National Laboratory activities in the field of catalyst development for hydrocarbon reforming, focussing on gasoline [246]. The patented technology [247] was utilised by the catalyst manufacturer Süd-Chemie for reforming catalyst production [246]. The catalysts were supported by an oxide ion conducting carrier, such as ceria, zirconia or lanthanum gallate, which was doped with small amounts of a non-reducible element, for example, gadolinium, samarium or zirconium. Nickel, cobalt, ruthenium, platinum or palladium could serve as active metals, the last three showing higher activity and hydrogen selectivity [246].

O'Connor *et al.* investigated the partial oxidation of gasoline and gasoline components, such as cyclohexane, n-hexane and isooctane over wash-coated ceramic monoliths carrying 5 wt.% rhodium supported by γ-alumina [248]. The stoichiometric O/C ratio of 1.0 revealed the best results with respect to activity and selectivity. Full conversion of cyclohexane was achieved at temperatures between 1050 and 900 °C along the foam. No by-products were observed despite the very high space velocity of 730 000 h^{-1}. Approximately the same results were obtained for n-hexane, while toluene could not be converted completely under the experimental conditions. The catalyst suffered from sulfur poisoning, when commercial gasoline containing 0.5 wt.% sulfur components was used as the feedstock [248].

Qi also investigated thoroughly the performance of a ruthenium catalyst for autothermal reforming of n-octane [249]. The catalyst was composed of 0.5 wt.% ruthenium stabilised by ceria and potassium on γ-alumina. It showed full conversion of n-octane for 800 h (see Figure 4.13), however, selectivities moved from carbon ioxide and methane towards carbon monoxide and light hydrocarbons. This has to be regarded as an indication of catalyst degradation during long-term tests performed at full conversion. After 800 h the catalyst showed incomplete

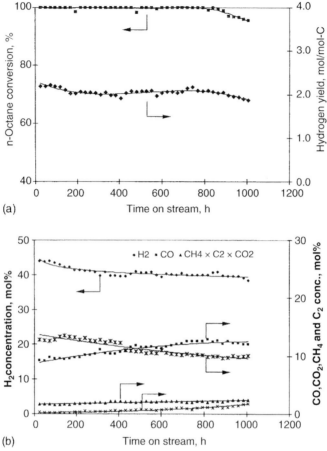

Figure 4.13 Conversion and selectivities as determined for a ruthenium catalyst during n-octane reforming for 1000-h test duration: S/C ratio 2.0; O/C ratio 0.76 [249].

conversion. Tests performed at the spent catalyst revealed losses of specific surface area and of 33 wt.% of the noble metal.

4.2.8
Catalysts for Diesel and Kerosene Reforming

Relatively few publications are yet available that deal with steam reforming of kerosene and diesel fuel.

The nickel/ceria/zirconia catalyst developed by Tadd et al. for gasoline reforming [238] (see Section 4.2.7) was also applied to autothermal reforming of diesel surrogates in a fixed bed [250]. When reforming began after pre-heating the catalyst to 550 °C in air, initial temperature fluctuations up to 1100 °C occurred. This originated from fuel combustion by nickel oxide. The temperature fluctuations caused significant sintering of the catalyst. The catalyst was then reduced *in situ* by the reformate hydrogen and was still active. At a S/C ratio of 2.0 and a gas hourly space velocity of 225 000 h^{-1}, full conversion of tetralin could be achieved at O/C ratios of 1.0 or higher. This was not the case for dodecane and a 50/50 mixture of dodecane and tetralin, because the temperature of the fixed bed was higher for the tetralin. The higher temperature also suppressed methane formation due to the thermodynamic equilibrium. Liquid species in the product of tetralin reforming were analysed and naphthalene, indene and other components were identified as by-products. Liquid by-products of dodecane were 4-dodecene, 1-undecene and oxygenated compounds.

Rosa *et al.* [251] investigated catalysts for autothermal reforming of diesel. Numerous systems were tested. Firstly, platinum, ruthenium, cobalt and nickel catalysts on an alumina carrier promoted with magnesium and lanthanides to improve thermal stability were investigated. The second set of samples was composed of cobalt perovskite catalysts. In addition, Rosa *et al.* investigated ruthenium and platinum catalysts on lanthanum/cobalt perovskites. The samples were tested at gas hourly space velocities of between 20 000 and 80 000 h^{-1} and S/C ratios from 3 to 5. The O/C ratio was set to between 0.4 and 1.4 and the temperature to 650–900 °C. Amongst the first set of samples, platinum stabilised by a dual set of lanthanum stabilisers, that were not disclosed, showed improved performance over nickel and ruthenium. The optimum conditions were 750 °C reaction temperature, S/C = 3 and O/C = 1. While diesel conversion was stable over the catalyst, a certain deactivation became obvious, owing to the decreasing content of hydrogen and the increasing content of carbon monoxide present in the reformate.

The LaCoO$_3$ perovskite catalyst alone showed similar degradation compared with the stabilised platinum/alumina catalyst, whereas the addition of 0.2 mol% ruthenium to the perovskite catalyst led to stable performance.

Dinka and Mukasyan prepared a large variety of perovskite catalysts for autothermal reforming of sulfur containing JP-8 fuel [252]. The perovskites were of LaFeO$_3$-type and lanthana was partially substituted by ceria, while iron was substituted by nickel, resulting in an overall formula of La$_{0.6}$Ce$_{0.4}$Fe$_{0.8}$Ni$_{0.2}$O$_3$. They were further doped with noble metals such as platinum, palladium, ruthenium and rhenium and alternatively with non-noble metals such as potassium, sodium, lithium, caesium, cobalt and molybdenum to improve stability in the presence of sulfur components. Activity tests were carried out at a reaction temperature of 800 °C, S/C ratio of 3.0 and O/C 0.7. The most promising results were achieved for 2 wt.% potassium and 1 wt.% ruthenium doping.

Liu and Krumpelt reported 100 h of stable operation for their perovskite catalyst LaCr$_{0.95}$Ru$_{0.05}$O$_3$ at 800 °C reaction temperature, low S/C 1.5 and O/C 1.0 for steam supported partial oxidation of dodecane [253].

Suzuki *et al.* reported catalyst development for kerosene steam reforming [254]. They prepared a 2 wt.% ruthenium catalyst supported by an alumina carrier, which was stabilised by 20 wt.% yttria, lanthana and ceria, respectively. At 800 °C reaction temperature, S/C 3.5 and 8-bar pressure the ceria stabilised sample showed the best performance in the medium term in the presence of 51 ppm sulfur in the feed. With hydrodesulfurised kerosene the catalyst showed a stable performance for 8000-h test duration, however, at an extremely low weight hourly space velocity of less than 5.0 L $(h\,g_{cat})^{-1}$.

Fukunaga et al. demonstrated stable performance for a time of 12 000 h on stream at full conversion for their self-developed ruthenium-based catalyst for steam reforming of kerosene at S/C ratio of 3 and 730 °C reaction temperature. The space velocity related to the liquid feed (LHSV) was low at 0.5 h^{-1} [255].

Gadolinium stabilised ceria was used as support for a platinum catalyst for autothermal reforming of gasoline and diesel. The catalyst contained only 0.5 wt. % platinum. It was operated at temperatures exceeding 800 °C [256]. However, stability data of the catalyst have not yet been provided.

Cheekatamarla and Lane [257] found higher hydrogen yield and a lower content of light hydrocarbons for autothermal reforming of synthetic diesel fuel over bimetallic platinum/palladium and platinum/nickel catalysts compared with the monometallic samples. The catalysts showed medium-term stability for 50-h test duration in the presence of sulfur in the feed [258].

Palm et al. [259] investigated the performance of a proprietary precious metal catalyst from the former dmc^2 company (now Umicore) in autothermal reforming of C_{13}–C_{19} hydrocarbon mixtures in a fixed bed. The temperature along the catalyst bed exceeded 700 °C at the inlet and decreased to 600 °C towards the outlet. The O/C ratio was set to 0.9, the S/C ratio to 2.2 and the gas hourly space velocity ranged between 13 000 and 18 000 h^{-1} [259]. From these experimental conditions provided by the authors, a rather low weight hourly space velocity of about 33 L $(h\,g_{cat})^{-1}$ could be calculated. The catalyst showed stable but still incomplete conversion (96%) for 20 h under these conditions. Changes to the hydrocarbon feed composition from a C_{13}–C_{15} mixture to a C_{13}–C_{19} mixture and addition of condensed cycloalkanes (cyclonaphthalenes) did not impair the catalyst performance significantly.

Kopasz et al. investigated the performance of commercial noble-metal based catalysts deposited onto three honeycomb monoliths switched in series for the autothermal reforming of diesel [260]. The main target of the work was to identify the effect of paraffinic and aromatic diesel compounds on the reforming process. N-Dodecane, 1-methylnaphthalene and decaline were used as feedstock components. The measurements were performed at a S/C ratio of 2.0 and an O/C ratio of 0.9. Addition of 10 vol.% decaline or 10 vol.% methylnaphthalene to the dodecane feed decreased the hydrogen and carbon monoxide contents of the reformate considerably, while the carbon dioxide content remained almost unchanged and light hydrocarbons appeared in the product. Unfortunately, only methane and ethylene were reported by the authors, however, more ethylene than methane was detected and it is probable that higher hydrocarbons such as propylene were also formed. The addition of decaline and methylnaphthalene in parallel impaired the catalyst performance even more, as shown in Figure 4.14. The initial catalytic combustion reactions were obviously not affected by the additives, but their conversion by steam reforming was then incomplete. Consequently, the temperature of the second and third monolith increased, when the additives were added to the feedstock, because less energy was consumed by steam reforming reactions. The investigations revealed that single- and three-ring aromatics were produced by the catalyst. This indicated a

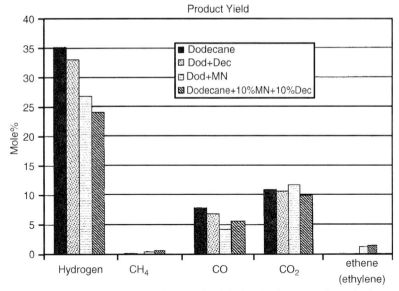

Figure 4.14 Dry gas composition of various diesel feedstocks during autothermal reforming over precious metal-based catalyst as measured by Kopasz et al. [260]: base feedstock, dodecane; 10 vol.% decaline; 10 vol.% 1-methylnaphthalene; and both additives were added to the dodecane.

tendency towards coke formation, because multi-ring aromatics are known to be coke precursors [260]. When increasing the O/C ratio of the feed from 0.75 to 1.05, paraffinic, olefinic and naphthenic by-products could be suppressed, but not the aromatic by-products. The last were formed downstream of the first monolith, where all the oxygen had been consumed. The optimum hydrogen yield was obtained for an O/C ratio of 0.9. When varying the S/C ratio from 2.0 to 3.0, a minimum of aromatic fragments and by-products was observed at an S/C ratio of 2.5 [260]. This is a surprising result, because steam excess is expected to suppress coke formation and improve catalyst performance for diesel reforming, as described by other workers (see below).

Krummenacher et al. investigated the partial oxidation of diesel and diesel components such as n-decane and hexadecane over wash-coated ceramic monoliths carrying 5 wt.% rhodium supported by γ-alumina [261]. A very high O/C ratio of 1.6 was required to achieve full conversion of n-decane at temperatures exceeding 1000 °C, while no by-products were observed apart from carbon dioxide and water due to the excessive oxygen surplus in the feed. When the O/C ratio was lowered, significant formation of light olefins, mainly ethylene, but also propylene, butylene and pentylene was observed. The same was the case for hexadecane, but the O/C ratio required to suppress hydrocarbon by-products was as high as 2.5. At lower O/C ratios, significant formation of C_{6+} olefins dominated the hydrocarbon breakthrough. This trend was followed when the feedstock was switched to low sulfur diesel fuel. Even at an O/C ratio of 2.8, the formation of olefins could not be suppressed, however, conversion still exceeded 98%. The addition of naphthalene to n-decane revealed that naphthalene was more difficult to convert than n-decane, but easier than hexadecane [262]. When partial oxidation of n-decane was performed over a platinum catalyst, complete conversion could not be achieved and olefins were produced in the main, instead of hydrogen [263]. The preparation of a bimetallic catalyst containing rhodium and platinum in a weight ratio of 1:10 did improve the performance, but not significantly.

The steam was then added to the diesel surrogates by Dreyer et al. [264]. However, the conversion of n-decane decreased with increasing S/C ratio, mainly because the reaction temperature dropped significantly [264]. The beneficial effect of steam addition on catalyst activity could not be observed, probably because the feed was only pre-heated to about 250 °C, which is not sufficient to maintain high reaction temperatures during autothermal reforming. However, selectivities developed in a manner similar to the observations of many other groups. Hydrogen and carbon dioxide selectivity increased with increasing S/C ratio, while carbon monoxide selectivity decreased owing to the water–gas shift equilibrium. Ethylene and propylene formation could be completely suppressed at S/C ratios higher than 2.0, even when the O/C ratio was as low as 0.66. The results were further improved by external heating [264]. Hexadecane also turned out to be more difficult to reform than n-decane in the presence of steam. An S/C higher than 2.0 was required to completely suppress light olefins at O/C ratios below 1.0. JP-8 fuel was fed to the rhodium catalyst at a high O/C ratio of 1.4 and an S/C of 1.0. Conversion at 98% was achieved and olefin by-products could be suppressed, but the catalyst suffered from carbon formation [264], probably because the steam addition was not sufficient.

4.2.9
Cracking Catalysts

Catalysts suitable for cracking reactions will not be discussed in depth as they are not within the scope of this book. Poirier et al. identified palladium catalysts as being the most suitable for methane cracking [80]. Because significant amounts of carbon are formed during the cracking process, the catalysts require regeneration through carbon combustion at regular time intervals [79]. Some aspects of this process will be discussed in Section 5.1.4. After reduction, the catalysts release significant amounts of carbon monoxide, which frequently exceed 10 vol.% in the effluent [79].

Choudhary et al. investigated the catalytic cracking of methane over 10 wt.% nickel catalysts supported by silica, zeolite Y and ZSM-5 zeolite [265]. Experiments were performed at a temperature of 550 °C and 20 L (h g_{cat})$^{-1}$ weight hourly space velocity. While the ZSM-5 supported catalyst deactivated completely within a time of 1 h on stream, the silica and zeolite Y supported catalysts were active for 12 h, which was attributed to the formation of filamentous carbon over the latter catalysts, while graphitic carbon was formed over ZSM-5. The most active was the silica-based catalyst, which also had the lowest selectivity towards carbon monoxide (about 50 ppm in the product after an initial period of higher selectivity, see Figure 4.15) [265]. However, the carbon monoxide content in a practical operation would be higher, because the methane feed had been diluted by 80 vol.% inert gas. Zhang et al.

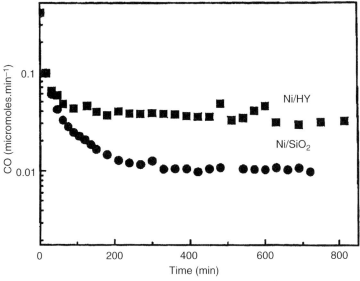

Figure 4.15 Carbon monoxide content in the hydrogen product of methane cracking over nickel/silica and nickel/zeolite Y; weight hourly space velocity 20 L $(h\, g_{cat})^{-1}$; temperature 550°C [265].

had also reported good activity of nickel-based catalysts containing 16.4 wt.% nickel for cracking. They demonstrated that the catalyst could be regenerated to full initial activity by treatment in air or steam at 550 °C. During steam regeneration, the catalyst produced a decreasing amount of additional hydrogen in the course of time.

Takenaka *et al.* applied nickel, nickel/copper and nickel/palladium catalysts supported on silica for the cracking of liquefied petroleum gas and kerosene components [266]. The catalysts contained 20 wt.% nickel while 17 mol% of the second metal were added to the nickel for the bimetallic catalysts. At a 550 °C reaction temperature, undiluted propane was fed to the reactor, which served as a surrogate for LPG. About 20% propane conversion could be achieved at a relatively high weight hourly space velocity of 80 L $(h\, g_{cat})^{-1}$ at 100 vol.% propane in the feed. The nickel/palladium catalyst was most active under these conditions for 180 min, conversion decreased from 20 to 10% in this period. Methane formation was observed, which lowered the hydrogen selectivity, but it still exceeded 90% for all catalysts under investigation [266]. Increasing the palladium content of the nickel/palladium catalyst did impair the catalyst performance. The cracking of n-hexane, cyclohexane and n-octane revealed a hydrogen generation comparable to propane for 90 min over the nickel/palladium catalyst. However, the catalyst deactivated rapidly within 10 min when toluene was used as feedstock. The activity of the nickel/palladium catalyst was not affected when 20 ppm benzothiophene were added to the feed. Higher amounts of sulfur lead to rapid deactivation. Hydrogen generation was even possible from kerosene for 60 min, however, the weight hourly space velocity had to be reduced to 30 L $(h\, g_{cat})^{-1}$ and the kerosene feed was diluted to 20 vol.%. Takenaka *et al.*

did not report carbon monoxide formation over their catalyst, but it probably did occur, taking into consideration the other work discussed above.

Pyrolysis oil is a mixture of 15–25 wt.% acids such as acetic acid, 1–3 wt.% esters such as methylformate butyrolactone, 2–6 wt.% alcohols such as methanol and ethanol, 2–4 wt.% ketones such as acetone, 10–20 wt.% aldehydes, 2–6 wt.% phenols, substituted aromatics and 2–6 wt.% furfurals along with about 30 wt.% water [267]. It was converted into a gas mixture containing hydrogen, methane and carbon oxides by thermal cracking over platinum and rhodium catalysts at 700 °C by Iojoiu et al. [267]. The samples contained 1 wt.% of the noble metal and were supported by ceria/zirconia mixed in a 1:1 molar ratio [267]. Owing to the water content of the pyrolysis oil, the cracking was actually a steam reforming at a very low S/C of 0.5 [267]. After cracking intervals of 5–15 min duration the catalyst had to be regenerated in air for the same duration. The rhodium catalyst outperformed platinum with respect to stability and hydrogen productivity. Coating the catalyst onto ceramic monoliths decreased the carbon dioxide/carbon monoxide ratio but improved the catalyst stability. However, the cracking/regeneration sequences decreased the surface area of the catalysts to about 43% of the original value. The regenerated catalysts still contained 0.3 wt.% carbon.

4.2.10
Deactivation of Reforming Catalysts by Sintering

Cheekatamarla and Lane [268] observed 50% loss of specific surface area for their 1 wt.% platinum/ceria catalyst after 56 h of autothermal reforming of synthetic low sulfur (10 ppm) diesel fuel. In parallel, the dispersion of the platinum decreased from 51 to 41%. These workers did not attribute a sintering process of the platinum to the latter effect. However, platinum in particular is known to suffer from dispersion losses when exposed to an oxidising atmosphere even at temperatures exceeding not more than 400 °C [269]. Thus, both the carrier material and the active species suffered from the elevated temperature of the reforming process.

4.2.11
Deactivation of Reforming Catalysts by Coke Formation

Hydrocarbon cracking [see Eq. (3.22), Section 3.2] may lead to coke formation. This reaction is well known from nickel catalysts, which form whisker-like coke deposits at high temperature [270]. This type of pyrolytic carbon formation takes place above a critical temperature [81], which is about 600 °C for methane [271]. Precious metal catalysts such as rhodium generally have a lower tendency towards coke formation [272] and are more active, which makes them attractive for hydrocarbon reforming in smaller-scale systems (see Section 4.2).

Another route towards coke formation originates from the polymerisation of unsaturated hydrocarbons such as ethylene being present as surface intermediates on the catalyst. These species may then loose more and more hydrogen by formation of aromatic (ring) structures and finally result in a carbon deposit, which then contains relatively little hydrogen. In general, a higher temperature suppresses coke formation according to the polymerisation route because of the thermodynamic equilibrium, and a minimum temperature exists to prevent the formation of this type of coke, which is also termed "gum" in industrial reforming processes [81]. However, coke may not just be formed at the catalyst itself. The higher the reaction temperature, the more likely is the formation of coke on the metal (steel) housing of the reactor also. In particular, nickel-containing steels and of course nickel-based high-temperature resistant alloys are known to have a tendency towards coke formation under certain conditions (see Section 3.12).

Paraffins and condensed aromatics such as naphthalene have a higher tendency towards coke formation compared with aromatic fuels [74]. This is probably explained by the spectrum of primary products, which has been determined by Flytzani-Stephanopoulos et al. [74,153] over nickel-based commercial catalysts (see Section 3.3, Figure 3.15. The logarithmic scaled graph shows more pronounced formation of unsaturated hydrocarbons such as ethylene, acetylene and propylene over the length of the catalyst bed for a tetradecane feed compared with a benzene feed. The concentration of these species passed through a maximum in most instances and decreased significantly at the reactor outlet, but unsaturated hydrocarbons are known to have a high tendency towards coke formation through polymerisation reactions as described above.

It has been shown by Rostrup-Nielsen et al. [270] that under the conditions of steam reforming (S/C = 2; 500 °C reaction temperature) the coking tendency increases in the following order:

> ethylene > benzene > n-hexane > n-heptane > cyclohexane > trimethyl butane
> > n-butane

Olefins tend to form carbon deposits by homogeneous pyrolysis reactions at temperatures greater than 700 °C.

Coke formation may also originate from carbon monoxide reacting with hydrogen to produce carbon and steam [see Eq. (3.20) in Section 3.2].

Generally, coke formation is less likely in the presence of oxygen and a steam excess. This is one reason for providing a certain excess of steam to steam reformer reactors. Calculations by Seo et al. [66] revealed that under conditions of methane steam reforming, carbon was no longer a stable species, when the reaction temperature was greater than 850 °C. Further calculations revealed that carbon was not stable at S/C ratios higher than 1.4 for reaction temperatures between 600 and 800 °C, as shown in Figure 4.16. Increasing the pressure also decreases coke formation [66].

Coke formation is most critical with respect to partial oxidation. Thermodynamic calculations of Seo et al. [66] performed for the partial oxidation of methane showed that a suprastoichiometric O/C ratio of 1.2 (equivalent to an air ratio λ of 0.3) is required to prevent carbon formation (see Figure 4.17). This means that a certain degree of total oxidation is required to prevent carbon formation. However, coke

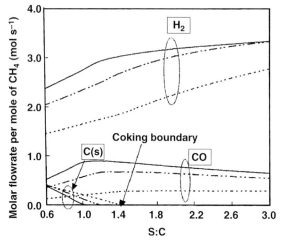

Figure 4.16 Effect of the S/C ratio on the equilibrium composition of methane steam reforming: reaction pressure, 1 bar; reaction temperature, 600 °C (dotted line), 700 °C (dashed line), 800 °C (straight line) [66].

formation may still occur despite the thermodynamics, especially with the higher hydrocarbon feed as discussed below.

Higher hydrocarbons may form stable coke at much higher S/C ratios than 1.4 as is known from many applications, because the polycondensed and polymeric clusters are not easily converted in the presence of steam.

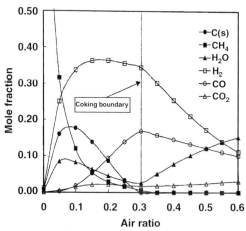

Figure 4.17 Effect of the O/C ratio [here expressed as air ratio $\lambda = (O/C)/4$] on the equilibrium composition of methane partial oxidation; reaction pressure, 1 bar; calculated for an adiabatic reactor with a 200 °C preheating temperature of the methane/air feed [66].

Carbon formed *in situ* in the monatomic state or in the form of small clusters can be removed by methanation:

$$C + 2H_2 \leftrightarrow CH_4 \quad \Delta H^0_{298} = -75 \text{ kJ mol}^{-1} \quad (4.12)$$

or gasification by steam:

$$C + H_2O \leftrightarrow CO + H_2 \quad \Delta H^0_{298} = 131 \text{ kJ mol}^{-1} \quad (4.13)$$

However, these reactions will not take place when the carbon has already formed a stable coke deposit phase.

Removal of the coke deposits is normally feasible through treatment with oxygen at an elevated temperature. However, it has the potential of damaging the catalyst, because the coke combustion is highly exothermic, thus easily over-heating the catalyst when significant amounts of coke are present.

Shamsi et al analysed coke deposits formed over a platinum/alumina catalyst during steam reforming, autothermal reforming and partial oxidation of various components of diesel, namely tetradecane, decaline and 1-methylnaphthalene [273]. The experimental conditions chosen are summarised in Table 4.1. For all types of reforming reactions, decaline produced the least amount of coke, followed by tetradecane and 1-methylnaphthalene. The coke was deposited both on the precious metal and the support. Temperature programmed oxidation experiments were carried out to determine the stability of the coke deposit. The coke could be removed easier from the precious metal surface by this oxidative treatment than from the support. The coke formed under conditions of steam reforming was easier to oxidise and could be removed completely by oxygen treatment at a temperature of 800 °C. The coke formed during autothermal reforming was more graphitic and could not be removed completely, even at 900 °C.

4.2.12
Deactivation of Reforming Catalysts by Sulfur Poisoning

Only a limited number of publications have dealt with the sulfur tolerance of reforming catalysts and even fewer with the reaction mechanisms. Very low levels of sulfur can even deactivate certain catalysts by irreversible adsorption [274].

Cheekatamarla *et al.* [8] investigated thoroughly the deactivation effects of sulfur components contained in JP-8 (American standard fuel containing about 1000 ppm sulfur) and synthetic diesel fuel (about 10 ppm sulfur) on a 1 wt.% platinum/ceria

Table 4.1 Experimental conditions chosen by Shamsi et al. for their investigations on coke formation by diesel components over a platinum/alumina catalyst [273].

Experimental conditions	ATR	SR	POX
O/C	0.6	0.0	1.0
S/C	1.5	3.0	0.0
L/(g_{cat} h) (tetradecane and decalin)	110	110	110
L/(g_{cat} h) (1-methylnaphtalene)	52	52	52
Temperature [°C]	825	800	800

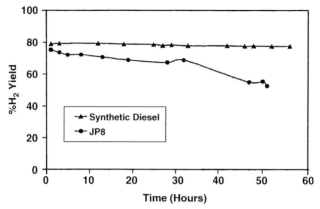

Figure 4.18 Deactivation of Pt/CeO$_2$ catalyst by synthetic diesel fuel (10 ppm S) and JP-8 diesel fuel (1000 ppm sulfur) measured via the hydrogen yield, experimental conditions: S/C = 2.5; O/C = 1; GHSV = 17 000 h^{-1} [8].

catalyst during autothermal reforming. The investigations were performed at S/C ratios between 1 and 3 and at O/C ratios from 1 to 4. Despite the fact that these O/C values are really too high for practical applications, some insight into the mechanisms of sulfur poisoning of platinum catalysts was gained. Figure 4.18 shows a comparison of the deactivation behaviour with time as observed for both fuels mentioned above on stream.

To study the effect of increasing sulfur content in the reformate, sulfur dioxide was added in small amounts of between 100 and 400 ppm to the feed mixture of the autothermal diesel reformer. Sulfur dioxide was chosen because under the conditions of autothermal reforming most sulfur components are converted into sulfur dioxide (see Section 3.5). Activity measured as hydrogen yield showed a drastic decrease from 75 to 40%, when 200 ppm sulfur dioxide were added to the feed. However, no further detrimental effects were recorded when more sulfur dioxide was added. A similar behaviour was observed when hydrogen sulfide was added. This is because the plateau had already been reached at 75 ppm hydrogen sulfide addition.

The most interesting observation of Cheekatamarla et al. [8] was that the poisoning by both sulfur dioxide and hydrogen sulfide was partially reversible (see Figure 4.19). This was demonstrated by long-term experiments where the poison addition was switched on and off in an alternating manner. The activity of the catalyst was partially recovered when the poison addition was stopped. Similar behaviour was observed for sulfur dioxide and hydrogen sulfide. The sulfur poisoning decreased the hydrogen yield and the concentration of carbon dioxide, while the concentration of carbon monoxide and methane in the product were increased. The reversibility of sulfur poisoning was confirmed under more realistic conditions by switching from sulfur rich JP-8 diesel to sulfur free synthetic diesel.

Palm et al. [259] also observed partially reversible, but severe, poisoning of a precious metal catalyst in medium-term experiments when adding between 10 and 30 ppm of sulfur to their C_{13}–C_{15} hydrocarbon feed.

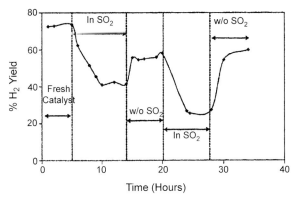

Figure 4.19 Reversible deactivation of Pt/CeO$_2$ catalyst by synthetic diesel fuel (10 ppm S) under temporary addition of 200 ppm sulfur dioxide, experimental conditions: S/C = 2.5; O/C = 1; GHSV = 17 000 h^{-1} [8].

The catalyst of Cheekatamarla and Lane lost 75% of its original surface area when exposed for more than 50 h to the sulfur containing feed, whereas only a 50% reduction in surface area was observed for the low sulfur feed [268]. In parallel, the dispersion of the platinum decreased from 51 to 35%. These workers explained these observations by two types of sulfur species being present on the catalyst surface: firstly, physically adsorbed sulfur species and, secondly, their oxidised products, which were then irreversibly bound to the catalyst sites by chemical adsorption. The activity of the catalyst could be completely re-gained when it was reduced in 20 vol.% hydrogen (with a balance of helium) for 30 min at 800 °C.

A 5% weight increase of the platinum/ceria catalyst was observed when exposed to sulfur dioxide and an excess of air at a 400 °C reaction temperature. Almost identical results were gained for a pure ceria carrier, which supports the assumption that ceria is extremely sensitive to sulfur poisoning.

The formation of sulfates by sulfur dioxide has been described for the ceria carrier [268]:

$$6CeO_2 + 3SO_2 \rightarrow Ce_2(SO_4)_3 + 2Ce_2O_3 \tag{4.14}$$

The mechanism is more complicated in the presence of platinum [268]:

$$Pt + SO_2 \rightarrow Pt - SO_2 \tag{4.15}$$

$$2Pt - SO_2 + O_2 \rightarrow Pt - SO_3 \tag{4.16}$$

$$4Pt - SO_3 + 2CeO_2 + 2.5O_2 \rightarrow Ce_2(SO_4)_3 + CeOSO_4 + 4\,PtO \tag{4.17}$$

The recovery of activity by diluted hydrogen treatment at 800 °C described above was explained by the following reactions:

$$CeOSO_4 + H_2 \rightarrow CeO_2 + SO_2 + H_2O \tag{4.18}$$

$$Ce_2(SO_4)_3 + 2H_2 \rightarrow 2CeO_2 + 3\,SO_2 + H_2O \tag{4.19}$$

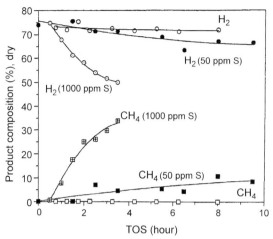

Figure 4.20 Deactivation of nickel-based catalyst by sulfur poisoning: S/C = 3.6; T = 800 °C; weight hourly space velocity of 7.7 L (min g$_{cat}$)$^{-1}$ [275].

A higher tolerance to sulfur poisoning was observed for autothermal reforming of synthetic diesel fuel over bimetallic platinum/palladium and platinum/nickel catalysts compared with monometallic samples [257].

Ming et al. demonstrated the effect of sulfur poisoning on a nickel catalyst formulation containing 12 wt.% nickel [275]; 50 and 1000 ppm of 1,4-thioxane were added to

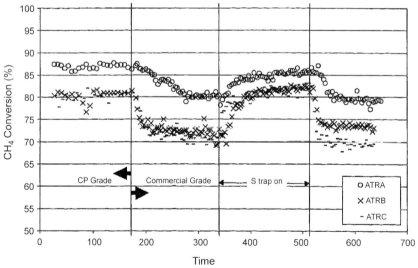

Figure 4.21 Partial and reversible poisoning of platinum/palladium/rhodium catalyst during autothermal reforming of methane; in the first and third phase of the experiment the feed was sulfur-free, but it contained 2 ppm sulfur in the second and fourth phases [57].

isooctane fed to a testing reactor operated at 800 °C, S/C ratio 3.6 and low weight hourly space velocity of 7.7 L (min g_{cat})$^{-1}$. Figure 4.20 shows the decreasing content of hydrogen and the increasing formation of methane due to the deactivation procedure.

Giroux et al. from Engelhard observed deactivation of their platinum/palladium/rhodium autothermal reforming catalyst by only 2 ppm sulfur under conditions of autothermal reforming of methane [57]. However, as shown in Figure 4.21, only partial deactivation occurred, which stopped after a certain loss of activity. The catalyst could be regenerated by switching back to a sulfur-free feed.

Lombard et al. reported full regeneration of nickel catalysts poisoned by sulfur through treatment with pure steam at 720 °C for 15 h [276].

When robust catalyst formulations are chosen for reforming, they are probably tolerant to low levels of sulfur in the region of 10 ppm [75].

4.3 Catalysts for Hydrogen Generation from Alternative Fuels

4.3.1 Dimethyl Ether

Steam reforming of dimethyl ether can be performed over methanol steam reforming catalysts mixed with an acidic catalyst [277]. In fact dimethyl ether is frequently a by-product of methanol steam reforming (see Section 3.1). However, the reforming of dimethyl ether is obviously more difficult than methanol reforming, as discussed below.

As described in Section 3.8, dimethyl ether steam reforming is two-step reaction, namely decomposition of the ether to methanol followed by methanol steam reforming. While the former reaction is catalysed by acidic catalysts, the latter reaction is commonly performed over copper/zinc oxide catalysts. Consequently, Semelsberger prepared numerous copper/zinc oxide catalysts on acidic carrier materials such as Y- and ZSM-5 type zeolites, alumina and zirconia [277]. The catalysts were tested at 0.78 bar pressure, S/C 1.5 and a high residence time of 1 s for reaction temperatures between 150 and 400 °C. However, the catalysts showed low selectivity towards hydrogen production. The best performing catalyst was copper/zinc oxide on a γ-alumina carrier, which required 400 °C reaction temperature to convert the fuel despite the high residence time applied. Owing to the high temperature, the carbon monoxide content of the reformate was high with 9 vol.% [277]. This is a high value especially when compared with values reported for methanol steam reforming over conventional copper/zinc oxide catalysts, which usually generate around 1 vol.% carbon monoxide in the feed.

Zhang et al investigated the partial oxidation of dimethyl ether at very high reaction temperatures of between 600 and 700 °C, because the intention was to utilise the reformate as fuel for a solid oxide fuel cell [278]. The O/C ratio was set to 1.5 (oxygen from dimethyl ether is included in this ratio), while the weight hourly space velocity of 38 L (h g_{cat})$^{-1}$ was low when taking into consideration the high reaction temperature

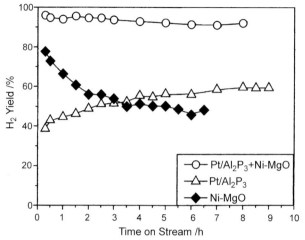

Figure 4.22 Hydrogen yields versus time over platinum/alumina and nickel/magnesia catalysts and a mixture of both: pressure, 1 bar; temperature, 700 °C; feed composition, 16 vol.% dimethyl ether, 17 vol.% oxygen, 67 vol°% argon [278].

and the relatively "fragile" fuel molecule. Platinum/alumina and nickel/magnesia catalysts were either switched in series or blended in a uniform catalyst bed. The platinum catalyst was active even at 600 °C but suffered from rapid deactivation, which was attributed to coke formation and sintering of the platinum particles. The nickel catalyst was more stable over the medium-term test duration of 10 h. However, the hydrogen yield did not exceed 60% even for the nickel catalyst, as shown in Figure 4.22. When both catalysts were blended, stable conversion was found for 9 h with high hydrogen yields. Methane and traces of ethane were formed as by-products, which was attributed to radical reactions [278]. However, the methane content of the product decreased with increasing reaction temperature according to the equilibrium of the methanation reaction (reverse methane steam reforming). Lower residence times suppressed methane formation. At a reactor inlet temperature of 725 °C, a hot spot at 800 °C was determined over the catalyst bed.

4.3.2
Methylcyclohexane

Kariya et al. performed dehydrogenation of methylcyclohexane and other cycloalkanes over platinum, palladium and rhodium monometallic and platinum/palladium, platinum/rhodium, platinum/molybdenum, platinum/tungsten, platinum/rhenium platinum/osmium and platinum/iridium catalysts supported on both petroleum coke active carbon and on alumina between 375 and 400 °C [279]. The platinum catalyst supported by petroleum active carbon showed the highest activity. While platinum was the most active monometallic catalyst, its activity could be increased by addition of molybdenum, tungsten and rhenium.

Roumanie et al. tested catalysts in a chip-like silicon microreactor for methylcyclohexane dehydrogenation [280]. A platinum/alumina catalyst achieved 88.5% conversion, while a platinum film sputtered onto "black silicon" showed only 2% conversion. Low activity is frequently observed for non-dispersed noble metal surfaces.

4.3.3
Sodium Borohydride

Ruthenium catalyst impregnated onto an ion exchange resin may serve as a catalyst for the hydrolysis of sodium borohydride [99]. Other catalyst formulations reported in the literature are metal halides such as nickel and cobalt chloride, colloidal platinum, active carbon, raney nickel, fluorinated particles of magnesium-based material [100] and nickel boride [281]. Zhang et al. reported that nickel powder is an active catalyst under conditions of a batch reactor [282]. With some time delay, the reaction started from ambient conditions.

Kojima et al. investigated a wide variety of catalyst carriers and active metals for sodium borohydride hydrolysis. They reported platinum precipitated onto $LiCoO_2$ to be the most active formulation [100].

4.3.4
Ammonia

Suitable catalysts for ammonia decomposition are nickel, ruthenium and iridium [12, 102].

Ammonia decomposition was investigated by Choudhary et al. [283] over nickel, iridium and ruthenium catalysts supported by various carrier materials such as ZSM-5 and Y-zeolites, alumina and silica. Ruthenium on silica was most active, followed by iridium and nickel [283].

Ganley et al. developed a ruthenium catalyst for ammonia decomposition [284]. The catalyst was deposited on small microstructured aluminium reactors of only 0.37-cm^3 volume. The reactors either carried channels 3-mm deep and between 140- and 260 μm wide or posts all fabricated by electro-discharge machining (see Section 10.2.3). The aluminium reactors were then treated by anodic oxidation to generate a porous alumina layer on their inner surface. Then 3.5 wt.% ruthenium was impregnated as ruthenium chloride onto the surface. Surprisingly the reactors were operated up to a temperature of 650 °C, which is rather high for aluminium. At this temperature, full conversion of 145 $cm^3 min^{-1}$ ammonia was achieved. Promotion of the catalyst with 0.8 wt.% potassium further improved its performance. Higher conversion was achieved in the channels compared with the posts, and the smaller channels showed higher conversion. This was attributed to mass transfer limitations of the reaction. At a 600 °C reaction temperature, 99% ammonia conversion could be achieved, which corresponded to about 12 $L h^{-1}$ hydrogen production by the small reactor.

4.4
Desulfurisation Catalysts/Adsorbents

Basically two strategies exist for the removal of sulfur compounds from hydrocarbons. They may be removed upstream or downstream of the reformer. For a higher sulfur content, as is present in mixtures of higher hydrocarbons such as heavy diesel or aviation fuel, the desulfurisation must be performed upstream of the reformer, otherwise the reforming catalyst suffers from sulfur poisoning. The operating temperature of the adsorption materials applied is usually around ambient or in a temperature range below 300 °C. Hydrodesulfurisation, which is applied on an industrial scale, is not suited to small mobile fuel processors, but could be used in stationary applications on the smaller scale.

Industrial hydrodesulfurisation of natural gas is performed over cobalt/molybdenum or nickel/molybdenum catalysts supported by alumina [109] in the presence of hydrogen under a high pressure of 20 bar and temperatures up to 500 °C. The catalysts contain high amounts of molybdenum oxide in the range of 16 to 20 wt.% in many instances [285]. A formulation containing 40 wt.% iron oxide, 5 wt.% zinc oxide, 24 wt.% silica, 7 wt.% alumina and chrome, calcium and manganese oxides was recently proposed by Zaki *et al.* for removal of dimethyl disulfide [286]. Hydrogen sulfide is a product of some of these processes, which is then adsorbed downstream by zinc oxide between 200 and 400 °C [107]. Hydro processing of gasoline and jet/diesel fuels is performed at even higher pressures of up to 100 bar and temperatures up to 400 °C. The heavy sulfur components of diesel fuel such as dimethyldibenzothiophene are more difficult to remove and therefore Richard *et al.* proposed the addition of acidic catalysts, such as zeolite Y, to nickel/molybdenum/alumina catalysts [287].

ConocoPhillips developed a special adsorbent S-Zorb, which is capable of removing hydrogen sulfide [288] but also the sulfur compounds of gasoline and diesel [109] in the presence of hydrogen. It released the sulfur-free organic compounds [109]. The process worked between 340 and 410 °C and in a pressure range of between 7 and 20 bar. The purified gasoline product contained only 10 ppm sulfur [109].

Techniques suitable for mobile fuel cell applications on the smaller scale are adsorption or catalysed adsorption techniques.

The catalysts/adsorbents are frequently nickel based [255], containing high amounts of nickel similar to hydrogenation catalysts. The operating temperature is in a range between 100 and 300 °C. The catalysts form nickel sulfides and their regeneration is possible by treatment in air. An alternative are the nickel/zinc oxide catalysts, which initially form nickel sulfide and subsequently zinc sulfide, while the nickel is regenerated [109].

Fukunaga *et al.* presented results from lifetime analysis of their self-developed nickel-based sulfur adsorbent for kerosene [255]. For 4000-h test duration, the sulfur components of commercial kerosene could be reduced from 48 to less than 0.05 ppm.

Desulfurisation at room temperature is a preferred solution for small-scale fuel processors [105]. It is performed by selective adsorption for removing sulfur (SARS) [1].

Activated carbons are possible adsorbents [289] as are carbon aero gels [290]. Activated carbons have a low adsorption capacity, for higher hydrocarbons this is in the region of only 1 g S per 100 g carbon with almost no capacity for carbon oxide sulfide [107]. Impregnating active carbons with alkali earth or transition metals such as potassium, copper, iron, chromium or combinations thereof improves their adsorption capacity for hydrogen sulfide and ethylmercaptane [5,107]. A zinc oxide/iron oxide/ceria/alumina adsorbent was proposed by Zhang *et al.* [291] for sulfur removal from gasoline. It contained 4.5 wt.% zinc oxide, 2.2 wt.% iron oxide and 2.5 wt.% ceria on alumina and was operated at a 60 °C adsorption temperature.

Trace amounts of sulfur compounds were removed over cerium exchanged Y-zeolite adsorbents in the work reported by Xue *et al.* [292].

Owing to the lower sulfur content of light hydrocarbon mixtures, such as low-sulfur gasoline and diesel, liquefied petroleum gas and natural gas, which usually only contain sulfur components in the region of 50 ppm or less, it is possible to convert all sulfur species in the reformer into hydrogen sulfide (see Section 4.2) and perform the desulfurisation downstream of the reformer. The corresponding adsorbents have an operating temperature well suited to being positioned downstream of the reformer, which works above 600 °C, and upstream of the water–gas shift, operated below 450 °C in most instances. Carbon monoxide clean-up catalysts for medium and low temperature water-gas shift, selective methanation and preferential oxidation are sensitive to sulfur poisoning (see Section 4.5).

Metal oxides of copper, zinc, calcium, iron and mixtures thereof [293] and active carbons are the preferred adsorbents for hydrogen sulfide. Owing to the operating temperature range of between 600 (reformer outlet) and 400–450 °C (water–gas shift inlet) as discussed above, the sulfides need to be resistant towards hydrolysis by the steam that is present in the reformate.

At temperatures below 300 °C, hydrogen sulfide can be reduced to below 100 ppb by zinc oxide in the absence of carbon monoxide and steam [294]. At higher temperatures, a mixture of zinc oxide and titania avoids the partial reduction of zinc oxide by the hydrogen contained in the reformate [293]. While zinc oxide has a high adsorption capacity of around 10 g S per g zinc oxide in dry gas, this value decreases considerably in the presence of steam to 0.1 g S per g zinc oxide [107]. According to Li an King, the addition of 1 vol.% carbon monoxide lowered the sulfur removal capability of their zinc oxide adsorbent [294]. The formation of carbon oxide sulfide by the reaction:

$$H_2S + CO \rightarrow H_2 + COS \qquad (4.20)$$

was observed and suspicions actually arose that the carbon oxide sulfide was formed from carbon dioxide via the reaction:

$$H_2S + CO_2 \rightarrow H_2O + COS \qquad (4.21)$$

However, the formation of carbon oxide sulfide is suppressed by binding the zinc oxide with acidic alumina [295]. Sulfate formation can be suppressed by suitable composition of the oxide adsorbent.

Rosso et al. prepared various zinc oxides and investigated their adsorption capacity in dry hydrogen sulfate containing gas at 250 °C [293]. The sulfur adsorption capacity of the zinc oxides decreased when calcination temperature exceeded 400 °C. However, sulfur uptake did not exceed 5 wt.%. The adsorbents could be partially regenerated in air at 625 °C.

Lampert presented a catalytic partial oxidation technique for sulfur compounds that was developed by the former Engelhard (now BASF) corporation [296]. The sulfur compounds of natural gas or liquefied petroleum gas were converted into sulfur oxides at a low O/C ratio of 0.03 in a ceramic monolith over a precious metal catalyst. These sulfur oxides were then adsorbed downstream by a fixed adsorber bed, which contained adsorption material specific to sulfur trioxide and sulfur dioxide, which could trap up to 6.7 g sulfur per 100 g adsorbent. The partial oxidation was performed at a 250 °C monolith inlet temperature, the adiabatic temperature rise in the monolith amounted to 20 K. Light sulfur compounds usually present in natural gas and liquefied petroleum gas, such as carbon oxide sulfide, ethylmercaptane, dimethyl sulfide and methylethyl sulfide, could be removed to well below the 1 ppm level. Exposure of the monolith to an air rich fuel/air mixture at temperatures exceeding 150 °C had to be avoided. The same applied for contact with fuel in the absence of air regardless of the temperature.

Catalytic partial oxidation of hydrogen sulfide by activated carbon is another alternative [297]. Coal-derived synthesis gas containing 36 vol.% carbon monoxide, 27 vol.% hydrogen, 12 vol.% carbon dioxide, 18 vol.% steam and 1000 ppm hydrogen sulfide with a balance of nitrogen was used as feed, as was natural gas containing 500 ppm hydrogen sulphide also. From the coal gas, hydrogen sulfide could be removed to below 1.2 ppm at an O/H_2S ratio of 10 in the temperature range between 135 and 155 °C. An oxygen excess was required to reduce carbon oxide sulfide formation to below 30 ppm. Formation of sulfur dioxide occurred at temperatures exceeding 155 °C. The natural gas was purified to less than 1 ppm hydrogen sulfide in the same temperature range.

Hydrogen sulfide removal from oxygen containing gas mixtures at ambient temperature was performed with various adsorbents by Meeyoo and Trimm [298]. Their adsorption capacity increased in the following order:

> alumina, magnesia, molecular sieve 13X $\ll$ charcoal < activated carbon fibre < activated carbon

The activated carbons could adsorb up to 18 wt.% sulfur by oxidation of hydrogen sulfide to give elementary sulfur. This value was even improved in the presence of less than 2.5 vol.% steam. However, the adsorption capacity of the activated carbon decreased dramatically when the operating temperature was slightly increased from 25 to 40 °C because the oxidation reaction was suppressed at elevated temperatures [298].

Recent work by Flytzani-Stephanopoulos et al. dealt with lanthana and ceria adsorbents, which were operated at high temperatures of up to 800 °C for solid oxide fuel cell applications [299]. However, their adsorption capacity was much lower compared with zinc oxide, in the region of 0.001 g S per g adsorbent.

4.5
Carbon Monoxide Clean-Up Catalysts

4.5.1
Catalysts for Water–Gas Shift

In large-scale industrial processes, water–gas shift is usually performed in two consecutive adiabatic beds. The first high-temperature stage contains an iron oxide catalyst stabilised by chromium oxide, its feed temperature ranges between 350 and 450 °C. The product temperature then rises to 400–500 °C because of the exothermic shift reaction in the adiabatic bed [57]. The reformate is then cooled and fed at about 200 °C to the low temperature shift stage. The temperature rise over the copper/zinc oxide catalyst is much lower at 20–30 K because less carbon monoxide is converted. The temperature resistance of the latter catalyst is limited below 280 °C [57]. In the industrial processes catalyst utilisation is poor for both stages [57]. Early fuel processors utilised fixed beds of commercial water–gas shift catalysts, and the shift stages dominated the overall system volume and weight by up to 50% due to the low catalyst utilisation [164].

Iron/chromium catalysts and even more copper/zinc oxide water–gas shift catalysts are pyrophoric, which means that they heat up significantly through oxidation reactions at the catalyst surface when exposed to air. Additionally, copper/zinc oxide catalysts deactivate when exposed to air [57], which makes them not particularly suitable for smaller scale fuel processors, where air exposure may happen both during start-up and during longer periods of shut-down. Oxygen assisted water–gas shift over copper/zinc oxide catalysts as proposed by Utaka et al. [300] does not seem to be a viable option owing to the sensitivity to air exposure. Copper/zinc oxide catalysts are very sensitive to poisons, such as sulfur species. Other copper catalysts, for example, copper/ceria and copper/alumina catalysts are less active than copper/zinc oxide systems and are also sensitive to sulfur poisoning [107].

Industrial iron/chromium catalyst is insensitive to sulfur and could be used to remove sulfur components from reformate by adsorption [57].

Non precious metal low-temperature water-gas shift catalysts, such as molybdenum carbide, have been studied [301], but they are also sensitive to the presence of oxygen [164], which makes them not very suitable for practical systems.

Cobalt catalysts such as cobalt/manganese and cobalt/chromium show higher activity than iron/chromium catalysts at temperatures exceeding 300 °C and are highly sulfur tolerant [107]. However, their activity is certainly lower than that of the precious metal catalysts discussed below. Additionally, they are not suitable for low-temperature applications due to their low activity in this temperature range. Ruettinger et al. reported on proprietary base-metal water–gas shift catalyst development. The catalysts were claimed to have lower pyrophoricity than copper/zinc oxide catalysts, and to be stable towards air exposure at 150 °C and even to liquid water [302].

As an alternative to industrial scale catalysts, precious metal catalysts coatings show much higher activity.

Figure 4.23 shows the comparison of a fixed-bed reactor filled with iron/chromium catalyst with a monolith coated with precious metal water–gas shift catalyst. At

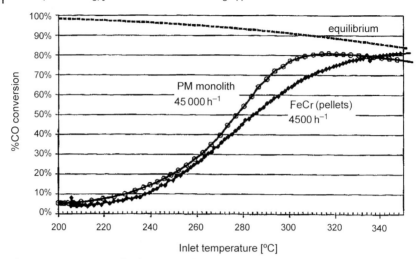

Figure 4.23 Comparison of carbon monoxide conversion over of a fixed bed of iron/chromium catalyst operated at 4500 h^{-1} gas hourly space velocity and a monolith coated with precious metal catalyst operated at 45 000 h^{-1} for reformate containing 5.9% carbon monoxide, 7.4% carbon dioxide, 31.7% hydrogen, 28.8% nitrogen and 26% steam [57].

ten times higher gas hourly space velocity, the activity of the monolith is still superior to the fixed bed. This means that the reactor would actually be ten times smaller in this instance.

Precious metals such as platinum, palladium, rhodium and ruthenium on ceria supports are reported in the literature as formulations for a water–gas shift [164]. Certainly the most prominent formulation is platinum/ceria, which has the drawback of a particular activity towards methane formation [164] at temperatures exceeding 375–425 °C, depending on the catalyst formulation. Its long-term stability is usually limited to a maximum temperature of between 425 and 450 °C due to the sintering effects of platinum.

NexTech Materials Company have developed a platinum/ceria catalyst, which is highly active and long-term stable in the temperature range between 300 and 400 °C [164]. Johnson Matthey have developed platinum-containing precious metal catalysts for water–gas shift, which are long-term stable, show high activity even at 250 °C and no selectivity towards methanation up to an operating temperature of 450 °C [164].

Precious metal based water–gas shift catalysts have zero reaction order for carbon monoxide below 300 °C, which means that the rate of conversion is not affected by the carbon monoxide concentration in the low temperature range [57,303]. This is not the situation for the copper/zinc oxide catalysts described above [304]. However, the products carbon dioxide and hydrogen have an inhibiting effect on the reaction in the low temperature range for both types of catalysts [305, 304]. Many publications in the field of water–gas shift catalysts do not take these effects into consideration, which impairs the applicability of the results considerably. Frequently only carbon monoxide and steam are fed to the catalyst samples and the activity is determined while ignoring the effects of product inhibition.

Phatak et al. performed kinetic measurements over platinum/alumina and platinum/ceria catalysts for a water–gas shift [306]. They observed similar activation energy in the range between 70 and 80 kJ mol^{-1}, but a 30 times higher turnover rate for 1 wt.% platinum/ceria at 200 °C than for platinum/alumina at 285 °C. A power-law expression was chosen for the kinetics:

$$r_{WGS} = k_f [CO]^a [CO_2]^b [H_2]^c [H_2O]^d (1 - \beta) \tag{4.22}$$

$$\beta = \frac{[CO_2][H_2]}{K_{eq}[CO][H_2O]} \tag{4.23}$$

The reaction order a for carbon monoxide was -0.03, while steam had a positive reaction order b of 0.44. Carbon dioxide and hydrogen showed significant inhibition, as indicated by the reaction order of -0.1 for c and -0.44 for d, respectively [306].

The effect of the support on the activity of platinum catalysts was investigated by Rosa et al. [251]. Alumina, ceria, titania and mixed ceria/titania were compared, all containing 0.5 wt.% of platinum. As shown in Figure 4.24, the activity ranking was as follows:

$$\text{ceria/titania} > \text{titania} > \text{ceria} > \text{alumina}$$

The ceria/titania based catalyst was also more stable. The catalyst showed 60-h durability under the conditions provided in Figure 4.24. However, the weight hourly space velocity of 21.2 L (h g$_{cat}$)$^{-1}$ was relatively low. About ten-times higher yields can be achieved with platinum/ceria systems. The activity of the platinum/ceria catalyst is approximately 15-times higher compared with a platinum/alumina catalyst for the

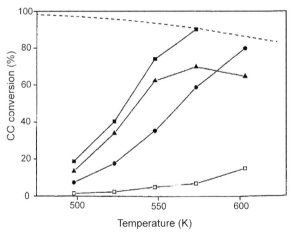

Figure 4.24 Conversion of carbon monoxide by a water–gas shift as determined for platinum catalysts made from different carrier materials: □, alumina; ●, ceria; ▲, titania; ■, titania/ceria; experimental conditions, 28 vol.% H$_2$, 0.1 vol.% CH$_4$, 4.4 vol.% CO, 8.7 vol.% CO$_2$, 29.2 vol.% N$_2$, 29.6 vol.% H$_2$O; WHSV = 21.2 L (h g$_{cat}$)$^{-1}$ [251].

water–gas shift [107]. Similar results were gained for palladium catalysts by Gorte et al. when comparing palladium/alumina, neat ceria and palladium/ceria catalysts. Only the last catalyst showed significant activity for the water–gas shift [307].

Germani et al. prepared platinum/ceria water–gas shift catalysts in microchannels containing between 0.8 and 1.4 wt.% platinum and between 8 and 20 wt.% ceria [308]. The sample containing 1.4 wt.% platinum and 8 wt.% ceria showed the highest conversion, which was decreased when the catalyst was reduced prior to the activity test. Kolb et al. varied the platinum content of their platinum/ceria washcoated catalysts between 1 and 5 wt.%, while the ceria content ranged between 6 and 40 wt.% [269]. The optimum platinum content was determined in the range between 3 and 5 wt.%, while the optimum ceria content was between 12 and 24 wt.%.

The stability of a proprietary water–gas shift catalyst was investigated by Pasel et al., who applied real reformate from an autothermal reformer [309]. No degradation of the catalyst was observed according to these workers for more than 80-h duration. It is difficult to judge this result, because no concentrations for the hydrocarbons in the water–gas shift feed were provided and nor were any weight hourly space velocities reported either.

However, it is questionable whether the water–gas shift catalysts can withstand the presence of higher hydrocarbons in the feed in the longer term. The group working with the author of this book has performed tests over a precious metal based water–gas shift catalyst, which had already passed a 1000-h stability test in the presence of 5 vol.% methane without apparent deactivation. It degraded significantly within less than 200 h when exposed to 5 vol.% propane in the feed.

The origin of deactivation of platinum/ceria water–gas shift catalysts is an issue that is still under discussion in the scientific community [310]. It may originate from sintering of the platinum particles, which impairs their dispersion, growth of the ceria crystallites, which may occlude platinum particles and formation of carbonate-like species, which cover the surface [164]. Zalc et al. stated that the stability of platinum/ceria catalysts is not sufficient for fuel processing applications [311]. However, Zalc et al. operated their platinum/ceria sample at a weight hourly space velocity of 90 L $(h\,g_{cat})^{-1}$ and found rapid degradation within 50 h, while Kolb et al. have reported stable performance of their self-developed platinum/ceria catalyst coatings in microchannels at a weight hourly space velocity of 144 L $(h\,g_{cat})^{-1}$ for more than 1000 h at 325 and 400 °C (see Figure 4.25) [312]. As usual, the preparation method and additives of the catalyst formulation obviously play a dominant role in the catalysis.

Choung et al. [313] reported significant activity and stability gains for their platinum/ceria/zirconia catalysts, which were supported on a mixture of 46 atomic% of ceria and 54 atomic% of zirconia when adding 1–2 wt.% rhenium to the catalyst formulation.

Liu et al. investigated the performance of a platinum/ceria catalyst exposed to reformate at a low temperature of 60 °C [310]. They found significant aging of the catalyst after just 5-min exposure to a mixture containing 3% carbon monoxide and 15% carbon dioxide as shown in Figure 4.26, which was attributed to the formation of

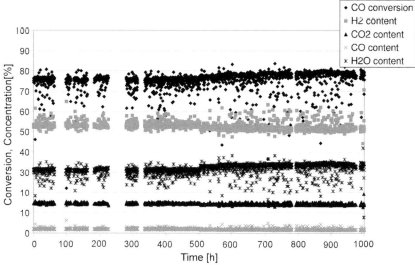

Figure 4.25 1000-h stability test of a platinum/ceria catalyst coating in microchannels for water–gas shift; weight hourly space velocity 72 L (h g_{cat}) first 500 h, 144 L (h g_{cat}) second 500 h; temperature 400 °C feed composition 50 vol.% hydrogen, 9.4 vol.% carbon dioxide, 8.0 vol.% carbon monoxide, 32.7 vol.% steam (source: IMM).

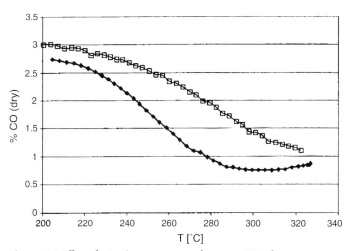

Figure 4.26 Effect of aging by exposure to reformate at 60 °C for 5 min on the performance of a platinum/ceria catalyst; reformate composition 2.2 vol.% carbon monoxide, 11 vol.% carbon dioxide, 36 vol.% hydrogen, 29 vol.% nitrogen, 26 vol.% steam WHSV 20000 (h^{-1}); ♦, fresh catalyst; □, aged catalyst [310].

carbonates under these conditions. Regeneration of the catalyst in reformate at 450 °C was not possible, but treatment with air at 450 °C recovered the catalyst activity partially through oxidation of the carbonates [310].

Care has to be taken that the precious metal water–gas shift catalyst does not overheat during regeneration treatment in an oxidative atmosphere, because this impairs the dispersion of the active metals [314]. It is difficult to judge the actual temperature at the catalyst surface and thus the overall temperature should be kept well below the maximum operating temperature of the catalyst during regeneration.

Wheeler et al. investigated the methanation selectivity of ruthenium, rhodium, nickel, platinum and palladium catalysts on alumina and ceria carriers for short contact times in monolithic reactors in the temperature range between 300 and 800 °C. The platinum and palladium catalysts showed the lowest tendency towards methane formation [315]. Utaka et al. investigated the methanation activity of ruthenium catalysts supported by different carriers [316]. While the ceria supported catalyst showed methanation above a reaction temperature of only 250 °C, vanadium oxide supported ruthenium showed lower activity, but also no significant methanation up to a 400 °C reaction temperature. Kolb et al. investigated the performance of platinum/ceria, platinum/palladium/ceria, platinum/rhodium/ceria and platinum/ruthenium catalysts for a water–gas shift in microchannels. Similar to the results of Wheeler, the samples, which contained only platinum, palladium and ceria, showed much lower tendency towards methane formation compared with rhodium or ruthenium containing samples [269]. Tonkovich also observed methane formation over a ruthenium/zirconia catalyst applied for water–gas shift [317]. Catalysts containing ruthenium are good candidates for the selective methanation of carbon monoxide, as discussed in Section 4.5.3.

Surprisingly, Basiniska et al. found little methanation activity over their ruthenium catalyst supported by α-alumina [318].

Gold catalysts supported on ceria, titania and iron oxide are reported in the literature as candidates for a low-temperature water–gas shift [164, 319], which may well be significantly active in the temperature range below 200 °C [320]. They are, however, very sensitive to the preparation conditions and gold particles generally suffer from sintering issues [164, 321], which makes them not particularly applicable for practical applications to-date.

4.5.2
Catalysts for the Preferential Oxidation of Carbon Monoxide

To understand the behaviour of preferential oxidation catalysts, the operating principle requires explanation. The common feature of these catalysts is the preferential adsorption of carbon monoxide at low temperature. When the reaction temperature increases, the carbon monoxide coverage decreases and reaction with oxygen (when it is present in the gas phase) takes place. At even higher temperatures and lower coverage of active sites with carbon monoxide, hydrogen oxidation occurs in parallel. Thus, an operating window exists for preferential oxidation catalysts.

Where full conversion is achieved over a preferential oxidation catalyst under certain feed conditions (O/C ratio set), a further increase of reaction temperature or over sizing of the reactor may well initiate side reactions, specifically a reverse water–gas shift (see Section 3.10.2).

Preferential oxidation catalysts usually consist of precious metals such as platinum, ruthenium, palladium, rhodium, gold and alloys of platinum with tin, ruthenium [164] or rhodium. Typical carrier materials are alumina and zeolites [164], such as zeolite A, mordenite and zeolite X. Other possible carriers are cobalt oxide, ceria, tin oxide, zirconia, titania and iron oxide [214]. A high precious metal loading usually improves catalyst performance [164].

The catalyst most frequently applied for the preferential oxidation of carbon monoxide is platinum on alumina, which contains low amounts of platinum in the range below 1 wt.% [102, 322], to reduce the cost of the catalyst.

The presence of steam in the feed has a beneficial effect on catalyst performance, at least for precious metal catalysts as demonstrated by several workers (see Figure 4.27) [323–326], although it may deactivate gold catalysts rapidly [107].

The group working with the author of the present book quantified that for a platinum/rhodium catalyst the beneficial effect of water was no longer relevant, when the temperature exceeded 160 °C over platinum/rhodium catalyst. Similar results were obtained by Zhou et al. [322].

Son et al. reported increased activity of their platinum/alumina catalysts by exposure to liquid water and subsequent evaporation [326]. This effect could be helpful in practical applications, because catalysts might well be exposed to condensing steam during start-up procedures in fuel processors. However, when the catalyst was operated below 100 °C, deactivation occurred, which was attributed to accumulation of the condensed water on the catalyst. Activity could be re-gained when the temperature was increased above 100 °C, clearly this was because the water was evaporated.

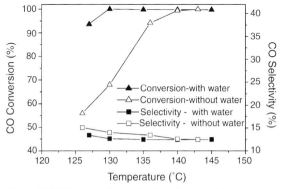

Figure 4.27 CO conversion and selectivity over platinum/rhodium/γ-alumina with and without steam present in the feed versus temperature; O/CO ratio of 8 [324].

Table 4.2 Effect of over-heated catalyst surface on oxygen conversion [325].

Inlet concentrations [Vol.%]					Reactor temperature [°C]		Conversion [%]	
O_2	H_2	CO	CO_2	H_2O	Inlet	Surface	O_2	CO
1.69	43.46	1.41	18.08	7.30	91.8	160.1	30.12	22.15
1.68	44.40	1.41	18.10	7.30	90.0	253.5	101.34	23.01

The heat removal from the catalyst surface is an important issue. As shown in Table 4.2, local overheating of the catalyst leads to completely different, but not desirable, performance of a reactor. For identical feed compositions, the oxygen is completely consumed, but carbon monoxide conversion is not affected [325].

Catalyst development work on catalyst formulations as alternatives to conventional platinum/alumina catalyst, for the preferential oxidation of carbon monoxide, will be discussed below.

An early study by Oh et al. described an extensive catalyst screening. The following activity ranking was obtained [327]:

$$Ru/Al_2O_3 > Rh/Al_2O_3 > Pt/Al_2O_3 > Pd/Al_2O_3 > Co/Cu/Al_2O_3$$
$$> Ni/Co/Fe/Al_2O_3 > Ag/Al_2O_3 > Cr/Al_2O_3 > Fe/Al_2O_3 > Mn/SiO_2$$

Zhou et al. tested various noble metal based catalysts containing 0.5 wt.% of the active species on γ-alumina carrier and found the following activity ranking: Pt > Rh > Ru > Pd [322]. The catalysts had been coated onto small monolithic supports. For a 1000-h test duration, conversion of carbon monoxide below 150 ppm was obtained over a 1 wt.% platinum/alumina sample at an O/CO ratio of 2.0.

Dudfield et al. [328] investigated various catalysts for preferential oxidation in a test reactor that had a design similar to a macroscopic shell and tube heat-exchanger. The catalysts tested are summarised in Table 4.3. The feed, at $10\,\mathrm{L\,min^{-1}}$, composed of

Table 4.3 Catalysts evaluated for preferential oxidation of carbon monoxide in a shell and tube heat-exchanger, by Dudfield et al. [328].

Catalyst Carrier	Active Component 1 [wt.%]	Active Component 2 [wt.%]
Hopcalite	–	–
Aluminium Stannate	Cu 5	La 3
Ferric Oxide	Cu 3	Ce 2
Silica + 4 wt.% Ce	Mn 8	Cu 2
Silica	Pd 2.5	–
Silica	Ru 2	–
Silica	Ru/2.25 Pt 2.25	–
Alumina	Ru/3.5 Pt 3.5	–
Hopcalite	Ru/2.5 Pt 2.5	–

75.0 vol.% H_2, 0.7 vol.% CO, with a balance of CO_2 and air for oxidation, was fed into a relatively large test reactor carrying 230 cm^3 catalyst micro-spheres of 1-mm diameter. Of the non-precious metal catalysts, the sample composed of pure hopcalite showed the highest activity achieving almost full conversion in the temperature window between 130 and 160 °C. The minimum CO concentration achieved was 40 ppm. Of the precious-metal catalysts, the platinum/ruthenium samples showed the highest activity, namely more than 99.8% conversion. In particular, the platinum/ruthenium sample based upon the hopcalite carrier showed even higher conversion in the wide temperature window between 90 and 160 °C; 7 ppm carbon monoxide were detected in the purified reformate [328]. However, hopcalite is not stable towards moisture, which would generate problems in practical applications [329].

However, ceria/zirconia supports may improve the activity of platinum catalysts especially in a temperature range below 100 °C [329].

Ayastuy *et al.* reported higher activity at a much lower temperature for their platinum/ceria catalysts [330]. While the platinum/alumina catalysts showed full conversion at 175 °C, the ceria supported samples showed full conversion by 80 °C. However, it is questionable whether platinum/ceria is a suitable catalyst for preferential oxidation, because it has excellent activity for the water–gas shift. This will impair the performance of the catalyst under partial load in a fuel processor environment due to the reverse water–gas shift taking place, as discussed in Sections 3.10.2 and 5.2.2.

Bimetallic catalysts, mostly combinations of platinum with other metals such as tin [331] and rhodium [324], are promising candidates for the preferential oxidation reaction. Platinum/tin oxide catalysts showed significant reaction rates even at 0 °C [214]. Similar with platinum/rhodium catalysts, an alloy is formed from both metals, which changes the properties of both source metals [214]. Other additive metals, which may improve the activity and selectivity of platinum catalysts, may well be ruthenium and cobalt [329].

Cominos *et al.* [324] tested platinum, ruthenium and rhodium catalysts supported by alumina and ceria and bimetallic platinum catalysts, namely platinum/rhodium, platinum/cobalt and platinum/ruthenium all supported by γ-alumina. The samples were tested as coatings in microchannels. The feed was composed of 58 vol.% hydrogen, 21 vol.% carbon dioxide, 1.12 vol.% carbon monoxide, 4.6 vol.% oxygen and 15 vol.% nitrogen, which corresponded to an O/CO ratio of 8. Over ruthenium/platinum, rhodium/platinum and rhodium conversion close to 100% was achieved. Less than 4 ppm carbon monoxide were detected at 126, 140 and 144 °C reaction temperatures, respectively. The platinum/rhodium catalyst containing 2.5 wt.% of each precious metal was most stable with time, maintaining its high activity at 120 L $(h\,g_{cat})^{-1}$ weight hourly space velocity for 50 h. Subsequently, the catalyst was improved further and passed 1000-h durability tests, as reported by Kolb *et al.* [312]. Conversion to less than 100 ppm carbon monoxide was achieved at a 160 °C reaction temperature, much lower O/CO ratio of 3.0 and higher weight hourly space velocity of 180 L $(h\,g_{cat})^{-1}$.

The performance of a platinum/cobalt/α-alumina catalyst was investigated in microstructures by Delsman *et al.* [332], with 0.5 vol.% carbon monoxide, 1.6 vol.% oxygen, 56 vol.% hydrogen and 18 vol.% carbon dioxide, and a balance of helium, being fed into the integrated device at a temperature of 300 °C. Nitrogen was used as coolant. At a weight hourly space velocity of 120 L (h g$_{cat}$)$^{-1}$, the carbon monoxide could be reduced to 7 ppm at an O/CO ratio around 3.0, but the catalyst deactivated and 300 ppm CO were released after 4 h.

Maier and Saalfrank developed novel catalyst formulations for preferential oxidation by combinatorial chemistry [333]. The catalysts were supported by cobalt and the formulations were Pt$_{0.5}$Al$_1$Mn$_{6.7}$Co and Al$_1$Mn$_{6.7}$Co. Full conversion of 1.2 vol.% carbon monoxide could be achieved at a 100 °C reaction temperature and low O/CO ratio of 1.6 for the former catalyst at a weight hourly space velocity of 30 L (h g$_{cat}$)$^{-1}$.

Rhodium/alumina and rhodium/potassium alumina catalysts were prepared by wash-coating onto microchannels by Chen et al. [334]. The feed was composed of 40 vol.% hydrogen, 20 vol.% carbon dioxide, 0.2–1.0 vol.% carbon monoxide and 0.2–1.5 vol.% oxygen on a dry basis; 10 vol.% water was added to the feed. The doped catalyst outperformed its potassium-free counterpart when tested in ceramic monoliths. The highest conversion achieved was 99.82% for a CO concentration of 5000 ppm, which corresponds to less than 10 ppm at the reactor outlet. Above 250 °C reaction temperature, significant methane formation was observed, which points to the suitability of rhodium catalysts for methanation of carbon monoxide in this temperature range. When the catalyst was tested in the microstructured reactor at a very high gas hourly space velocity of 500 000 h^{-1}, 230 °C reaction temperature and an O/CO ratio of 2.0, high conversion of carbon monoxide in the region of 95% was still achieved. Less than 30 ppm methane were detected under these conditions.

According to Kawatsu, from Toyota [30], a ruthenium catalyst was used in the fuel processor of the Toyota methanol-fuelled fuel cell vehicle (see Section 9.1). It was claimed to show less reverse water–gas shift and higher selectivity compared with a platinum catalyst. Giroux et al. also proposed the application of ruthenium- or copper-based catalysts, especially for a second stage preferential oxidation reactor [57]. Farrauto et al. even mention the possibility of three preferential oxidation stages [107], which seems too complicated for a practical application.

Recent investigations have dealt with the application of gold for the preferential oxidation of carbon monoxide, which is active even at ambient temperature or at least below 100 °C, the operating temperature of regular low temperature PEM fuel cells. Gold catalysts were prepared on various carrier materials, such as titanium oxide, magnesia, iron oxide and zinc oxide [329, 335]. However, it is unclear to the author of this book what the advantage of this low operating temperature is, because the reformate has to be cooled upstream of the fuel cell anyway and it does not matter if this cooling is performed upstream or downstream of the preferential oxidation reactor. In addition, the preferential oxidation reaction suffers from heat removal problems due to its exothermicity, which are not lowered at lower operating temperatures. One possible advantage could be that the reactor could heat itself up during start-up. However, gold catalysts generally suffer from stability problems [102], which basically originate from agglomeration of the gold nanoparticles.

The deactivation of gold catalysts in the presence of steam was explained by the formation of formate species on the catalyst surface in the presence of water. These species were formed about 300-times faster in the presence of water. However, addition of smaller amounts of water (2.5 vol.%) had no negative effect on a gold/ceria catalyst in another study [333]. Luengnaruemitchai et al. also found no significant deactivation of their gold/ceria catalyst in the presence of 10 vol.% steam [336]. Carbon dioxide did reduce the activity of gold catalysts more significantly, which was also observed by Panzera et al. [337]. Nevertheless 3000-h durability was proven for the catalyst in the presence of 2 vol.% carbon dioxide and 2.6 vol.% steam. The catalyst was prepared by a co-precipitation method, which seemed to be the preferred route, leading to more stable gold/ceria catalysts, also according to Panzera et al. [337].

Silver/silica catalysts are other systems that have been under investigation more recently.

Wakita et al. investigated the effect of poisons such as sulfur dioxide and hydrogen sulfide on the performance of their ruthenium and platinum catalysts supported by alumina [338]. The catalysts contained between 1.6 and 3.8 wt.% ruthenium or between 3.1 and 3.8 wt.% platinum, respectively. Because a water–gas shift reactor is usually positioned upstream of a preferential oxidation reactor, low temperature water–gas shift product was fed to the testing reactor. The feed composition was 51 vol.% hydrogen, 0.3 vol.% carbon monoxide, 13 vol.% carbon dioxide 0.5 vol.% oxygen, 6.0 vol.% nitrogen, 28.8 vol.% steam and 21 ppm sulfur dioxide. The ruthenium catalyst gradually lost its carbon monoxide oxidation capability within 5 h as shown in Figure 4.28, until only 33% conversion could be achieved. Interestingly, the selectivity towards carbon monoxide remained approximately the same. Initially the catalyst had some methanation activity, which immediately vanished in the presence of sulfur. The sulfur was trapped on the ruthenium and the alumina support. This was assumed, because the contaminant content in the product was much lower compared with its content in the feed. After stopping the addition of sulfur dioxide, the catalyst recovered only slightly; 4.3 ppm hydrogen sulfide had a less dramatic effect on the ruthenium catalyst. The carbon monoxide content in the product increased slightly

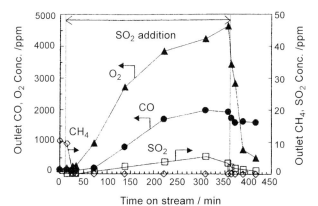

Figure 4.28 Effect of 21 ppm sulfur dioxide on the activity of a ruthenium/alumina catalyst for the preferential oxidation of carbon monoxide; the catalyst contained 1.6 wt.% ruthenium on alumina; reaction temperature 150°C [338].

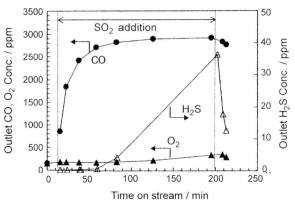

Figure 4.29 Effect of 21 ppm sulfur dioxide on the activity of a platinum/alumina catalyst for the preferential oxidation of carbon monoxide; the catalyst contained 3.1 wt.% platinum on alumina; reaction temperature 175°C [338].

from 150 to 180 ppm and the methane selectivity gradually vanished within 5 h. The platinum catalyst showed different behaviour. As shown in Figure 4.29, it lost its oxidation capability almost completely within 2 h in the presence of 21 ppm sulfur dioxide. After 1 h the catalyst started to release hydrogen sulfide. The effect of 2.1 ppm sulfur dioxide and 2.1 ppm hydrogen sulfide was almost the same for the platinum catalyst. It lost about 50% of its initial activity within 2 h. The catalyst could be regenerated by oxidation treatment for 10 min at 175 °C. Further investigations revealed that sulfate from sulfur dioxide and sulfide from hydrogen sulfide were adsorbed on the ruthenium catalyst, while only sulfide was found on the platinum surface, which was oxidised to sulfate by the oxidation treatment. The sulfate then obviously migrated to the alumina carrier forming aluminium sulphate, and thus the activity could be recovered.

Non-noble metal based catalysts are also reported in literature for the preferential oxidation of carbon monoxide.

Amongst many others not cited here, Wang et al. found a high conversion of carbon monoxide in the temperature range between 140 and 160 °C over copper oxide catalysts supported by samaria-doped ceria [339]. Park et al. investigated the performance of copper/ceria catalysts partially doped by cobalt and supported on γ-alumina [340]. The catalysts contained a total of 10 wt.% of copper and ceria on an alumina support. Initially the feed used for the tests was only composed of 1 vol.% carbon monoxide, 1 vol.% oxygen and 60 vol.% hydrogen, with a balance of nitrogen. The maximum carbon monoxide conversion was achieved with a copper content of 5 wt.%, a lower copper content had a negative effect on activity. In general, the catalysts required a temperature greater than 160 °C to achieve full conversion despite the relatively low weight hourly space velocity of $60 L\,(h\,g_{cat})^{-1}$. However, the O/CO ratio of 2.0 was rather low. Doping with 0.2 wt.% cobalt further decreased the minimum temperature required to achieve full conversion to 150 °C and the O/CO ratio could be reduced to 1.5, which is a rather low value. However, addition of 13 vol.% carbon dioxide to the feed had a very negative effect on the performance of the catalyst. Both the activity decreased and a reverse water–gas shift was observed for temperatures higher than 200 °C, which narrowed the operating window of the catalyst to about 20 K. The same effect was observed when steam was added to the feed. This observation is obviously a specific feature of the present catalyst formulation because the opposite is usually observed for noble metal based catalysts (see above). The steam effect was less pronounced for the samples not doped with cobalt. Snytnikov *et al.* even reported similar results for their catalyst containing 15 wt.% copper on ceria [341]. The activity of the catalysts was lower in the presence of carbon dioxide, steam and both reactants. Avgouropoulos *et al.* also observed this effect for copper/ceria catalysts containing only 3 wt.% copper [327].

4.5.3
Methanation Catalysts

Nickel and ruthenium [342,343] catalysts are those most frequently under investigation for the selective methanation of carbon monoxide. The main issue of concern is that the concentration of carbon dioxide is much higher in the reformate compared with carbon monoxide and thus the catalyst used for carbon monoxide methanation has to be very selective. Normally, the operating window of methanation catalysts is relatively small, around 250 °C, because a trade-off is required between sufficient activity and selectivity. Well above 250 °C all methanation catalysts tend to be selective for carbon dioxide methanation.

Takenaka *et al.* studied the activity of various catalysts for carbon monoxide methanation in the absence of carbon dioxide [342]. From the different active species on a silica carrier, 5 wt.% ruthenium, 10 wt.% nickel and 10 wt.% cobalt were significantly more active than iron, palladium or platinum, each prepared with an active species content of 10 wt.%. Then Takenaka tested nickel, ruthenium and cobalt catalysts on different carrier materials, namely, alumina, silica, titania and zirconia. The formulations most active were nickel/zirconia and ruthenium/titania catalysts. The best performing catalyst was the 5 wt.% ruthenium/titania, which converted the carbon monoxide apart from less than 20 ppm from a feed mixture containing 60 vol.% hydrogen, 15 vol.% carbon dioxide, 0.9 vol.% steam, 0.5 vol.% carbon monoxide, with a balance of helium at 220 °C. The space velocity was rather high at 300 L (h g_{cat})$^{-1}$.

Men *et al.* investigated selective methanation over nickel and ruthenium catalysts supported by different carriers [344]. A nickel catalyst containing 43 wt.% nickel, which was doped with 6 wt.% calcium oxide and supported by alumina, turned out to be the most active sample. The catalyst produced methane exclusively, from both carbon monoxide and carbon dioxide in the presence of hydrogen. 90% conversion of carbon monoxide could be achieved at a 300 °C reaction temperature with 35% selectivity. The presence of steam reduced the activity of the catalyst but improved its selectivity towards carbon monoxide. When oxygen was added to the feed, the catalyst exclusively converted carbon monoxide into carbon dioxide, and methane formation did not start until all the oxygen had been consumed.

The selectivity towards carbon monoxide is of course no longer an issue when methanation is applied for the removal of traces of carbon monoxide from purified hydrogen permeate downstream of a membrane separation device. The carbon dioxide concentration is then as low as the carbon monoxide concentration and the parallel conversion of both carbon oxides becomes viable (see Section 5.2.4).

A particular issue is the deactivation of methanation catalysts by carbon formation. Kuijpers *et al.* [345] observed significant carbon formation over a nickel/kieselgur catalyst containing 54 wt.% nickel when exposed to a mixture of 10 vol.% carbon monoxide, 15 wt.% hydrogen, with a balance of nitrogen at 0.6 bar pressure and a 250 °C reaction temperature. Carbon filaments were found, which contributed to 10 wt.% of the catalyst mass at the inlet of the fixed bed. A nickel/silica catalyst showed practically no coke formation for 1000 h duration under the same conditions.

4.6
Combustion Catalysts

Catalytic combustion of fuels may be performed in catalytic burners for fuel processing applications with the advantage of a significantly lower operating temperature, absence of NO_x emissions and, with hydrogen and methanol, even light-off of the burner under ambient conditions, which releases the fuel processor from the electrical pre-heating demand. In general, platinum is a good catalyst for catalytic combustion. Alcohol fuels are oxidised fairly easily over precious metal catalysts.

Hydrogen and carbon monoxide are readily oxidised over a platinum catalyst at room temperature [214]. The ignition temperature rises with increasing hydrogen concentration, because the hydrogen is preferentially adsorbed at the platinum surface and a higher temperature is required to initiate the reaction. Thus, Rinnemo *et al.* determined that a temperature of 100 °C is required to ignite a mixture of 55 vol.% hydrogen and 45 vol.% oxygen over a platinum wire [346,347], while at a hydrogen concentration below 10 vol.% the reaction even occurred at room temperature. Dilution with nitrogen reduces the flammability limits. Fernandes *et al.* found that a mixture of 16 vol.% hydrogen, 3.3 vol.% oxygen and 80 vol.% nitrogen could be heated up to 200 °C without ignition of the reaction mixture [348].

Methanol starts light-off even at room temperature over noble metal catalysts [349]. Over platinum wire, conversion is initiated at only 100 °C. Lindström and Pettersen evaluated a wide variety of catalysts for the cold light-off by dispersing cold methanol in air over the catalysts [349]. Only noble metal catalysts showed light-off from room temperature and palladium suffered from different drawbacks, amongst them rapid deactivation, while silver was even less suitable. Best suited was platinum, which was more active the higher the metal loading was. Addition of manganese enhanced the activity of platinum [168]. A catalyst containing 10 wt.% manganese and 0.3 wt.% platinum was identified as a suitable formulation for the cold start of a methanol fuel processor [349].

The catalytic combustion of hydrocarbon fuels becomes more difficult the lower the number of carbon atoms that are present in the fuel molecule. Platinum and palladium are the noble metals most commonly applied [350], while gold is less suited for alkane combustion.

Platinum catalysts are superior for hydrocarbon oxidation [350]. Platinum is usually present as metallic platinum at temperatures exceeding 400 °C [351]. Besides alumina, zirconia, molecular sieves and metal oxides are applied as supports, amongst others. However, an inhibition effect of oxygen was observed over platinum/alumina for light hydrocarbons [350,352] and other platinum-containing catalysts for methane [353] and propane oxidation [354]. This effect becomes problematic, in particular because full hydrocarbon conversion is usually required for a catalytic burner. The temperature control of catalytic burners also becomes difficult, because this is frequently carried out by regulation of the air feed flow rate to suprastoichiometric values.

Platinum combustion catalysts are probably more tolerant to sulfur poisoning than palladium catalysts [350,355]. Corro *et al.* reported that sulfur dioxide may even have a promoting effect on propane oxidation [356]. They claimed that aluminium sulfate needs to be present on the catalyst surface to promote the reaction in the low temperature range below 300 °C. The sulfate formation was assumed to start at temperatures exceeding 500 °C. Therefore, the catalyst must have previously been exposed to such a temperature in the presence of sulfur dioxide. Over a pre-sulfated platinum/alumina catalyst, 50% methane conversion was achieved by 530 °C, while 560 °C was required for the sulfur-free counterpart [357]. However, no promotion effect is to be expected over non-sulfating carrier materials, such as silica, according to Gelin and Primet [351].

Addition of 1000 ppm propane to the methane feed decreased the temperature required to achieve 50% conversion from 560 to 380 °C over a pre-sulfated catalyst [357]. Combustion of the 1000 ppm propane occurred by 150 °C over the platinum catalyst. A 50% propane conversion was achieved for the sulfur-free catalyst at 260 °C, while only 210 °C was required for the pre-sulfated sample. However, the weight hourly space velocity was rather low at 30 L $(h\,g_{cat})^{-1}$ for the experiments of Corro *et al.* [357].

Yazawa *et al.* investigated the effect of additives on the performance of platinum catalysts in propane combustion [358]. While sodium, lithium, potassium, rubidium, caesium, magnesium, calcium, strontium, barium, yttrium, zirconium, niobium and phosphorus did not improve or even decrease the activity, molybdenum, tungsten and vanadium improved the catalyst activity significantly. The electronegativity of the additives prevented oxidation of the platinum in the oxidising atmosphere of combustion [359]. More acidic supports, such as alumina, silica or silica/alumina, showed higher activity than magnesia, lanthana and zirconia for similar reasons [359,360]. In particular, the inhibition effect is no longer observed over platinum/tungsten and platinum/molybdenum catalysts supported by alumina according to the experience of the author of this book. Guan *et al.* reported high catalytic activity of a molybdenum promoted platinum catalyst supported by alumina inverse opals [361]. Light-off occurred at 250 °C and complete conversion of propane was achieved at 300 °C for a high weight hourly space velocity of 300 L $(h\,g_{cat})^{-1}$. The catalyst exhibited stable performance for 170 h.

An activation procedure is frequently observed for the catalytic hydrocarbon combustion over noble metal catalysts [362], which is assumed to originate from an inhibiting effect of residual chlorine originating from the catalyst preparation procedure [351]. However, Guerrero et al. observed an activation effect for their palladium/zirconia catalysts, which was believed not to have originated from the release of chloride from the catalyst surface [363]. As compared with most other heterogeneously catalysed reactions, increasing the noble metal particle size increases the rate of reaction for catalytic combustion [351]. This holds at least up to a certain limit (15 nm for methane combustion catalysed by palladium [350]). Steam has an inhibiting effect on the combustion reaction over noble metal catalysts, although this is not the situation for carbon dioxide [351].

With palladium, palladium oxide is usually regarded as the active species, which probably covers the palladium particles [350]. Palladium in its oxidised form (PdO) may well be a suitable catalyst for the catalytic combustion of methane [107] and the preferable solution compared with platinum catalysts [214]. Alumina is usually the preferred support for palladium combustion catalysts, but silica, zirconia, ceria and molecular sieves are also under investigation [350]. Palladium was more active for methane oxidation compared with rhodium and iridium, according to investigations of Deng and Nevell [364]. The stability of palladium/alumina catalysts in the high temperature range is improved by addition of ceria, zirconia and rare earth metals [365]. Silicon nitride (α-Si_3N_4) is an alternative support, which is stable up to 1200 °C. The activity of palladium/zirconia catalysts with various palladium loadings was analysed for methane oxidation. It revealed that increasing the palladium content up to 10 wt.% significantly increased the activity.

Sulfur species may form sulfates and sulfites, which lower the catalyst surface area and cause deactivation for palladium [350]. An alumina support can trap or adsorb [351] sulfate groups. Deng and Nevell reported deactivation of alumina supported palladium, rhodium and iridium catalysts by hydrogen sulfide during methane oxidation [364]. Sulfate species were formed on the catalyst surface under the oxidising conditions.

Light-off of methane may occur even at 200 °C over palladium when the space velocity is low enough. Trimm et al. reported that the light-off temperature decreased with decreasing O/C ratio and was significantly lower under partial oxidation conditions [214].

Kiwi-Minsker et al. prepared platinum and palladium catalysts from commercial aluminoborosilicate glass fibres [354]. The glass fibres were partially modified with titania, zirconia and alumina to increase their thermostability. The noble metal content of the fibres was very low, between 0.1 and 0.3 wt.%. The platinum samples were more active than the palladium catalysts and titania and zirconia modification of the glass increased its activity for propane combustion. Light-off occurred at 200 °C, 50% conversion was achieved at 280 °C but full conversion was not achieved even by 450 °C, owing to the inhibition effects discussed above for a sample containing 0.3 wt.% platinum. However, the space velocity was rather low at 15 L $(h\,g_{cat})^{-1}$.

Bimetallic catalyst formulations of palladium and platinum have been applied for the catalytic combustion of light hydrocarbons, which showed higher stability compared with monometallic samples.

Metal oxide catalysts, such as copper, manganese, cobalt and chromium oxide, are economically favourable alternatives to noble metals for catalytic combustion, but they suffer from lower activity and higher light-off temperature [350,366]. Cobalt oxide deposited on sintered inconel fibres [367] and on nickel and copper grids [366] showed high activity for total oxidation of propane. However, the catalysts suffered from deactivation under a surplus of oxygen [366,367]. Full conversion of methane was achieved at a reaction temperature of 400 °C over cobalt oxide (Co_3O_4)/alumina catalysts [368]. Another low-cost alternative to noble metals are perovskite catalysts (see Section 4.2) such as lanthanum/manganese oxide and lanthanum/cobalt oxide systems ($LaMnO_3$ or $LaCoO_3$), either doped with other metals or not [350]. Zhao *et al.* report promising results obtained with ceria catalysts for the catalytic combustion of C_1–C_4 alkanes [369].

Addition of methanol to hydrocarbon mixtures allows their light-off even at room temperature, because the alcohol oxidation is obviously not impaired by the presence of other fuels [214].

5
Fuel Processor Design Concepts

5.1
Design of the Reforming Process

Table 5.1 shows the energy requirements for steam reforming and partial oxidation of various fuels [370]. Generally, partial oxidation creates a surplus of energy, and while steam reforming requires energy, alcohol fuels require less than hydrocarbons in the latter case.

A comparison of the reformate composition, hydrogen mass flow and efficiency was performed for steam reforming and autothermal reforming by Specchia *et al.* [371]. For both cases a total fuel feed flow rate of $1 \, \text{kg h}^{-1}$ was assumed. For steam reforming the energy supply was assumed to stem from combustion of part of the fuel in a separate unit or flow path of the reformer. As shown in Tables 5.2 and 5.3, the hydrogen flow rate and thus the efficiency of the steam reforming process is higher compared with autothermal reforming. $(m_{H_2})_{\text{th,R}}$ in the tables stands for the hydrogen mass flow contained in the fuel itself. It is obvious that model fuels such as isooctane for gasoline or cetane for diesel fuel overestimate the performance of the fuel processor due to the lower hydrogen content of the real fuels [371] (for fuel compositions see Section 2.1).

5.1.1
Steam Reforming

When designing the steam reforming process, the reaction temperature needs to be chosen sufficiently high to achieve full conversion for all hydrocarbon and alcohol fuels, except for methanol. A certain amount of unconverted methanol can be tolerated with high temperature PEM and phosphoric acid fuel cells. However, carbon formation needs to be minimised and with the high temperatures required for hydrocarbon steam reforming, the temperature becomes one viable measure for preventing carbon formation. As discussed in Section 4.2.11, coke formation, especially with methane steam reforming, is suppressed by high reaction temperatures. The second measure to prevent coke formation is a sufficiently high S/C ratio. However, the steam excess required becomes higher with an increasing chain

Fuel Processing for Fuel Cells. Gunther Kolb
Copyright © 2008 WILEY-VCH Verlag GmbH & Co. KGaA, Weinheim
ISBN: 978-3-527-31581-9

Table 5.1 Theoretical process energy requirements for steam reforming and partial oxidation of various fuels [370].

Process	Energy needs during processing (kJ/mol of usable H_2)	Energy needed to vaporize and heat feed (kJ/mol of usable H_2)	Total theoretical process energy need (kJ/mol of usable H_2)
Steam-reforming of			
Methane	40	36	76
Methanol	17	39	56
Ethanol	20	40	60
Gasoline	41	49	90
Diesel fuel	40	49	89
Jet fuel	42	49	91
Partial oxidation of			
Methane	−43	26	−17
Gasoline	−144	50	−94
Diesel fuel	−145	49	−96
Jet fuel	−151	49	−102

Table 5.2 Mass flow rates m, reformer outlet temperature T_{out} and reformer efficiency η_R for autothermal reforming of various fuels: S/C ratio 3.0; O/C ratio 0.7; feed temperature 700 °C [371].

m_{fuel} (1 kg/h)	$(m_{H_2})_{th}$ (kg/h)	T_{out} (°C)	m_{H_2} (kg/h)	m_{CO} (kg/h)	m_{H_2O} (kg/h)	m_{CO_2} (kg/h)	m_{N_2} (kg/h)	η_R (%)
Gasoline	0.135	892	0.245	1.074	2.896	1.484	2.656	67.8
Iso-octane	0.158	764	0.276	0.895	2.718	1.676	2.581	75.6
Light diesel	0.130	788	0.253	0.918	2.808	1.747	2.670	72.4
Heavy diesel	0.124	771	0.251	0.885	2.803	1.822	2.690	73.6
Cetane	0.150	794	0.265	0.960	2.779	1.601	2.604	73.3
Bio-diesel	0.119	876	0.208	0.896	2.664	1.421	2.369	68.2

Table 5.3 Fraction of the fuel fed to the steam reformer β_R, mass flow rates m, reformer outlet temperature T_{out} and reformer efficiency η_R for steam reforming of various fuels; S/C ratio 3.0; O/C ratio of the catalytic afterburner 0.77; the fraction $(1-\beta_R)$ of the fuel is fed to the afterburner [371].

m_{fuel} (1 kg/h)	β_R (kg/h)	S/C	$(m_{H_2})_{th,R}$ (kg/h)	T_{out} (°C)	m_{H_2} (kg/h)	m_{CO} (kg/h)	m_{H_2O} (kg/h)	m_{CO_2} (kg/h)	m_{N_2} (kg/h)	η_R (%)
Gasoline	0.762	3.94	0.103	897	0.258	0.896	2.495	1.003	0	71.2
Iso-octane	0.776	3.87	0.123	769	0.285	0.773	2.322	1.175	0	78.0
Light diesel	0.759	3.95	0.099	794	0.263	0.776	2.432	1.201	0	75.3
Heavy diesel	0.754	3.98	0.093	778	0.261	0.745	2.439	1.250	0	76.5
Cetane	0.771	3.89	0.116	801	0.276	0.806	2.372	1.131	0	76.3
Bio-diesel	0.751	3.99	0.089	882	0.219	0.748	2.306	0.995	0	71.8

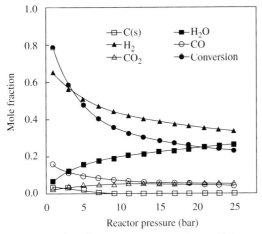

Figure 5.1 Effect of reaction pressure on the equilibrium conversion of methane steam reforming: reaction temperature, 700 °C; S/C ratio 1 [66].

length of the hydrocarbon. Finally, an elevated system pressure also helps to prevent carbon formation but in addition reduces conversion, as shown for methane steam reforming in Figure 5.1. Higher pressure also increases energy consumption by the compressor.

Because steam reforming is endothermic, it requires a heat source for the energy supply. On an industrial scale, this is performed by heating tubular steam reforming reactors externally via homogeneous combustion. For stationary fuel cell systems on the scale of large power plants this technology is clearly still applicable. However, the smaller the energy supply system becomes, the less viable this solution is, because heat losses, integration and space demand become more stringent factors. Because fuel cell systems running on reformate do not consume the hydrogen contained in the reformate completely, a significant amount of hydrogen, usually about 20%, leaves the fuel cell anode unconverted (see Section 2.3). It may be fed back to the fuel processor and provide energy to the reforming process, which is not necessary when autothermal reforming is applied. Integrated heat-exchangers/reactors open the door to such integrated processing concepts. The endothermic steam reforming reaction may be coupled to an exothermic catalytic combustion in such reactors.

Ma *et al.* investigated autothermal methanol reforming in five different reactor configurations, as shown in Figure 5.2. A dual bed arrangement, where the combustion catalyst bed was followed by the steam reforming bed, and two concentric tubular arrangements, where the combustion zone was either in the cylindrical centre of the reactor or in the annular outer zone, were compared. In both latter examples the combustion zone was first entered by the feed and the steam reforming zone followed downstream [372]. In addition, concentric spherical arrangements were investigated. The spherical reactor with the combustion zone

(a) dual fixed bed reactor

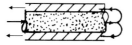

(b) Coaxial cylindrical reactor I.

(c) Coaxial cylindrical reactor II.

(d) Concentric spherical reactor I.

(e) Concentric spherical reactor II.

▨ Oxidation catalyst

▨ Steam reforming catalyst

Figure 5.2 Five different fixed bed reactor configurations as investigated by numerical calculations of Ma et al. for methanol steam reforming [372].

in the centre showed three to four times higher efficiency compared with the other spherical and with the tubular arrangements. The optimum S/C ratio for the spherical reactors [reaction I, (d) in Figure 5.2] was in the range of 0.7, while the calculations for the tubular reactors revealed much higher values of around four [372], which is of course outside the scope for a practical application, because water consumption would be too high by far.

It is evident that both coaxial arrangements investigated by Ma et al., as discussed above, are counter-flow configurations, which are not favourable for optimum heat management when combining exothermic combustion with endothermic steam reforming, as will be discussed below.

Avci et al. investigated combined catalytic combustion and steam reforming but also quasi-autothermal reforming of methane in fixed catalyst beds. Two different catalyst formulations were used to combine the combustion reaction with steam reforming, namely a nickel catalyst for steam reforming and a platinum catalyst for methane combustion. The two catalysts were assumed to be placed into two serial beds, where the combustion catalyst bed was followed by the steam reforming bed, similar to the arrangement assumed by Ma *et al.* [372] discussed above. The third option was a physical mixture of both catalysts. It is not surprising that the serial catalyst arrangement led to higher temperature fluctuations compared with the catalyst blend, which was also verified by experimental work performed by Avci *et al.* [373]. Intraparticle mass transfer limitations reduced catalyst activity for both catalyst bed configurations [373].

The results of Avci *et al.* are interesting because oxidation and steam reforming reactions are separated through the assumption that they take place over different catalysts. However, in a practical fuel processor system, a single catalyst formulation would be used for autothermal reforming at least in the smaller scale.

For small- to medium-scale stationary methane steam reforming applications, Johnston and Haynes [374] proposed a series of adiabatic reactors with heat-exchangers switched in between rather than directly removing the heat in a single heat-exchanger/reactor. Thus, a saw-tooth type of temperature profile resulted, as is familiar from conventional large-scale industrial processes. The proposed flow-paths were on the meso-scale and small catalyst particles instead of coatings were proposed for their application. The study showed a complicated network of 27 catalyst beds (5 for pre-reforming, 9 for reforming, 9 for catalytic combustion to supply the heat, 2 for water–gas shift and 2 for preferential oxidation) and 20 heat-exchangers. The power supply stemmed from the anode off-gas. However, many of the reactors were integrated into a single device by the Printed Circuit Heat-Exchanger (PCHE) technology, which is fabricated by a wet chemical etching procedure of stainless steel metal foils. A reforming temperature below $800\,°C$ and a S/C ratio of 2.6 was regarded as sufficient to run the system. A theoretical efficiency of 89.2% was calculated for the system. A prototype fuel processor was presented, which will be discussed in Section 9.3.

From the group of Eigenberger, one of the first papers dealing with the simulation of coupled steam reforming and catalytic combustion is the paper by Frauhammer *et al.* [375]. Methane combustion and methane steam reforming were the subject of the calculations. The reactor concept was a ceramic monolith, which was modified by specially designed reactor heads to achieve heat-exchanger functionality, a design that will be described in detail in Section 6.3.1. Surprisingly, a counter-current concept was assumed for coupling both reactions in this early work. The concept suffered from the formation of temperature peaks in most instances and required homogeneous reactions on the combustion side to achieve higher degrees of conversion in the second flow-path, where methane steam reforming took place. Because the temperature inside the reactor decreased sharply close to the exit of the steam reforming flow path, the conversion of methane steam reforming decreased again towards the reactor exit, owing to the equilibrium of the reaction (see Figure 5.3). Very high maximum temperatures exceeding $1300\,°C$ were calculated by Frauhammer *et al.*, which, however, would damage the structure of the catalyst in a practical system. Figure 5.3 shows the temperature profile for different ratios of the heat capacity flows:

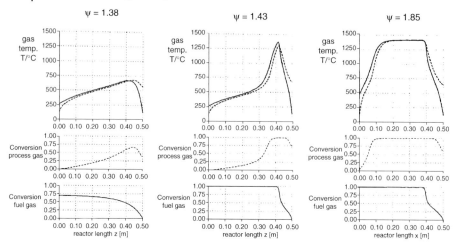

Figure 5.3 Coupling of methane combustion (feed enters the reactor at the right side of the reactor) and methane steam reforming (feed enters the reactor at the left side of the reactor) in a ceramic monolith, modified as a counter-current heat-exchanger. Top graphs, gas temperature of the combustion side (straight line) and of the reformate (dotted line); centre graphs, conversion of methane steam reforming; bottom graphs, conversion of methane combustion; the graphs from left to right represent results for different ratios, ψ, of heat generation by combustion to heat required for steam reforming [375].

$$\frac{(\dot{m}\,C_p)_{\text{combustion feed}}}{(\dot{m}\,C_p)_{\text{steam reforming feed}}} \tag{5.1}$$

of 2.29 (left), 2.37 (centre) and 3.06 (right). The ratio, ψ, of heat generation by the combustion reaction to the heat requirement by steam reforming is shown on the top of the graphs. When a part of the reactor was not coated with catalyst, as shown in Figure 5.4, conversion of methane steam reforming could be increased. Not more than 75% conversion of steam reforming could be achieved for a reactor coated with catalyst over the whole length for both reactions, as shown at the left side of Figure 5.4. Full conversion could be achieved, when the exit section of the steam reformer side of the reactor was not coated with catalyst (centre of Figure 5.4). This is obvious owing to the equilibrium of steam reforming as mentioned above and the sharp temperature drop in this area. When both flow ducts were not coated with catalyst on the right side of the reactor, full conversion of both reactions could also be achieved. However, this means that the right part of the reactor worked only as a heat-exchanger. The peak temperature increased to 1500 °C under these conditions.

Summarising the calculations of Frauhammer *et al.*, a counter-current concept of coupling steam reforming with combustion reactions leads to significant temperature gradients and consequently thermal stress within the reactor and is not a suitable solution. Rather, co-current operation would be preferred, as will be discussed below.

Kolios *et al.* [376], from the same group, performed an extensive study revealing that coupling of exothermic and endothermic reactions is only possible under safe and stable operation conditions in catalytic wall reactors and not in coupled packed beds.

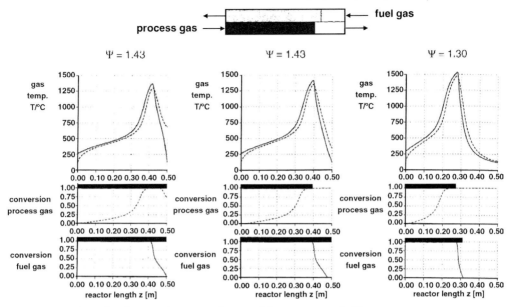

Figure 5.4 Coupling of methane combustion (feed enters the reactor at the right side of the reactor) and methane steam reforming (feed enters the reactor at the left side of the reactor) in a ceramic monolith, modified as a counter-current heat-exchanger. Top graphs, gas temperature of the combustion side (straight line) and of the reformate (dotted line); centre graphs, conversion of methane steam reforming; the part of the reactor coated with steam reforming catalyst is indicated by the black bar; bottom graphs, conversion of methane combustion; the part of the reactor coated with combustion catalyst is indicated by the black bar; the graphs from left to right represent results for different ratios, ψ, of heat generation by combustion to heat required for steam reforming [375].

In the latter, instability and thermal runaway of the reactor may occur. The two reactions taking place at the two sides of the same reactor wall need to be coupled through a highly heat conductive material, according to Kolios *et al.* These workers proposed a folded wall reactor operated in the counter-current mode as the optimum solution to the problem. This design was successfully applied for methanol steam reforming. Corrugated metal foils, similar to the foils used for automotive exhaust gas systems (see Section 6.2), were incorporated into the reactor and formed the flow ducts. The feed for the catalytic combustion reaction entered the ducts from the side through a spacer over the whole length of the reactor, which made optimum process control possible.

Later Gritsch *et al.*, from the same group [377], presented a new concept of coupling endothermic and exothermic reactions. The distributed feed addition along the reactor channel was regarded as too complicated. Therefore, a co-current flow arrangement was proposed for the two reactions, including two heat-exchangers heating the feed of each flow path by its hot product gases. The folded wall reactor concept was used again. The feasibility of this concept was then investigated in a single combustion layer combined with two adjacent reforming layers, which will be described in Section 7.1.3.

A comparison between a tubular fixed-bed steam reformer, a fixed-bed plate reformer and a plate reformer coated with catalyst was carried out by Zalc and Löffler [378]. The application motivating these investigations was a 50 kW$_{el}$ automotive fuel processor/fuel cell system. Assuming a commercial nickel catalyst was used for steam reforming, a catalyst weight of 10.8 kg and catalyst volume of 7.2 L was calculated for the conventional tubular design. The fixed catalyst bed suffered from heat transfer limitations. For a plate heat-exchanger filled with catalyst particles, significantly less catalyst volume of 2.2 L was required. For the plate heat-exchanger coated with catalyst, only 20 g of catalyst were required, owing to the tremendously improved utilisation of the catalyst, which originated from the elimination of mass transfer limitations. The reformer volume was calculated to be 1.2 L in the latter, which corresponded to 17% of volume calculated for the tubular fixed-bed reactor. The space demand for heat supply to the different reactor types was not taken into consideration in these calculations.

Zanfir and Gavriilidis [379] studied in detail the feasibility of combined methane oxidation/steam reforming in an integrated heat-exchanger with micro- and mesoscaled reaction channels using numerical simulations based upon a two-dimensional model. Kinetic expressions determined for nickel-catalysts in industrial scale processes extracted from the literature were used. First-order kinetics of typical precious metal combustion catalysts were chosen to describe the combustion process. Co-current flow of the reformer and combustor gases was assumed for the reactor model. Pressure close to ambient, S/C ratio of 3.4 and a 520 °C feed temperature were assumed for the process. The last two values were close to the conditions of large-scale industrial processes. Figure 5.5 shows the single plate that was simulated.

An almost 260 K temperature rise was calculated for the reactor wall temperature along the feed flow path due to the faster kinetics of the combustion reaction. The axial temperature gradient increased when the half-height of the channels was

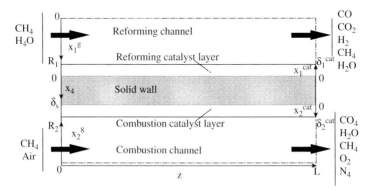

Figure 5.5 Schematic of the single plate coated with catalyst from both sides for methane steam reforming and methane combustion as used by Zanfir and Gavriilidis for their calculations [379].

increased from 0.5 to 2 mm at constant space velocity. Larger channel dimensions resulted in less efficient heat transfer and higher temperature gradients in the gas phase. Across the reactor wall very low temperature gradients below 0.5 K were calculated. Temperature gradients from the wall to the gas phase ranged between 20 and 70 K, depending on the channel half-height of 0.5 and 2 mm, respectively. Methane conversion increased with increasing channel height in the first half of the reactor due to the higher temperature in both flow paths. This effect was equalled after 70% of the reactor length. At the end of the reactor, methane conversion was lower for higher channels mainly due to mass transport limitations in the higher channels. Increasing the channel half-height from 0.5 to 2 mm decreased the methane conversion from 92 to 95%. This effect was much more pronounced when the inlet flow velocities were kept constant rather than the inlet flow rates. With constant flow velocities, the outlet conversion dropped from 100% with the 0.5-mm channel half-height to 70% with 2-mm half-height (see Figure 5.6). Full conversion was actually achieved after just 45% of the reactor total length for the smaller

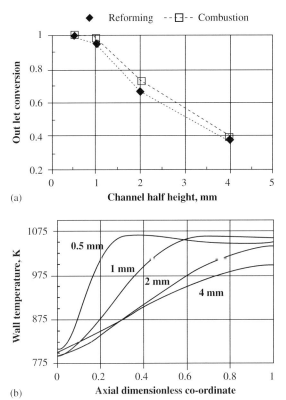

Figure 5.6 Results from numerical calculations for combustion assisted methane steam reforming: (a) outlet conversion as a function of channel half-height; (b) wall temperature as a function of dimensionless reactor length; calculation results determined at constant inlet velocity [379].

channels. The temperature rise was fast for the small channels approaching a maximum after 30% of the reactor length, thus leading to higher conversion. In the large channels, mass transfer limitations occurred. Consequently, the catalyst mass present in the reactor was then too low for the particular space velocity. In a third case study, the thickness of the catalyst layer was increased at a constant weight hourly space velocity by increasing the inlet flow rate. When the catalyst layer thickness was increased from 10 to 60 μm a decrease of conversion from 100% to below 70% was calculated for steam reforming and combustion reactions. The axial temperature rise was steeper with thinner coatings.

The calculations of Zanfir and Gavriilidis highlighted the benefits of catalyst coatings in general and also indicated the basic feasibility of the combination of steam reforming and fuel combustion from a theoretical point of view, even at the elevated temperatures of hydrocarbon reforming/combustion.

At this point an interesting experiment performed by Giroux et al. [57] will be described, because it partially contradicts some of the calculated results of Zanfir and Gavriilidis dealing with the thickness of the catalyst coating. Giroux et al. coated two metallic monoliths with catalyst for methane steam reforming. One monolith had 200 cells per square inch (cpsi) the other had 400 cpsi, which corresponds to channel sizes of 1.1 and 1.4 mm, respectively. The mass of the catalyst wash-coat was kept constant for both monoliths by increasing the wash-coat thickness. As shown in Figure 5.7, the conversion was essentially the same within the range of accuracy of the experiments. However, the variation in the channel size as assumed by Zanfir and Gavriilidis was more drastic.

Robbins et al. performed transient modelling of methane steam reforming over a 1.7 wt.% rhodium/γ-alumina catalyst combined with hydrogen or methane combustion over a 0.6 wt.% palladium/γ-alumina catalyst in a co-current plate heat-exchanger

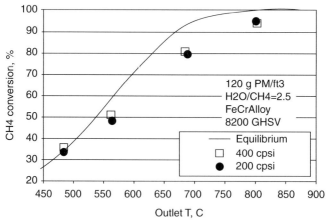

Figure 5.7 Methane conversion for two monoliths with 1.1- and 1.4-mm channel size; catalyst loading 120 g of precious metal catalyst in both instances; gas hourly space velocity 8200 h^{-1} for both monoliths [57].

[380]. A one-dimensional isobaric model was used for the gas phases and a single pair of combustion and steam reforming channels was considered. The wash-coat was discretised from the gas phase. However, a kinetic model for partial oxidation of methane over rhodium was used to simulate methane steam reforming. Results were also verified experimentally in a reactor composed of a Fecralloy plate coated with 25 μm of each of the combustion and steam reforming catalysts from both sides. The channels had 20-mm width and 3.2-mm height and were 100-mm long. The S/C ratios ranged between 0.8 and 2.0. Only 40% methane conversion could be achieved in the steam reformer part, while full conversion of the combustion reactions was always achieved regardless of whether methane or hydrogen was the fuel. A temperature profile from 700 °C at the reactor inlet to 600 °C at the outlet developed in the reactor when hydrogen was combusted and the feed temperatures were set to 500 °C. Self-sustaining conditions were feasible for feed temperatures as low as 400 °C. When methane was combusted, the temperature profile was more even over the reactor length [380]. This was probably due to the slower kinetics of the methane combustion.

Kirillov et al. modelled a combined hydrogen catalytic burner and methane steam reformer in plate heat-exchanger technology [381]. The model considered temperature differences between gas phase and catalyst and axial heat conductivity along the reactor wall, which is an important issue. The reactor under investigation had been built previously and evaluated by experiments. While a platinum catalyst was used for hydrogen oxidation, a nickel catalyst served for steam reforming. The plates of the reactor were of 1-mm thickness, while the thickness of the platinum catalyst porous metal supports was 1 mm. The thickness of the nickel catalyst porous metal support was even higher at 10 mm [381]. The inherent low activity of nickel catalysts probably made the higher thickness necessary. The nickel catalyst was prepared by sintering 84 wt.% powdered nickel with industrial nickel catalyst [381]. Owing to the rather low catalyst activity, the feed rate to the steam reformer side of the reactor was fairly low and the reactor had a relatively high volume of 0.5 L for a maximum electrical power equivalent of about 800 W_{el}. The reactor was operated in co-current mode. An increasing temperature profile was calculated for the reactor and verified by experiments, which originated from the fact that the feedstock was introduced cold into the reactor, which is very unusual. The temperature increased from 550 to 800 °C along the reactor length axis. The steam reforming catalyst block showed a significant transverse temperature profile due to the endothermic reaction and its high thickness. The temperature increased from 485 to 600 °C over the transverse axis. Start-up of the reactor was performed with hydrogen and air [381], which is not feasible in a practical system, because neither hydrogen nor anode off-gas is available.

Pan and Wang performed simulations of a plate-fin reformer for methanol steam reforming [382]. The system, which was not only investigated numerically but also built and tested (see Section 7.1.3), was an integrated device containing functionalities for fuel and water evaporation, steam reforming and catalytic combustion, as shown in Figure 5.8. Because a cross-flow arrangement had been chosen and catalyst fixed-bed technology was used, significant mal-distribution of the temperature and consequently also of the carbon monoxide selectivity were calculated over the reformer catalyst bed.

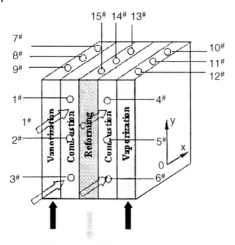

Figure 5.8 Integrated plate-fin reformer (PFR) developed by Pan and Wang [382]; the device incorporated plates for evaporation, steam reforming and catalytic combustion.

Petrachi et al. compared the performance of two microstructured isooctane steam reformers, which were supplied with energy either by catalytic combustion by integrated combustion channels, which were coated with catalyst (case A) or by integrated heating channels heated by combustion gases from an external burner (case B) [383]. The microchannels were assumed to be of 5.5-mm height, 550-µm width and 190-mm length. The thickness of the catalyst layer was set to 50 µm, while the thickness of the steel wall amounted to 500 µm. A S/C 3.0 was used for the calculations in both cases. For case A, the inlet temperature of the combustion gases into the reactor were 1000 °C, while the feed temperature of the integrated burner was 580 °C. It is important to mention that the air excess fed to the external burner of case B was 94%, while it was only 20% for the internal catalytic burner of case A [383]. This different air excess was obviously necessary because the energy had to be transferred from the combustion gases in case B, while the chemical energy was generated inside the integrated steam reformer in case A.

Figure 5.9 shows the isooctane conversion profiles for both cases. Surprisingly, these workers came to the conclusion that steam reforming was faster than catalytic combustion within the integrated reactor of case A, which was attributed to mass transfer limitations of isooctane in the microchannel burner, while the steam reforming reaction was controlled by kinetics, according to their calculations [383].

This result is, however, not in line with many other observations and investigations in the field of reforming and catalytic combustion (see Sections 4.2.7 and 4.6).

As a consequence of the mass transfer limitations in the burner, the temperature profile had a maximum close to the centre of the reactor for case A, while the temperature was declining due to the heat transfer from the heating gas in case B [383] (see Figure 5.10).

According to the practical experience of the author of this book, temperature profiles are declining in both cases. However, this is not a practical problem because

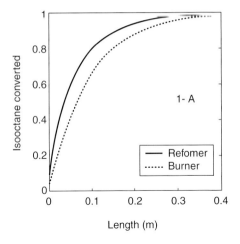

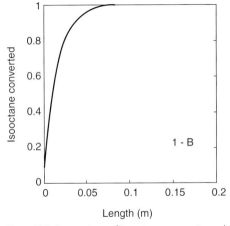

Figure 5.9 Comparison of isooctane conversion achieved in a microchannel steam reformer with integrated catalytic burner for isooctane combustion (case A), and a microchannel steam reformer supplied with energy from hot combustion gases fed into heating channels (case B) [383]; S/C ratio was 3.0 in both cases; the air excess was 20% in case A and 94% in case B.

the heavy hydrocarbon feedstock is more or less completely converted into lighter hydrocarbons, and also carbon oxides and hydrogen at the inlet section of the reactor. A lower temperature does not impair catalyst stability at the reactor centre and outlet section, because conversion of light hydrocarbons is taking place there. Lower temperatures at the exit are even beneficial because they reduce the carbon monoxide concentration of the product due to the equilibrium of the water–gas shift reaction. Too low temperatures, however, favour methane formation (see Section 3.1).

Dynamic simulations for a single couple of microchannels performed by Petrachi *et al.* revealed a fast response to load changes of the microchannels due

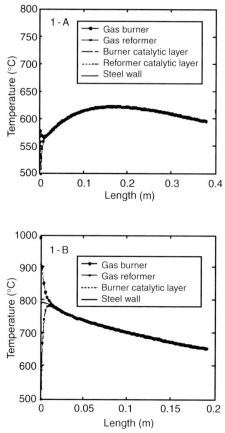

Figure 5.10 Comparison of temperature profiles along the reactor length axis as calculated for a microchannel steam reformer with integrated catalytic burner for isooctane combustion (case A) and a microchannel steam reformer supplied with energy from hot combustion gases fed into heating channels (case B) [383]; S/C ratio was 3.0 in both cases; the air excess was 20% in case A and 94% in case B.

to the direct coupling of heat input and heat consumption through only 500 μm of metal foil [383].

Cao et al. performed numerical calculations for methane steam reforming in microchannels [384]. They compared the performance of catalysts in straight wide gaps and in a corrugated "ruffle" design as shown in Figure 5.11. The latter geometry showed lower conversion during the experiments, which was explained by the less isothermal conditions and uneven concentration profiles through the numerical calculations. Then Tonkovich et al. performed numerical calculations and practical experiments of a combination of a similar gap for methane steam reforming, which was 11.4-mm long, and 10.7-mm wide but only 356-μm deep, with combustion

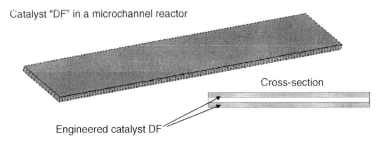

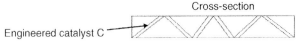

Figure 5.11 Different catalyst geometries as investigated by Cao et al. [384].

channels [385]. The catalyst coating for steam reforming, which was composed of 10 wt.% rhodium, 4.5 wt.% magnesia on an alumina carrier, had a high thickness of 280 μm. Thus, the remaining free flow path of the gap was only 76 μm. Steam reforming was performed at a S/C ratio 3.0. In the cylindrical combustion channels, hydrogen was converted. The channels had a surprisingly high diameter of 2.54 mm. The hydrogen was introduced into each channel through a small whole of 254-μm diameter and then mixed with the air by a static mixer. The latter was composed of spiral twisted Inconel strips. The combustion catalyst was palladium simply impregnated, without any carrier material, onto the Inconel walls, which had been pretreated at 950 °C. Table 5.4 shows the results that were determined experimentally for methane steam reforming at very short contact times in the range between 900 and 90 μs. Conversion close to the thermodynamic equilibrium could be achieved at a 900-μs residence time. Then Tonkovich et al. [385] performed calculations, unfortunately, for a different geometry of the gap. It was 4.06-mm wide and 177.8-mm long. Figure 5.12 shows the ratio of calculated conversion to equilibrium conversion for different space velocity and thickness of the catalyst coating for a 50-μm height of the remaining free gap. A coating thickness of about 100 μm was required to achieve 94% of the equilibrium conversion even at a 5-ms residence time according to Tonkovich et al.

Delsman et al. investigated the advantages of a microstructured methanol reformer coupled with a catalytic burner for anode off-gas over a conventional shell and tube design [386]. The conventional reformer was a shell and tube heat-exchanger, the combustion was performed over a catalyst coated to the inner surface of the tubes while a fixed-bed in the shell served for steam reforming. Two ranges of electrical power output of the corresponding fuel processor/fuel cell system were considered, namely 100 W and 5 kW. The specifications of the conventional devices are provided

Table 5.4 Experimental results of methane steam reforming in a gap of 76-μm height over a rhodium/magnesia/alumina catalyst coating at different contact time [385].

	1	2
SMR contact time (μs)	900	90
Time on stream for sample (h)	73	10
Molar steam to carbon ratio	3.0	3.0
Percent excess combustion air (%)	450	260
Inlet flows and compositions		
SMR CH_4 flow rate (SLPM)	0.153	1.55
SMR steam flow rate (SLPM)	0.461	4.64
Air flow rate (SLPM)	5.4	5.0
Fuel H_2 flow rate (SLPM)	5.508	0.81
Gas stream temperatures		
SMR inlet gas temperature [°C]	837	788
SMR outlet gas temperature [°C]	802	754
Air inlet gas temperature [°C]	806	732
Exhaust gas temperature [°C]	912	862
Gas stream pressures and pressure drops		
SMR inlet pressure [bar]	13.0	13.0
SMR outlet pressure [bar]	12.9	12.1
SMR pressure drop [bar]	0.1	0.9
Air inlet pressure [bar]	1.47	1.43
Air pressure drop [bar]	0.13	0.1
SMR performance		
SMR CH_4 conversion [GC, Vol.%]	88.2	17
Selectivity: CO [%]	38.3	43
Average reactor web temperature [°C]	837	811
Equilibrium conversion at temperature [°C]	89.1	86.4
Approach to equilibrium [%]	99	19.7
Average heat flux [W/cm^2]	18.9	21.3
Combustion performance		
Combustion H_2 conversion [%]	100	100

in Table 5.5, while the specification of the microstructured catalyst coated reformer/burner heat-exchanger reactor are provided in Table 5.6. The calculations revealed more than 50% lower reactor size and that more than 30% less catalyst mass was required for the microreactor for the 100-W system (see Figure 5.13). For the 5-kW system, the reactor volume was only 30% lower, but the catalyst mass required decreased to about 50% (see Figure 5.14). Scale-up issues for larger scale devices were also discussed by Tonkovich et al. [385].

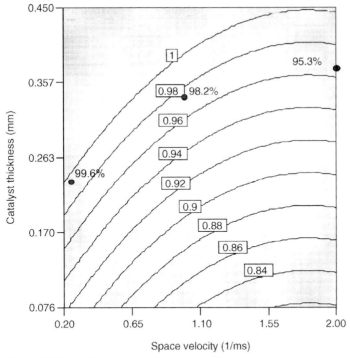

Figure 5.12 Ratio of calculated conversion of methane steam reforming to equilibrium conversion for various space velocities and thicknesses of the catalyst layer; the gas flow was assumed to take place in a 50-μm high gap above the catalyst [385].

Table 5.5 Specifications of the conventional shell and tube methanol reformer/burner reactors of 100 W_{el} and 5 kW_{el} power equivalent as assumed by Delsman et al. for their calculations [386].

	RefBurn reactor	
Design Parameter	100 W_e	5 kW_e
Construction material	Steel	Steel
Pellet diameter [mm]	0.4	1.0
Tube diameter [mm] (wall thickness [mm])	2 (0.5)	2 (0.5)
Tube pitch	2	2.5
Fixed-bed length [mm]	52	190
Number of tubes	56	974
Shell diameter [mm] (wall thickness [mm])	28 (3)	150 (4)
Reactor volume [dm³]	0.091	7.0
Insulation volume [dm³]	0.28	1.1
Catalysis mass [g]	30	3500
Total mass [g]	530	17,000
Mean bed temperature [°C]	235	222
Mean cat. Effectiveness [%]	79	57

Table 5.6 Specifications of the microstructured reformer/burner heat-exchanger reactors of 100 W_{el} and 5 kW_{el} power equivalent as assumed by Delsman et al. for their calculations [386].

	100 W_e		5 kW_e	
Design Parameter	ref[a]	burn[a]	ref	burn
Number of plates	96	48	348	174
Units in parallel	4	4	4	4
Channels per plate	30	30	143	143
Open channel height [µm]	40	70	91	160
Channel length [mm]	23	23	120	120
Coating thickness [µm]	100	–	100	–
Reactor material	Carbon steel		Carbon steel	
Reactor size [w × h × l cm³]	3.5 × 3.5 × 3.5		16 × 16 × 16	
Reactor volume [dm³]	0.043		4.3	
Insulation volume [dm³]	0.098		0.56	
Catalyst mass [g]	19		1700	
Total mass [g]	210		12,000	
Mean bed temperature [°C]	247		228	
Mean cat. Effectiveness [%]	75		86	

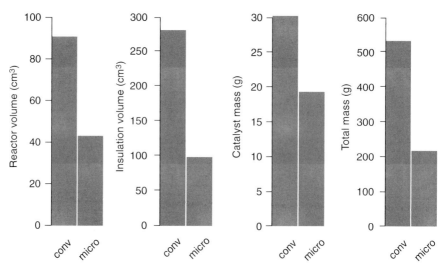

Figure 5.13 Comparison between conventional fixed-bed technology (conv) and a combined micro-structured plate heat-exchanger/catalytic afterburner (micro) according to Delsman et al. [386]; power range, 100 W_{el}.

5.1.2
Partial Oxidation

Running a fuel processor under conditions of partial oxidation is most favourable with methane, because it has the highest atomic hydrogen content of all hydrocarbons. While the theoretical hydrogen concentration of the reformate still reaches

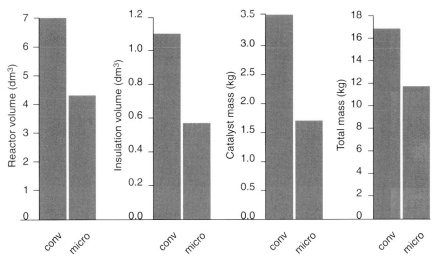

Figure 5.14 Comparison between conventional fixed-bed technology and a combined microstructured plate heat-exchanger/catalytic afterburner according to Delsman et al. [386]; power range, 5 kW$_{el}$.

values around 34 vol.% at an O/C 1.0 with methane, it rapidly decreases for higher hydrocarbons to 25.7 vol.% with propane and then levels off to about 23 vol.% for C$_{8+}$ species. The hydrogen concentration of the reformate is still lower in practical applications, because partial oxidation catalysts have a certain selectivity towards total oxidation, even under stoichiometric conditions of partial oxidation (O/C ratio 1.0), as discussed in Section 3.2. Usually a surplus of air is required to avoid coke formation under conditions of partial oxidation. This excess of air reduces the hydrogen content additionally as shown in Figure 5.15 for O/C 1.2. Not only the hydrogen content, but also the total amount of hydrogen and carbon monoxide is decreasing and thus the overall heating value of the reformate is lower for higher hydrocarbons.

Despite the dilution of hydrogen, partial oxidation is an attractive option due to the simplicity of the system, because water evaporation is not required. One favourable application is the pre-reforming for high temperature fuel cells, which consume carbon monoxide. When PEM fuel cells are the consumers of the reformate, either a membrane separation is required to provide undiluted hydrogen or substantial amounts of steam need to be added downstream of the reformer to remove carbon monoxide and increase the hydrogen content via a water–gas shift.

Thermodynamic calculations as described in Section 3.2 help to explain the optimisation problem of a partial oxidation reactor. The feed pre-heating temperature must not be too low, because otherwise the reaction may suffer from carbon formation at the reactor inlet, especially at high feed flow rates. On the other hand, an O/C ratio of about 1.2 is required to avoid coke formation. Finally, the reaction temperature should be kept below 800 or 850 °C to avoid damage to the catalyst structure (unless high temperature resistant catalysts are applied, which are, however, less active than a precious metal based system, see Section 4.2).

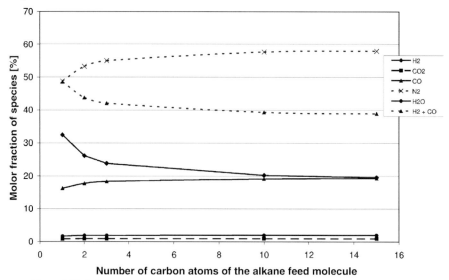

Figure 5.15 Reformate composition versus carbon number of alkane feed as calculated for an O/C ratio of 1.2.

Figure 5.16 shows three different modes of partial oxidation. The first one is the so-called "Spaltvergaser", developed by Daimler Benz in the 1970s to improve emissions of internal combustion engines. The reactor worked with homogeneous partial oxidation (without a catalyst). Part of the product gases is fed back into the reactor

Thermal partial oxiation with product feed back

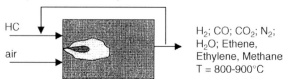

Thermal partial oxidation with water injection

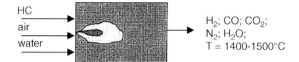

Thermal partial oxidation combined with steam reforming

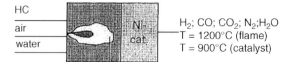

Figure 5.16 Design concepts of partial oxidation reformers [7].

inlet, which prevents coke formation. However, the operating temperature of the process was low, between 800 and 900 °C and lead to formation of numerous light hydrocarbons, especially methane. Thus this concept was *per se* not applicable in fuel cell systems. The second system shown in Figure 5.16 also worked with homogeneous partial oxidation but at temperatures between 1400 and 1500 °C, and thus no light hydrocarbons were stable any longer. Finally, the third system worked with two zones, namely a zone for homogeneous reaction operated at 1200 °C and a second zone for steam reforming at 900 °C over a nickel catalyst. The last system leads to autothermal reforming on an industrial scale, as discussed below.

Stutz *et al.* performed thorough numerical calculations for the partial oxidation of methane in a single microchannel of 10-mm length and 1-mm diameter as a model of an adiabatic ceramic monolith [387]. The heat conductivity of the wall was taken into consideration and kinetics of 38 elementary reaction steps were used to model the chemical reaction network in a two-dimensional model. The feed temperature was set to 580 °C and ambient pressure was assumed. Hot spots between 1100 and 1200 K were calculated for the channel, which were situated 1-mm downstream the reactor inlet. The oxygen was almost completely consumed after the first 2 mm of the channel. Surprisingly, Stutz *et al.* found steam as the main product in the hot spot region followed by carbon monoxide, particularly at high residence times. This result is in line with the experimental results from several workers, which were described in Section 3.2. The steam then reacted downstream in the channel with methane or carbon monoxide. Consequently, hydrogen was only a minor product of the reaction system provided oxygen was still present [387]. Increasing the site density of the catalyst increased the hydrogen production in the oxygen rich region and also increased steam conversion into hydrogen downstream of the reactor. Stutz *et al.* then varied the channel diameter from 0.67 to 3 mm. Methane conversion and hydrogen yield showed a maximum at a 1.7-mm channel diameter. However, oxygen was not completely converted in the largest channels of 3 mm diameter (see Figure 5.17). This lead to a uniform and high temperature of 1130 °C over the whole reactor length, because the exothermic reactions occurred along the whole channel. For the smallest channels, a hot spot temperature of 1000 °C was calculated. The temperature decreased to 830 °C towards the outlet as shown in Figure 5.17.

5.1.3
Autothermal Reforming

Autothermal reforming, also termed oxidative steam reforming, on an industrial scale is usually comprised of a combination of homogeneous combustion at flame temperatures exceeding 2000 °C and catalytic steam reforming [388]. The steam reforming catalyst is frequently exposed to temperatures exceeding 1100–1400 °C [388].

For processes on a smaller scale, the entire autothermal reforming process is usually performed in a catalyst bed. Alternative concepts are the catalytic pre-reforming of higher hydrocarbons or cool flames (see Section 3.5 and 5.1.5).

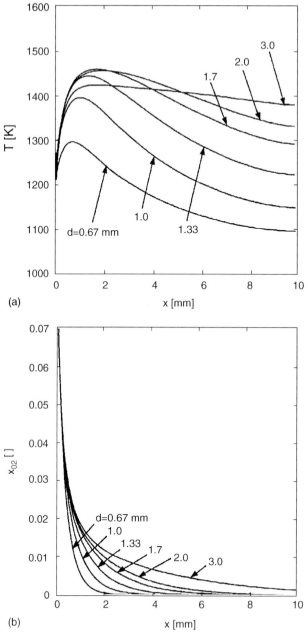

Figure 5.17 Temperature (a) and oxygen mole fraction (b) for various channel diameters for partial oxidation of methane in a ceramic monolith [387].

Chan et al. calculated the optimum O/C ratio for autothermal natural gas reforming to be 0.72, while a high S/C ratio of between 2.5 and 3.0 was regarded as optimum. However, these calculations only focussed on a maximum hydrogen yield of the reformer [76]. The subsequent water–gas shift stages and the energy efficiency of the system were not taken into consideration. Additionally, adiabatic conditions were assumed for the calculations, which make the evaluation of the data difficult because different temperatures have to be compared.

Seo et al. calculated optimum ratios for S/C of 0.35 and O/C of 1.16 for maximum hydrogen yield. However, these conditions are a long way from being autothermal but, rather, are in the range of partial oxidation supported by steam addition. The corresponding carbon monoxide content of the reformate amounted to about 13 vol. %, which is a very high number and puts a substantial load onto the subsequent water–gas shift stages. For high temperature fuel cells, the reformate quality obtained at S/C 0.35 and O/C 1.16 would of course be sufficient, because carbon monoxide also serves as the fuel.

Table 5.7 shows the equilibrium reformate composition at four different O/C and S/C ratios as calculated from the data provided by Seo et al. [66]. The equilibrium composition of the reformate after water addition (provided as mol water per mol of

Table 5.7 Equilibrium reformate composition as achieved by steam supported partial oxidation (O/C = 1.2) and autothermal reforming (O/C = 0.88) of methane; gas compositions are provided after the reformer and after water addition plus water–gas shift equilibrium at 250 °C.

Feed				
S/C	0.35	1.20	0.80	1.20
O/C	1.2	1.2	0.9	0.9
CH_4 [Vol.%]	20.8	17.7	23.1	21.1
O_2 [Vol.%]	12.5	10.6	10.1	9.3
N_2 [Vol.%]	59.4	50.5	48.3	44.2
H_2O [Vol.%]	7.3	21.2	18.5	25.3
Equilibrium after reformer				
H_2 [Vol.%]	30.6	31.2	30.5	33.4
CO [Vol.%]	12.9	8.5	8.5	6.4
CO_2 [Vol.%]	3.2	5.7	8.5	9.5
H_2O [Vol.%]	7.2	14.2	17.0	17.5
N_2 [Vol.%]	46.0	40.5	35.5	33.3
H_2O addition [mol/mol CH_4]	2.0	1.2	0.8	0.7
Equilibrium Water–gas shift after water addition (250 °C)				
H_2 [Vol.%]	32.8	33.5	34.0	35.6
CO [Vol.%]	0.2	0.2	0.3	0.3
CO_2 [Vol.%]	12.0	11.8	14.6	14.1
H_2O [Vol.%]	20.0	20.0	20.0	20.0
N_2 [Vol.%]	34.9	34.4	31.2	30.1
CO Conversion [%]	97.6	96.7	96.0	94.8

methane feed) and subsequent water–gas shift at a 250 °C reaction temperature is also shown. The water addition was adjusted such that the water content of the reformate after the water–gas shift amounted to 20 vol.%, which is the least amount required to prevent a low temperature PEM fuel cell from dry-out. This comparison demonstrates that excessive amounts of steam need to be fed to the system downstream of the reformer to achieve sufficient water content. This steam addition is approximately six times higher than the steam fed to the reformer for an O/C ratio of 1.16 and an S/C ratio of 0.35. The reformate generated by partial oxidation contains 1–2 vol.% less hydrogen compared with autothermal reforming. One aspect that needs to be taken into consideration is that the selectivity towards total oxidation is usually higher than that calculated from the thermodynamic equilibrium under conditions of partial oxidation due to the catalyst properties. This reduces the hydrogen content to even lower values.

Springmann et al. [152] simulated a 10-kW$_{el}$ autothermal reformer for gasoline with a one-dimensional, dynamic model. The reformer was a metallic monolith coated with precious metal catalyst. The reformer consisted of two parts both of 70-mm diameter, a small electrically heated monolith only 5-mm long (400 cells per square inch or about 1300-μm channel size) and a second, about 100-mm long monolith (1600 cells per square inch or about 630-μm channel size). Both monoliths were coated with the same catalyst. Kinetic data were determined prior to the simulations [56] and heat losses were estimated for the reactors, which is a critical issue. The reformer was operated at a S/C ratio of 2 and an O/C ratio of 0.75. The pressure was 4 bar and the feed inlet temperature 500 °C.

The partial oxidation reaction occurred almost completely in the first part of the small electrically heated reformer, which consequently was exposed to a temperature of 1000 °C.

The calculations revealed that full conversion of the fuel was no longer possible with more than 10% load of the reactor due to the increasing effect of heat losses and an increasing contribution of axial heat conduction along the axis of the reactor length, which flattened its temperature profile. Increasing the O/C ratio of the feed from 0.75 to 1 compensated for these effects, as shown in Figure 5.18.

Start-up was simulated assuming a three-step procedure: Initially the first reformer was heated by electricity to 500 °C in 5 s with 2.5-kW electrical energy input. Then gasoline and air were fed in at a temperature of 200 °C, which was assumed to be sufficient for light-off of partial oxidation or catalytic combustion reactions. After 1 min, superheated steam was assumed to be available from the external evaporators, which were not included into the calculations. However, gasoline conversion was incomplete during the second phase under conditions of partial oxidation. About 10 vol.% methane was then present in the product [389]. Thus, as an alternative catalytic combustion of the gasoline fuel was used in the second phase at a high O/C ratio of 9.1. This value corresponded to approximately a three-fold air excess for complete combustion. The air excess should prevent temperature fluctuations of the reformer. After 60 s at 8-kW thermal power generation, the reformer was almost uniformly pre-heated to 1000 °C, according to the calculations.

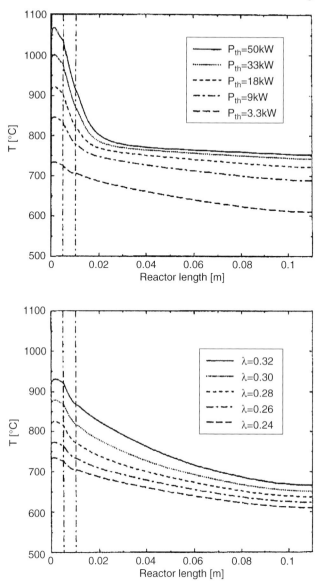

Figure 5.18 Effect of load changes (top) and variation of the O/C ratio at 3.3-kW thermal power of the reactor (in this instance the O/C ratio is expressed as λ; $\lambda = 1$ corresponds to complete combustion and an O/C ratio of 3.12); S/C = 2; feed temperature 500 °C; pressure 4 bar [152].

5.1.4
Catalytic Cracking

An alternative to the oxidation of the carbon product generated during the cracking step as described in Section 3.4 is its transformation into carbon dioxide and hydrogen by steam treatment [102].

Choudhary and Goodman performed methane cracking over nickel/zirconia catalysts [102]. Two reactors had be switched in parallel for this process. Both the methane and the steam feed had to be switched between the two reactors, as shown in Figure 5.19. The optimum switching time between the processes was determined to

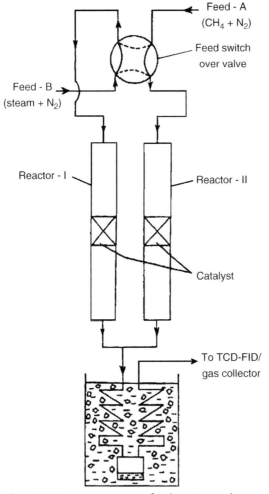

Figure 5.19 Process arrangement for alternating methane decomposition and coke gasification by steam as proposed by Choudhary and Goodman [390].

be 10 min. The advantage of the cracking procedure was the low operating temperature of 500 °C for both reactors, the disadvantage was the valves that were required for changing the feedstock. Continuous hydrogen production was achieved by Choudhary and Goodman a test duration of more than 6 h. However, methane conversion did not exceed 25% despite the very low gas hourly space velocity that was applied, which was in the range of $3\,L\,(h\,g_{cat})^{-1}$ for the cracking reaction [390]. Another alternative cracking concept was presented by Muradov, which worked with carbon-based catalysts. The major part of the carbon product was removed discontinuously during the cracking process [6]. This last option seems least viable for mobile and portable systems. Thus, Muradov proposed this system as an idea for niche application fields that are emission sensitive, such as hospitals and tunnels.

5.1.5
Pre-Reforming

An application of pre-reforming is the feed pre-conditioning for high temperature fuel cells, such as solid oxide fuel cells. When steam reforming is applied, the steam generation of the fuel cell anode may balance the steam consumption by steam reforming [38]. To utilise the anode off-gas, a certain amount needs to be split and re-fed to the pre-reformer [38].

A typical gas composition of pre-reformer feed blended with anode off-gas amounted to 17 vol.% natural gas, 43 vol.% steam, 20 vol.% carbon dioxide, 7 vol.% carbon monoxide and 13 vol.% hydrogen. The gas composition downstream of the pre-reformer then changed to 11 vol.% methane, 32 vol.% steam, 21 vol.% carbon dioxide, 8 vol.% carbon monoxide and 28 vol.% hydrogen [38].

Figure 5.20 shows the set-up of a cool flame evaporator. The diesel fuel was evaporated by an atomiser pulsing with 50 Hz, which was water-cooled to 120 °C. The air was pre-heated by electricity, which could be taken over by a heat-exchanger in a fuel processing system. The evaporator required pre-heating for start-up, which could be done with a start-up burner in a fuel processor system [69].

5.2
Design of the Carbon Monoxide Clean-Up Devices

5.2.1
Water–Gas Shift

The water–gas shift reaction usually suffers from mass transfer limitations similar to the reforming process when fixed-bed catalysts are applied. Levent found catalyst effectiveness factors for high temperature iron/chromium water–gas shift catalysts in the range between 10 and 20% [391]. Giunta et al. calculated catalyst effectiveness factors lower than 10% for a fixed low temperature water–gas shift catalyst bed [392].

The dominant effect on the size and performance of water–gas shift reactors of the S/C ratio fed to the reformer reactor was emphasised by Castaldi et al. [325]. As a rule

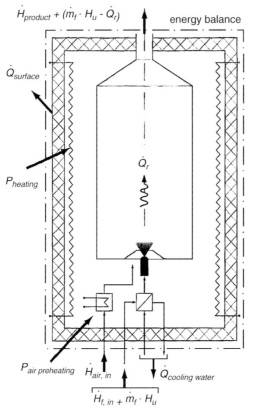

Figure 5.20 Principle of the set-up of a cool flame vaporiser [69].

of thumb, these workers estimated a 10% reduction in size for each additional vol.% of water fed to the reformer. Figure 5.21 shows the increasing carbon monoxide conversion as a result of the higher S/C ratio despite the higher space velocity, which was calculated and verified by experiments.

Zalc and Löffler proposed a plate heat-exchanger design for the water–gas shift [378]. By adjusting an optimum temperature profile in the plate heat-exchanger, the carbon monoxide conversion could be significantly improved compared with isothermal and adiabatic (uncooled) operation of the reactor, according to the calculations, as shown in Figure 5.22. The main advantage of the optimum temperature profile was that high reaction rates were achieved at the reactor inlet, where the carbon monoxide concentration was still high and the reaction far from equilibrium. By counter-current cooling of the reactor, a declining temperature profile was achieved, which shifted the thermodynamic equilibrium in a favourable direction at the reactor outlet. However, these workers then calculated that a catalyst mass of 17 kg was required for a water–gas shift reactor dedicated for a 50-kW$_{el}$ fuel cell. The weight hourly space velocity assumed for these calculations amounted to 3.7 L (h g$_{cat}$)$^{-1}$.

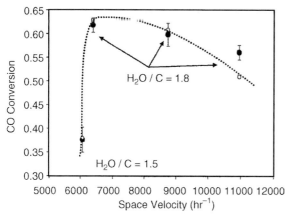

Figure 5.21 Effect of higher S/C ratio of the reformer on carbon monoxide conversion as calculated (dotted line) and determined experimentally by Castaldi et al. [325].

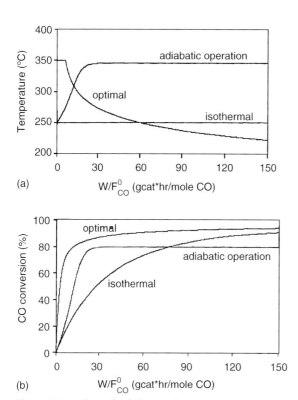

Figure 5.22 Isothermal, adiabatic and optimum temperature profiles and carbon monoxide concentration versus modified residence time W/F^0_{CO} [378].

This value, assumed by Zalc and Löffler, was rather low. Recent catalyst development work revealed operation of water–gas shift catalysts at 180 L (h g_{cat})$^{-1}$ and thus only 350 g catalyst would be required. However, catalyst coatings need to be applied to achieve such improved performance, which increases the utilisation of the porous catalyst structure.

TeGrotenhuis *et al.* simulated the water–gas shift reaction in a microchannel plate heat-exchanger [393] over a precious metal catalyst. Diffusion effects were taken into consideration both for the gas phase and for the catalyst. For a given feed composition of 9 vol.% carbon monoxide, 9 vol.% carbon dioxide, 36 vol.% steam and 45 vol.% hydrogen, which corresponds to the product of an octane steam reformer at a S/C ratio 3, an optimum temperature profile was calculated for the reactor, at which the maximum reaction rate was achieved. 82% conversion was calculated for first-third of the reactor, while the remaining two-thirds of the reactor were required to achieve a further 8% conversion. As the reaction temperature had to be limited to about 400 °C due to stability issues of the noble metal catalyst (see Section 4.5.1), the optimum profile was limited accordingly. A comparison was performed between a reactor running with an optimum temperature profile and two stage adiabatic reactors; 2.3-times more catalyst was necessary to achieve the same degree of conversion for the two adiabatic reactors according to the calculations. Thus, significantly less catalyst, only a single reactor and no intermediate heat-exchanger or water injection were required for the integrated heat-exchanger. The ratio of reformate to cooling gas flow had to remain within a certain range. Too low a cooling gas flow created quasi-adiabatic conditions, while the reaction was quenched by too high a cooling gas flow. Finally, it was demonstrated that the reformate inlet temperature had only a minor effect on the maximum conversion achieved. The inlet temperature of the coolant gas was more crucial (see Figure 5.23).

Then Kim *et al.* performed calculations for a microchannel heat-exchanger for the water–gas shift reaction [394]. They applied one- and two-dimensional models for their calculations. However, fixed-bed reactors were chosen for the calculations and the kinetics of a copper/zinc oxide catalyst were used. The one-dimensional model did not take the heat conductivity of the reactor wall into consideration, which reduced its applicability to values of the wall heat conductivity of less than 1 W (m K)$^{-1}$. The reactor inlet temperatures assumed by Kim *et al.* exceeded the maximum operating temperature of a copper/zinc oxide catalyst by far. Thus, the kinetic expressions were applied outside their range of applicability. The effect of wall thermal conductivity was investigated by Kim *et al.* using a two-dimensional model for both co-current and counter-current flow of the heat-exchanger [394]. As shown in Figure 5.24, high carbon monoxide conversion could be achieved only for a very low wall conductivity of around 10^{-2} W (m K)$^{-1}$ in the co-current mode, while the counter-current flow allowed high conversion for a broad range of wall conductivities, which covered the values for stainless steel [between 10 and 20 W (m K)$^{-1}$]. Kim *et al.* observed that the counter-current operation was less sensitive to the inlet temperature and flow rate of the coolant gas and to the inlet temperature of the reformate feed [394]. Cooling by steam generation was ruled out, because high conversion could only be achieved when the heat conductivity of the wall was around 10^{-2} W (m K)$^{-1}$ independent of the flow

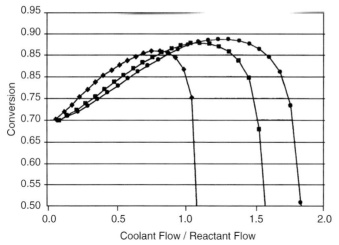

Figure 5.23 Effect of coolant flow rate and temperature on CO conversion for the water–gas shift reaction in a microchannel reactor at constant reformate feed inlet temperature of 350 °C: ◆, 125 °C coolant temperature; , 200 °C; •, 225 °C [393].

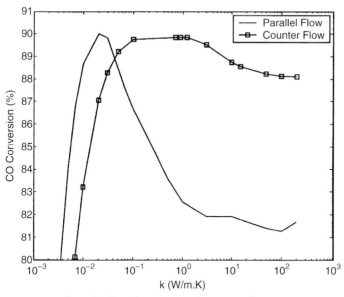

Figure 5.24 Effect of wall conductivity on carbon monoxide conversion for co-current and counter-current operation of a heat-exchanger reactor for water–gas shift, feed composition: 6 vol.% carbon monoxide; 10 vol.% carbon monoxide; 25 vol.% steam; 29 vol.% hydrogen; 30 vol.% nitrogen [394].

arrangement. The wall thickness had no effect on the reactor performance in the range between 200 and 1000 μm. Kim et al. calculated that a two-fold catalyst mass was required for an adiabatic reactor compared with the heat-exchanger arrangement.

Giroux et al. performed a comparison of an adiabatic water–gas shift reactor, which had a temperature rise from 300 to 360 °C, isothermal operation at 350 °C achieved by integrated heat-exchange and a reactor with a declining temperature profile from 550 to 300 °C [57]. In each case the same degree of carbon monoxide conversion was achieved in the reactors, but the space velocity increased from 35 000 h^{-1} for the adiabatic reactor to 50 000 h^{-1} for the isothermal reactor and even up to 70 000 h^{-1} for the reactor with the declining temperature profile, which means that the last reactor would require only half the size of the adiabatic counterpart.

Baier and Kolb analysed the water–gas shift reaction in a microstructured heat-exchanger by one- and two-dimensional models [395]. Kinetic expressions of conventional water–gas shift catalysts were modified to ten-times higher activity to simulate the higher activity of precious metal catalysts. The performance of two adiabatic reactors was compared with the heat-exchanger, which required 30% less length. Reducing the flow rates (turn-down) did show improved performance by the heat-exchanger due to the higher residence time. Another advantage of the heat-exchanger design was that no water addition was required to achieve equivalent performance compared with the two adiabatic reactors, on top of the 30% reduced length. The effect of diffusion limitations in channels of different size was analysed. For a channel height of 200 μm, the absence of diffusion limitations was observed (see Figure 5.25). Severe diffusion limitations were observed for a channel height of 800 μm. Thus practical applications should stay below this value taking into consideration the required limitation of a pressure drop in a practical system. However, increasing the channel height to 800 μm required a reactor twice as long in order to

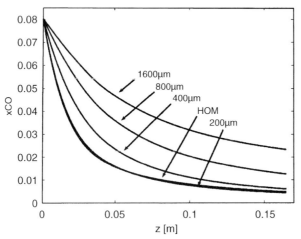

Figure 5.25 Carbon monoxide molar fraction versus channel length for various channel heights in a water–gas shift heat-exchanger reactor; HOM corresponds to the results from a homogeneous model (no diffusion limitation) [395].

achieve the same degree of carbon monoxide conversion. Results of this work were successfully applied for the construction of a water–gas shift heat-exchanger reactor on a kW scale [312].

Earlier experimental work performed by Pasel *et al.* [396] confirmed the results of Baier *et al.* [395]. Lower conversion for the water–gas shift in microchannels of 800-µm diameter was observed compared with channels of 400-µm diameter in the temperature range between 200 and 350 °C under otherwise identical conditions [396].

5.2.2
Preferential Oxidation of Carbon Monoxide

Both the preferential oxidation reaction and the hydrogen oxidation that takes place in parallel are highly exothermic (see Section 3.10.2). The local overheating of fixed catalyst beds was demonstrated by Ouyang and Besser [397], as shown in Figure 5.26. The small catalyst bed was overheated by almost 70 K for a 180 °C reaction temperature in the centre.

Kahlich *et al.* performed preferential oxidation of carbon monoxide over platinum/alumina and gold/iron oxide catalysts. Because their platinum catalyst was operated at a very high reaction temperature of 200 °C, incomplete conversion was obtained at lower space velocities. Reverse water–gas shift occurred as soon as the oxygen had been consumed, over a surplus of catalyst [114]. These conditions correspond to a partial load of a fuel processor. Reverse water–gas shift is favoured at higher temperatures. Thus, noble metal catalysts need to be optimised for operation at low temperatures, as close to 100 °C as possible to minimise the reverse water–gas shift. Kahlich *et al.* solved this issue by switching to a gold/α-alumina catalyst. Less reverse

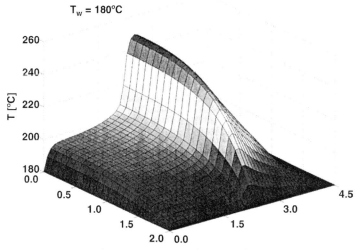

Figure 5.26 3D plot of reactor temperature for a small fixed catalyst bed of 2 mm × 4 mm size; wall temperature 180 °C; weight hourly space velocity 1500 h^{-1} [397].

water–gas shift was observed. Kahlich et al. then operated a two-stage preferential oxidation system with small fixed beds for more than 100 h. The first stage worked with a platinum/alumina catalyst, the second stage carried a gold/iron oxide catalyst, which was operated at 80 °C. The carbon monoxide present could be successfully reduced from 1000 ppm to less than 10 ppm. The O/CO ratios applied were rather high at 4.8 and 9.4, respectively.

Giroux et al. proposed a two-stage approach with inter-stage cooling for the preferential oxidation of carbon monoxide for monolithic reactors. According to these workers, the addition of the small air flows required for the second oxidation stage was difficult. Therefore, the first stage was operated at a low O/CO ratio of 1.2 with incomplete conversion of carbon monoxide. This also limited the undesired oxidation of hydrogen to a minimum [57].

Different feed addition strategies for a porous reactor concept were compared by Schuessler et al. [398] (see Figure 5.27). The first arrangement was when both reformate and air were fed to the front of the reactor (case I). This scenario is valid for most common applications. However, in case II, the reformate was gradually added to the air feed and in case IV air was gradually added to the reformate. Case III covered the gradual addition of air through the porous channel walls, while the reformate was added through the porous catalyst itself.

The kinetics of the preferential oxidation reaction of carbon monoxide (PrOx), for the hydrogen oxidation reaction (H_2Ox) and for the reverse water–gas shift reaction (RWGS) were provided, which had been determined in the relevant parameter space [398]:

$$r_{\text{PrOx}} = \left[\frac{k_4 p_{\text{CO}}}{1 + K_{\text{CO}}\, p_{\text{CO}}} - k_{\text{inh}} p_{\text{CO}} \right] p_{O_2}^{0.5} \tag{5.2}$$

$$r_{H_2Ox} = k_5 p_{\text{CO}}^{-0.33} p_{O_2}^{0.5} \tag{5.3}$$

$$r_{\text{RWGS}} = k_5 p_{\text{CO}}^{-0.33} p_{O_2}^{0.5} \tag{5.4}$$

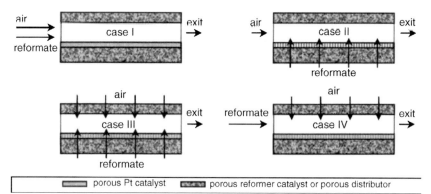

Figure 5.27 Four different feed addition strategies for air and reformate for a preferential oxidation reactor, as proposed by Schuessler et al. [398].

It is assumed that these kinetics were determined for the platinum catalyst incorporated in a compact methanol fuel processor (see Section 9.1). The expressions were valid for carbon monoxide and oxygen in the range between 10 and 50 000 ppm [398]. For hydrogen oxidation, the rate equation was valid for mole fractions exceeding 40%. The reaction rate did not depend on the hydrogen concentration, because hydrogen was present in a large excess. Thus, the reaction rate depended only on the oxygen concentration and was inhibited by carbon monoxide. The reverse water–gas shift reaction was also inhibited by carbon monoxide [398].

Simulations of the four cases described above were carried out. Mass transport limitations of oxygen and carbon monoxide were considered for the calculations. The pressure was set to 1.5 bar and reaction temperature to 300 °C, which is an extremely high value; 1 vol.% carbon monoxide and 2 vol.% oxygen were assumed as feed. In case II the carbon monoxide content was not reduced below 4 000 ppm, regardless of the load, owing to the massive hydrogen oxidation at the reactor inlet, which consumed too much oxygen. As indicated in Figure 5.28, case I had the narrowest operating window. The reactor had one optimum operating point and the carbon monoxide concentration rose dramatically at the lower load, because a reverse water–gas shift was taking place. However, this result was probably exaggerated by the high operating temperature. Case III had a wider operating window compared with case I, but the carbon monoxide concentration could not be reduced below 600 ppm. Case IV offered the widest dynamic range for load changes at a low carbon monoxide content of the product [398].

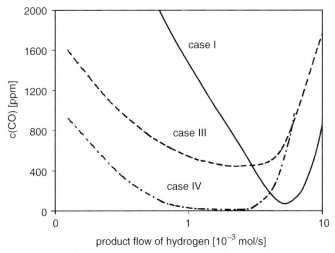

Figure 5.28 Carbon monoxide concentration as a function of reactor load for the various cases shown in Figure 5.27: feed composition, 1 vol.% carbon monoxide and 2 vol.% oxygen; reaction temperature 300 °C; pressure 1.5 bar [398].

Zalc and Löffler also proposed the application of plate heat-exchanger technology for the preferential oxidation of carbon monoxide to improve the heat-management of the highly exothermic process [378].

5.2.3
Selective Methanation of Carbon Monoxide

Xu et al. proposed temperature-staged methanation as an alternative to preferential oxidation [399]. The considerations were supported by experimental work performed with a 0.5 wt.% ruthenium catalyst on an alumina carrier. Complete conversion of the carbon monoxide by methanation with a selectivity of about 40% was achieved at around a 220 °C reaction temperature. However, the weight hourly space velocity of $2\,L\,(h\,g_{cat})^{-1}$ was extremely low. The selectivity of the reaction increased with decreasing reaction temperature but also the carbon monoxide content in the product increased, as shown in Figure 5.29. Thus Xu et al. proposed carrying out the methanation reaction in several reactors switched in series, which worked at decreasing temperatures from stage to stage. By decreasing the temperature, selectivity could not be maintained at high values, but this played a minor role, because the remaining carbon monoxide content was already low. Table 5.8 provides three example of triple-, dual- and single-staged methanation as proposed by Xu et al. for different feed compositions.

5.2.4
Membrane Separation

The membrane separation process involves several elementary steps, which include the solution of hydrogen and its diffusion as atomic hydrogen through the membrane bulk material. Polymeric membranes are frequently applied in industrial

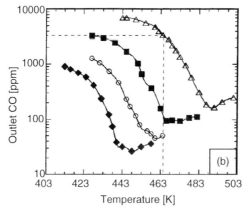

Figure 5.29 Carbon monoxide content at the outlet of the methanation reactor; the symbols correspond to the gas compositions provided [399].

Table 5.8 Three examples of process parameters proposed for temperature staged methanation as proposed by Xu et al. [399]; selectivity is defined as the ratio of moles carbon monoxide converted into moles of methane formed.

Inlet CO [mol%]	Parameter	Stage 1	Stage 2	Stage 3	Overall
1.00	Outlet CO [ppm]	3,500	1,000	~50	~50
	Temperature [K]	465	438	~420	–
	Formed CH_4 [mol%]	0.65	~0.25	~0.2	1.10
	Selectivity [–]	1.0	1.0	0.5	0.93
0.50	Outlet CO [ppm]	1500	~60	–	~60
	Temperature [K]	446	~425	–	–
	Formed CH_4 [mol%]	0.37	~0.30	–	0.67
	Selectivity [–]	1.0	0.5	–	0.75
0.20	Outlet CO [ppm]	75	–	–	75
	Temperature [K]	424	–	–	–
	Formed CH_4 [mol%]	0.40	–	–	0.40
	Selectivity [–]	0.5	–	–	0.5

processes for hydrogen separation, which seem to be less convenient on a smaller scale, because several separation steps are required due to the relatively low selectivity of the separation process. However, palladium and palladium alloy membranes have almost 100% selectivity and seem to be better suited for mobile fuel processors, despite their higher cost [105]. A palladium membrane of 20-mm thickness allows a hydrogen flux of 4 m^3 (m^2 h)$^{-1}$ at an operating temperature of 350 °C and a pressure gradient of 1 bar [107]. Decreasing the membrane thickness doubles the permeability and halves the weight of palladium, thus resulting in a cost reduction of 75%. Pure palladium membranes tend to become brittle during operation. Thus, palladium membranes should not be operated at temperatures below 300 °C and pressures lower than 20 bar, while the upper temperature range is between 500 and 900 °C [400]. Alloying the membrane with silver and copper extend their operating temperature down to room temperature [400]. The introduction of silver into the palladium membrane increases its lifetime, but also the cost, as compared with copper [401]. Han et al. reported the application of a membrane containing 40 wt.% of copper, which had a very high diffusivity at room temperature compared with other metal-hydrogen systems [402]. However, Wieland et al. compared the permeation rate of 25-µm thick palladium membranes containing 40 wt.% copper and 25 wt.% silver with a vanadium membrane coated with palladium [403]. Unfortunately the operating temperature was not provided by Wieland et al. As shown in Figure 5.30, the vanadium membrane failed by rupture at a hydrogen partial pressure of less than 4-bar.

Generally, the membranes need to be protected against rapid temperature and pressure changes and they are subject to poisoning by carbon monoxide, hydrogen

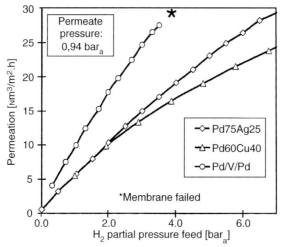

Figure 5.30 Permeation rate of different metal membranes versus hydrogen partial pressure: Pd75Ag25 is a 25-μm thick palladium membrane containing 25 wt.% silver; Pd60Cu40 is a 25-μm thick palladium membrane containing 40 wt.% copper; Pd/V/Pd is a 40-μm thick vanadium membrane coated with palladium from both sides [403].

sulfide and unsaturated hydrocarbons [401]. Figure 5.31 shows the permeation reduction of a 25-μm thick palladium membrane containing 40 wt.% copper at different carbon monoxide partial pressures [403].

Membrane separation of reformate is usually operated at elevated pressure, as the driving force for the permeation process. Thus, steam reforming is the preferred procedure, because only liquid pumps are required instead of large compressors for air pressurisation, which draw unacceptable parasitic losses [105], especially for small mobile systems with an electrical power equivalent of less than 10 kW$_{el}$.

Another measure to increase the hydrogen partial pressure difference between permeate and retenate is to use sweep gas on the retenate side. Because the hydrogen requires humidification for low temperature PEM fuel cells to prevent membrane dry-out, steam is the preferred sweep gas [405]. Oklany et al. highlighted the effect of steam as the sweep gas for the permeate in a methane steam reforming membrane reactor [406]. Higher methane conversion was observed, which originated from back-diffusion of steam from the permeate to the reaction side of the membrane, which increased the S/C ratio and consequently the conversion.

To minimise the pressure required for small-scale systems, the membrane thickness was reduced to a few μm in certain applications. Ceramic supports ensure the mechanical stability of these thin membranes. At such low membrane thickness, leakage of species other than hydrogen occurs through membrane imperfections.

A methanation reactor is then frequently switched downstream of the membrane separation to remove carbon monoxide present in the permeate.

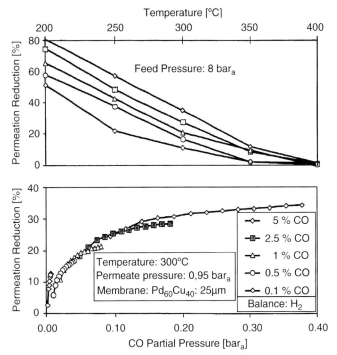

Figure 5.31 Permeation reduction of a 25-μm thick palladium membrane containing 40 wt.% copper at 8-bar feed pressure versus operating temperature (top) and at a 300 °C operating temperature versus carbon monoxide partial pressure (bottom); the permeate pressure was 0.95 bar [403].

Amphlett et al. developed a correlation for the permeability P of palladium membranes from literature data [407]:

$$P = 2.2 10^{-7} e^{\frac{-1600}{T}} \quad (5.5)$$

where T is the temperature in Kelvin and the permeability has the unit $[\text{mol H}_2 \, (\text{ms Pa}^{0.5})^{-1}]$.

Figure 5.32 shows Arrhenius plots of permeability as determined by various workers.

The hydrogen flux through the membrane J_{H_2} is proportional to the diffusion coefficient D_{H_2} of hydrogen in palladium and Sievert's solubility constant K_S of the hydrogen/palladium system:

$$J_{H_2} \propto D_{H_2} K_S \quad (5.6)$$

$$N_{H_2} = \frac{QA\left[(p_{H_2,1})^n - (p_{H_2,2})^n\right]}{l} \quad (5.7)$$

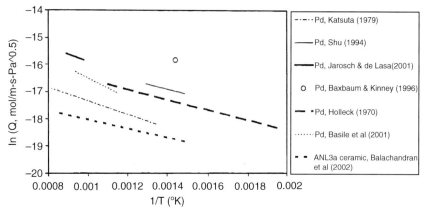

Figure 5.32 Arrhenius plot of permeability versus temperature as determined by various workers [405].

where Q is the permeability coefficient $[\mathrm{mol}\,(\mathrm{m}^2\,\mathrm{s}\,\mathrm{kPa}^n)^{-1}]$, A is the membrane surface area [m^2] and l the membrane thickness [m]. The exponent n for the partial pressure is 0.5 if the bulk phase diffusion of atomic hydrogen is the only rate limiting step. In this instance, Eq. (5.7) is known as Sievert's law. Otherwise, when the dissociative chemisorption and the reversible dissolution of hydrogen also play a role, the value of n is higher in the range between 0.62 [400] and 0.65 [408]. However, Li et al. claimed that an exponent other than 0.5 may only be observed for composite membranes [409]. They argued that the higher values of n originated from the combination of hydrogen diffusion through the membrane with the Knudsen diffusion through the ceramic support. For this situation, a coefficient of 0.66 was calculated [409].

The hydrogen separation factor is sometimes used to specify membrane quality. It is defined as:

$$s = \frac{\dot{n}_1 \Delta p_1}{\dot{n}_2 \Delta p_2} \tag{5.8}$$

where $\dot{n}_i$ stands for moles of species, i, transferred through the membrane and Δp_i stands for the partial pressure difference of species i through the membrane.

The reforming and water–gas shift steps may be combined with integrated membrane separation functionalities, which leads to the field of membrane reactors. In membrane reactors the equilibrium of the chemical reactions involved is shifted in a favourable direction, because the hydrogen product is removed from the reaction system. Product compositions exceeding the equilibrium of the original feed composition can thus be achieved. Joergensen et al. [410] calculated the stability of carbon according to the Boudouart reaction during methane steam reforming in their membrane reactor (see Section 7.1.4). At 6-bar pressure and a 500 °C reaction temperature, a S/C ratio of 2.5 was required to prevent carbon formation according to these calculations due to the removal of hydrogen from the product gas. This value is significantly higher than the S/C ratio close to 1.0 that is usually applied in industrial scale steam reformers.

A concept for a methanol (or ethanol) fuel processor based upon steam reforming and membrane separation was presented by Gepert *et al.* [400]. As shown in Figure 5.33, the alcohol/water mixture was evaporated and converted by steam reforming in a fixed-bed catalyst, into which palladium capillary membranes were inserted. The retenate then entered the combustion zone, which was positioned concentrically around the reformer bed at the reactor wall. Air was fed into the combustion zone and residual hydrogen, carbon monoxide and unconverted methanol combusted therein. The sealing of the membranes at the reactor top was an issue solved by air-cooled elastomers.

Barbieri and Di Maio modelled methane steam reforming in a tubular reactor over a nickel catalyst [411]. A 20-μm thick membrane positioned in the centre of the reactor tube was inserted to separate hydrogen product from the reformate. They compared the reactor performance for parallel- and counter-flow of the sweep-gas with the reformate flow. For low sweep gas flow rates the difference between both operation modes was marginal, whereas superior performance was calculated for the counter-flow at higher flow-rates of the sweep gas. This originated from the thermodynamic equilibrium of the steam reforming reaction. It shifted the hydrogen partial pressure in the permeate above the value of the reformate in the first part of the reactor for the counter-flow at low temperature, as shown in Figure 5.34.

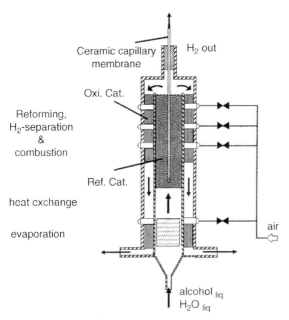

Figure 5.33 Alcohol fuel processor with integrated membrane separation as proposed by Gepert *et al.* [400].

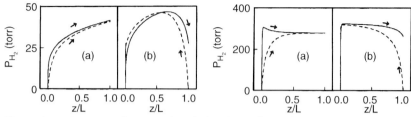

Figure 5.34 Hydrogen partial pressure along the length axis of a membrane tubular reactor operated in parallel (a) and counter-flow (b) arrangements: solid lines, retenate partial pressure; dashed lines, permeate partial pressure; reformate flow rate, 162 cm^3 min^{-1}; reformate pressure, 1.36 bar; sweep gas flow rate, 40 cm^3 min^{-1}; permeate pressure, 1.01 bar; left, reaction temperature 300 °C; right, reaction temperature 500 °C [411].

Gallucci and Basile analysed methanol steam reforming in a tubular membrane reactor operated in the co- and counter-current mode [412]. Methanol steam reforming was, of course, performed at a much lower temperature than methane steam reforming discussed above. The pressure difference between reaction side and permeate was much higher compared with the values assumed in the work of Barbieri and Di Maio discussed above. The target of the calculations by Gallucci and Basile was complete hydrogen separation. While methanol conversion was rather similar for co- and counter-current flow independent of reaction temperature and reaction pressure, this was not the case for the hydrogen separation. Because hydrogen partial pressures equalised inevitably at the end of a membrane reactor operated in co-current mode, some hydrogen was lost to the retenate. This was not the case for counter-current operation. Thus, hydrogen recovery was much higher for the counter-current mode. At 2-bar pressure on the reaction side, a 270 °C reaction temperature and a high S/C ratio of 3.0, only 52% of the hydrogen product could be separated with co-current operation, compared with 82% gained by the counter-current flow. This difference was reduced by increasing the reaction pressure, as shown in Figure 5.35.

Further calculations by Barbieri and Di Maio [411] revealed that the advantages of the membrane separation process were lost when the thickness of the membrane exceeded 30 µm, because the hydrogen flow through the membrane became too small. The maximum membrane thickness to achieve still more favourable results depended strongly on the operating conditions.

Barbieri et al. also performed experimental work on a methane steam reforming membrane reactor, which will be discussed in Section 7.1.4.

Marigliano et al. compared the performance of two different types of tubular methane steam reforming membrane reactors by numerical simulations [413]. In the first, the catalyst was packed into the palladium/silver membrane tube, and in the second it was positioned in the annular region surrounding the membrane tube. Both configurations were heated from the outside. Owing to the indirect heat transfer to the catalyst bed inside the membrane tube, methane conversion was lower in this instance, and under certain conditions it was even inferior to a conventional fixed bed

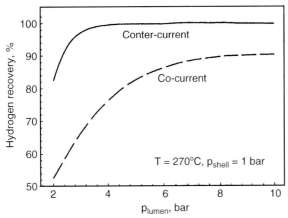

Figure 5.35 Hydrogen recovery versus pressure for co-current and counter-current operation of a membrane reactor [412].

without membrane separation. Decreasing the thickness of the membrane did not improve this inferior performance [414].

Marigliano et al. performed further modelling work for methane steam reforming and water–gas shift in membrane reactors [415]. They defined the sweep factor I as the ratio of flow rates of inert gas on the permeate side to the flow rate of methane on the reaction side of the membrane:

$$I = \frac{\dot{V}_{\text{inert}}}{\dot{V}_{CH_4}} \qquad (5.9)$$

Increasing the inert gas flow on the retenate side increased the driving force for the steam reforming and water–gas shift reactions due to the thermodynamic equilibrium of methane steam reforming and water–gas shift. Consequently, conversion increased. While increasing pressure decreases (equilibrium) conversion in conventional methane steam reformers, conversion could be increased at elevated pressure in a membrane reactor under certain conditions. In a similar way, the water–gas shift reaction is pushed in a favourable direction in a membrane reactor, while the reaction is pressure independent in a conventional reactor (see Section 3.10.1).

Simulations by Barbieri and Di Maio [411] and experimental work performed by Lin et al. [416] showed the opposite trend for the case of methane steam reforming under non-ideal conditions. With a palladium membrane of 20–25 μm thickness, which showed infinite selectivity for hydrogen separation, decreased methane conversion was obtained with increased reaction pressure [416].

A thorough analysis of the effect of pressure on methane conversion was performed by Gallucci et al. [417]. It revealed that the operating conditions will determine whether the methane conversion decreases or increases with reaction pressure in a membrane reactor. The dominance of either the thermodynamic effect of decreasing methane conversion with increasing pressure or increasing conversion

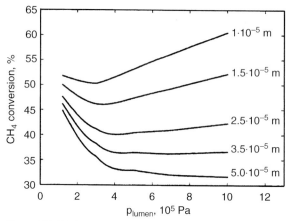

Figure 5.36 Methane conversion versus pressure in a membrane reactor with membranes of varying thickness [417].

by hydrogen removal from the reformate is determined by different factors. The conversion will increase with pressure for a high reactor length, high reaction temperature and low membrane thickness. In particular, in the latter case, a minimum of the conversion is to be expected for intermediate membrane thickness with increasing pressure, as shown in Figure 5.36.

Different mechanisms dominate the performance of dense and porous membranes. While Sievert's law describes the hydrogen dissolution and diffusion processes in dense membranes, Knudsen diffusion is the transport mechanism in porous membranes. This membrane type will always have a specific permeability for all gases, whereas the hydrogen permeability is, of course, the highest. Oklany et al. performed numerical simulations to describe the behaviour of both membrane types in a methane steam reforming membrane reactor [406]. A tubular fixed-bed reactor operated at a S/C ratio 3 was combined with a 20-µm thick dense palladium/silver membrane and alternatively with a micro-porous ceramic membrane with a pore diameter of 1 nm. The conversion could be increased compared with the thermodynamic equilibrium for both membrane types. A basic difference was the dependence of reaction pressure on conversion, as shown in Figure 5.37. While the conversion decreased with increasing pressure for the porous membrane, this was not the case for the dense membrane, because hydrogen was removed selectively.

Vogiatzis et al. modelled heat dispersion effects in future water–gas shift membrane reactors on an industrial scale [418]. A catalyst bed of 2-m length was assumed, which contained tubular membranes with a 4-mm outer diameter. The distance between the tubes, which was filled by the fixed catalyst bed, was only 23 mm and thus the model could be applied for systems of a smaller scale. Heat dispersion effects occurred in the reactor. Overheating of the catalyst bed by about 30 K occurred, which lead to increased conversion compared with an isothermal catalyst bed. However, the opposite effect is to be expected with an endothermic reaction, such as methane steam reforming.

5.2 Design of the Carbon Monoxide Clean-Up Devices | 173

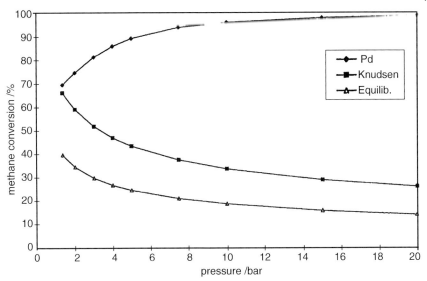

Figure 5.37 Effect of reaction pressure on methane conversion for methane steam reforming in membrane reactors with different types of membranes: reaction temperature 500 °C; S/C ratio 3; △, thermodynamic equilibrium; □, conversion in a membrane reactor with porous ceramic membrane; ◇, methane conversion in a membrane reactor with dense palladium membrane [406].

Figure 5.38 shows a complete integrated methanol steam reforming membrane fuel processor concept as designed by Edlund [419]. Methanol and steam entered the reformer at (30) and were evaporated in the coil (30a). Hydrogen was generated in the catalyst bed (62) by steam reforming and diffused partially through the membrane (54) along with some carbon oxides due to imperfections in the membrane. The permeate then moved through another annular gap between the membrane and a central tube (56) until it reached the open end of tube (56) and left the reactor through this central tube, which was filled with catalyst for carbon oxide removal

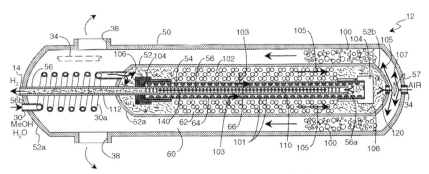

Figure 5.38 Membrane fuel processor according to a patent by Edlund [419].

through methanation. The retenate left the reactor through (120), air was added at (34) and hydrogen and carbon monoxide were combusted in the annular combustion catalyst bed (100), which surrounded the reformer and provided heat for steam reforming.

Examples of membrane separation systems and membrane reactors will be discussed in Section 7.4.

5.2.5
Pressure Swing Adsorption

Pressure swing adsorption (PSA) is a gas purification process, which consists of the removal of impurities on adsorbent beds. The usual adsorbents and gases adsorbed thereon are molecular sieves for carbon monoxide, activated carbon for carbon dioxide, activated alumina or silica gel. Iyuke et al. reported that the addition of tin onto activated carbon improved the efficiency of the adsorptive removal of carbon monoxide from hydrogen [420].

The adsorbents usually have a long lifetime. The impurities are adsorbed at high partial pressure. Then the adsorber is depressurised in two steps: firstly, co-current to the adsorption process down to a medium pressure level, followed by counter-current depressurisation to the low pressure level of the desorption step. While a minimum of two adsorbers is required to run the process, industrial pressure swing adsorption plants consist of up to 12 adsorbers and along with the number of valves required, this makes the systems complicated [421].

The pressure swing adsorption process is usually a repeating sequence of the following steps shown in Figure 5.39 [422]:

1. adsorption at feed pressure
2. co-current depressurisation to intermediate pressure
3. counter-current depressurisation to (sub-)atmospheric pressure usually starting at 10% to 70% of the feed pressure
4. counter-current purge with hydrogen enriched or product gas at ambient pressure
5. co- or counter-current pressure equalisation
6. co-current pressurisation with feed or secondary process gas (see below).

For hydrogen purification by pressure swing adsorption, hydrogen purity is high, but the amount of rejected hydrogen is also relatively high (10–35%). This is due to the fact that during the second stage of depressurisation (step 3 above) of the adsorber so-called dump gas is formed. It contains 30–70 vol.% hydrogen at the beginning and 15–25 vol.% hydrogen at the end when the methane steam reforming product is purified. Additionally, the adsorber is usually purged with low pressure purified hydrogen, the so-called purge gas (step 4). Both gas flows form the PSA off-gas of the pressure swing adsorption unit. It consists of 30–50 vol.% hydrogen, 30–60 vol.% carbon dioxide and 10–25 vol.% carbon monoxide and methane in case of methane being the feed fuel. A third gas species formed during the first step of depressurisation (step 2) is secondary product gas, which is

5.2 Design of the Carbon Monoxide Clean-Up Devices | 175

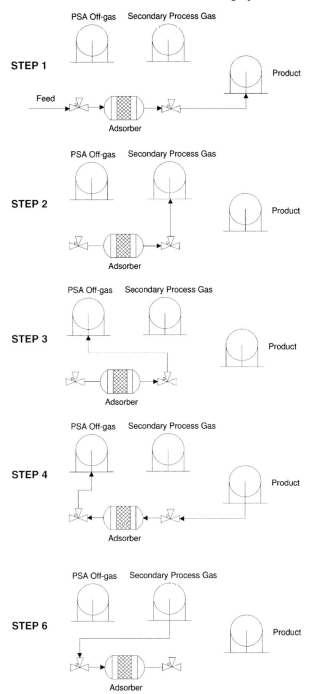

Figure 5.39 The main process steps of pressure swing adsorption.

frequently used for re-pressurisation of the adsorber (step 6). The hydrogen recovery from the reformate of a methane steam reformer was calculated to be only 77.6% by Sircar et al. [422]. This value could be improved by re-compressing the PSA off-gas, a subsequent membrane separation and re-cycling of the hydrogen enriched permeate to the PSA feed [422].

Because the efficiency of the process depends on the pressure difference during adsorption and desorption, it usually requires a high operating pressure [404]. This makes pressure swing adsorption not very attractive for mobile applications. It is certainly a viable option for distributed hydrogen and power generation in smaller scale stationary applications, such as hydrogen fuelling stations, where it shows lower parasitic losses compared with membrane separation processes.

Dalle Nogare et al. compared two natural gas fuel processors working with pressure swing adsorption and preferential oxidation for small-scale stationary applications [423]. They found a slightly higher efficiency for the pressure swing adsorption.

5.3
Aspects of Catalytic Combustion

Catalytic combustion or total oxidation is frequently applied in fuel processor systems. The motivations are two-fold:

- Provision of energy for start-up procedures; for most practical systems, the fuel converted in the fuel processor during normal operation is also supplied to a start-up burner, which works with either catalytic or homogeneous combustion.

- Combustion of anode off-gas in a catalytic afterburner or in a burner integrated into a steam reformer or evaporator, which could be achieved by plate heat-exchanger technology coated with combustion catalyst.

A large variety of conventional burners is commercially available and some of them may well be suited to supplying heat to fuel processors especially during start-up. However, these devices will not be discussed as they are not within the scope of this book.

Combustion of hydrogen contained in gas mixtures such as anode off-gas is a critical issue, because the explosion limits of the reaction range from 4 to 75% for hydrogen in air.

Microchannels offer unique possibilities for the safe operation of the reaction, because they act as flame arresters, which quench the flame. Veser et al. developed a single-channel microreactor and performed a thorough analysis of the reaction and related explosive regimes [424]. A 150-µm thick platinum wire was used as the catalyst, which was heated by electricity to ignite the reaction. The ignition of the reaction occurred at 100 °C and complete conversion was achieved at a stoichiometric feed ratio generating 72 W of thermal power. Later, Veser performed the reaction in a quartz glass microreactor with a diameter of 600 µm and length 20 mm [425]. The ceramic housing of the reactor and the reactor itself were stable for temperatures greater than 1100 °C. Again a platinum wire was used as the catalyst. Residence times as low as 50 µs could be achieved.

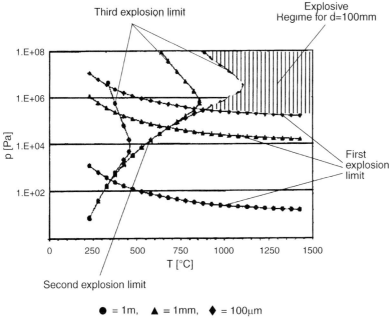

Figure 5.40 Explosion limits for a stoichiometric H_2/O_2 mixture and different reactor diameters determined by Veser [425].

The fact that no homogeneous reactions could be detected demonstrated the possibility of separating homogeneous and heterogeneous reactions in microreactors because of their higher surface area to volume ratio. A detailed mechanistic investigation of the explosion limits revealed that the first and the third explosion limits of the reaction are dependent on the reactor dimensions (see Figure 5.40). The first explosion limit is reached when the mean free path of the molecules becomes smaller than the reactor dimensions, which was already the situation at very low pressures in the microreactor (5 mbar). Thus the first explosion limit, which is given by the flat horizontal lines in Figure 5.40, strongly depends on the reactor dimensions. The flat lines move up significantly with decreasing reactor dimensions. The second explosion limit reduces the explosive regime in the direction of high pressure, because trimolecular reactions dominate, which slow down the self-acceleration of the explosion [425]. The second explosion limit, which is the first branch of the rising lines in Figure 5.40, is not affected by the reactor dimensions. The third explosion limit, which is the second branch of the rising lines, is related to both kinetic explosion by wall effects and thermal explosion by self-heating of the reaction. It is responsible for explosions at ambient and elevated pressure. With decreasing reactor dimensions, both the kinetic and the thermal explosion limit are increased and the third explosion limit moves into the upward right part of Figure 5.40. A core statement of the investigations was that none of the three explosion limits can be crossed at ambient pressure in a 100-μm microchannel regardless of the

Table 5.9 Flammability limits of hydrogen, carbon monoxide and methane in a stainless steel tube at different temperatures [427].

Fuel	Temperature [°C]	Flammability limits [Vol.%]	
		Lean	Rich
H_2	25	3.9	74.7
	200	2.8	80.3
	300	2.4	82.6
CO	25	13.6	66.5
	200	11.5	74.4
	300	10.3	76.1
CH_4	25	4.9	14.0
	200	4.1	15.4
	300	3.8	17.0

temperature, i.e., the reactor is "inherently" safe. Later, Chattopadhyay and Veser defined the limiting channel diameter for the absence of homogeneous reactions more precisely to 285 μm at ambient pressure. This value decreased almost linearly with increasing pressure, until at 9.4 bar the microchannels could no longer quench the homogeneous reactions, regardless of how small the channels were [426]. Platinum catalyst coated onto the wall of the microchannel reactor accelerated the ignition below a temperature of 730 °C. The ignition was slowed down above that temperature according to the calculations. Decreasing the hydrogen/air ratio increased the reaction rate in a platinum coated microchannel, because the hydrogen was preferentially adsorbed at the platinum surface, thus inhibiting the oxidation reaction [426].

Table 5.9 shows the flammability limits for hydrogen, carbon monoxide and methane in air [427]. The values were determined in a stainless steel tube of 50.8-mm diameter for a residence time of 30 s. The electric power was provided at 10 kV with a current of 23 mA.

Veser and Schmidt studied the catalytic and homogeneous ignition and the homogeneous extinction behaviour of lower alkanes and alkenes [428]. Platinum foil was used as the catalyst. The surface ignition temperature decreased with increasing number of carbon atoms in the fuel and with increasing modified equivalence ratio θ, which was derived from the equivalence ratio. The latter is defined as the ratio of vol.% air to vol.% fuel in the feed normalised by the ratio required for total combustion:

$$\phi = \frac{\frac{\text{vol.\% fuel}}{\text{vol.\% air}}}{\left(\frac{\text{vol.\% fuel}}{\text{vol.\% air}}\right)_{\text{stoichiometric}}} \tag{5.10}$$

$$\theta = \frac{\phi}{\phi + 1} \tag{5.11}$$

$\theta=0.5$ is the stoichiometric ratio for total combustion, $\theta=0$ for fuel free air, $0-1$ for air free fuel. Figure 5.41 shows the surface ignition temperatures for various alkanes. Below $\theta=0.2$, which corresponds to about 0.9 vol.% fuel in air for propane, surface ignition of the fuel was no longer possible. Methane showed a significantly different behaviour compared with the other alkanes, requiring a much higher temperature for surface ignition. Alkenes, namely ethylene and propene, required higher surface ignition temperatures, which increased with increasing θ. Surface ignition was possible for alkenes down to $\theta=0.1$. Veser et al. explained the different behaviour of alkanes and alkenes through the different surface coverage of platinum. Platinum seemed to be preferably covered with oxygen for the alkanes, while a hydrocarbon over layer was expected for the alkenes.

Homogenous ignition of C_1–C_4 alkanes and alkenes did not take place below 1000 °C according to Veser et al. (see Figure 5.42). Methane actually required at least 1200 °C. The flammability limits were wider for alkenes, especially for ethylene.

Raimondeau et al. modelled the homogeneous high temperature combustion of a pre-heated stoichiometric mixture of methane and air at $2\,\mathrm{m\,s^{-1}}$ flow rate and a 1000 °C reference temperature, which corresponded to the methane ignition

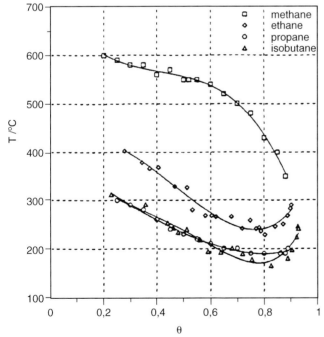

Figure 5.41 Surface ignition temperature of methane, ethane, propane and butane as determined by Veser and Schmidt over platinum foil [428].

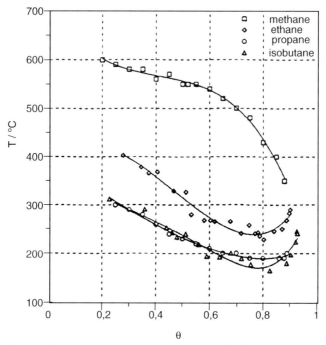

Figure 5.42 Homogeneous ignition temperature of methane, ethane, propane and butane as determined by Veser *et al.* over platinum foil [428].

temperature under these conditions according to their calculations [429]. For a channel of 100-μm diameter no concentration or temperature gradients were determined by the calculations. It was demonstrated that temperature losses through the wall lead to flame extinction. This effect was more pronounced for a decreasing channel diameter. Flame propagation was not possible in stainless steel channels due to the high heat conductivity and affinity for radicals of stainless steel. However, even for materials that do not adsorb radicals, a minimum channel width existed below which homogeneous combustion was no longer possible. Insulating materials such as silicon nitride and inert layers such as alumina were required to maintain the homogeneous reaction.

Wang and Wang reported on the initial flow distribution problems discovered for their plate and fin heat-exchanger, which was designed to combust hydrogen in order to supply the evaporation of a methanol/water mixture with energy [430]. The combustion chamber was filled with a fixed-bed catalyst and hydrogen addition to the air feed had to be spread over the combustion chamber to achieve even temperature distribution. The design was then scaled up and operated successfully in a 5-kW methanol fuel processing system.

5.4
Design of the Overall Fuel Processor

5.4.1
Overall Heat Balance of the Fuel Processor

The reforming process is composed of process engineering steps, which require energy such as fuel evaporation and steam generation. When catalytic carbon monoxide removal is chosen for reformate purification, the clean-up steps produce some energy (water–gas shift, preferential oxidation). The reforming process itself requires or produces energy depending on the technology applied (see Section 3).

Thus, the first design step of fuel processor development is an overall net heat balance of the system.

Figure 5.43 shows the design concept of a natural gas pre-reformer coupled to a 170 kW solid oxide fuel cell as developed by Riensche et al., which was also discussed by Peters et al. [431]. It was composed of an evaporator, a pre-reformer working with steam reforming and an air pre-heater for the cathode feed. All these components were supplied with energy by the fuel cell anode off-gas combustion. A S/C ratio of 2.5 was chosen. The efficiency of this system was calculated to be 43%. However, this value increased to 67%, when the residual heat contained in the anode off-gas was utilised for other purposes, such as hot water production. For a system size as being considered here, the recirculation of anode off-gas by hot-gas blowers could well be considered [431,432]. This measure has the consequence that external water supply is no longer required. As a consequence, the overall efficiency calculated for the system increased further from 67 to 81%. The feasibility of this concept was investigated on the smaller 10-kW scale by Peters et al. [431], see Section 7.1.3.

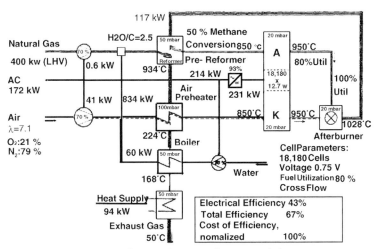

Figure 5.43 Basic concept of a natural gas pre-reformer coupled to a solid oxide fuel cell [432].

Steam reforming requires substantial energy input not only for the reforming process itself, but also for steam generation, and with liquid fuels for fuel evaporation. This energy may be gained by combustion of unconverted hydrogen contained in the fuel cell anode off-gas. Steam reforming of all common fuels apart from methanol requires lower anode hydrogen utilisation of the fuel cell anode below the value of 80%, which is usually assumed to be the maximum value. By these means more residual hydrogen may be re-circulated to the fuel processor and feed the fuel processor with energy. Alternatively, the hydrogen utilisation could be maintained at a high value and additional fuel then fed to the fuel processor for energy generation by catalytic or homogeneous combustion.

Amphlett *et al.* calculated that for a methanol steam reformer the hydrogen utilisation needs to be maintained around 80% to avoid methanol combustion in the burner of the fuel processor. The maximum overall system efficiency had a weak maximum at the lowest possible hydrogen utilisation [407].

For a methane steam reforming fuel processor, more than 15% higher fuel processor efficiency was determined experimentally by Mathiak *et al.* [433] when utilising fuel cell anode off-gas compared with combustion of additional methane. Doss *et al.* analysed an autothermal gasoline fuel processor and found improved efficiency by utilisation of anode off-gas [434].

Schmid *et al.* [16] compared steam reforming and autothermal reforming of methane. They assumed that the anode off-gas was burnt in an external burner. The heat was utilised for heating purposes of a combined heat and power system. The analysis revealed much lower practical efficiency for autothermal reforming compared with steam reforming at 73% hydrogen utilisation in the fuel cell anode, as shown in Figure 5.44.

It has already been mentioned, in Section 2.2, that anode off-gas re-circulation can increase the fuel processor efficiency to values exceeding 100% where efficiency is

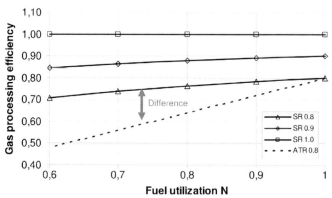

Figure 5.44 Fuel processor efficiency versus anode fuel utilisation for different fuel processor configurations: steam reforming with 100, 90 and 80% efficiency (SR 1.0, 0.9, 0.8) and autothermal reforming with 80% efficiency; the arrow compares steam reforming and autothermal reforming at 80% efficiency for a hydrogen utilisation of 73% [16].

calculated according to Eq. (2.1), Section 2.2. However, this only depends on the definition chosen for efficiency. Realistic calculations need to take the heat losses of the fuel processor into consideration. In particular, steam reformer/fuel processor systems on the smaller scale then have less than 100% efficiency – regardless which definition of efficiency is applied.

Full or almost complete conversion of the fuel should be targeted for the fuel processor. This is not mandatory with methane reforming, because methane does not damage the CO clean-up catalysts and nor even the PEM fuel cell. However, less than 90% conversion lowers the efficiency of the overall process, which was demonstrated by the calculations of Avci *et al.* for methanol processing [47].

Fuel processor concepts based upon autothermal reforming or partial oxidation usually generate a surplus of heat, which has to be removed from the system at the low temperature level, as discussed below.

The heat balance of an autothermal reformer is dictated by its S/C ratio, which affects the amount of air feed required for a net heat balance. Increasing the S/C ratio thus increases the O/C ratio, as shown in Table 5.10. From the overall chemical reaction of an autothermal fuel processor (assuming that carbon monoxide is completely converted into carbon dioxide) it follows that less moles of hydrogen are produced per mole of fuel at higher O/C ratios. This because less fuel is converted

Table 5.10 Effect of the S/C ratio on system parameters and water balance of an autothermal methane fuel processor [435].

Basis: 1 gmol/min of methane		Base case		
S/C ratio into FP, Ψ	1.2 (−20%)	1.5	1.8 (+20%)	3.0 (+100%)
Water feed into FP (gmol/min)	1.2	1.5	1.8	3.0
O_2/CH_4 molar ration into FP, χ	0.450	0.478	0.523	0.614
Air feed into FP (gmol/min)	2.14	2.27	2.40	2.92
Idealized FP products				
H_2 (gmol/min)	3.10	3.05	2.99	2.77
H_2 conc. in reformate (%-dry)	52.6	48.4	44.6	33.4
LHV of H_2 (kW$_{th}$)	12.5	12.3	12.1	11.2
Fuel processor efficiency (%)	93.4	91.8	90.1	83.5
H_2O in reformate (gmol/min)	0.10	0.46	0.81	2.23
H_2O conc. in reformate (%-wet)	1.71	7.2	12.1	26.8
Air into cathode (gmol/min)	14.8	14.5	14.2	13.2
H_2O in burner product (gmol/min)	3.2	3.5	3.8	5.0
H_2O concentration in burner product (%-wet)	16.8	18.2	19.06	24.9
Exhaust gas temperature, $T_{exhaust} = T_{air} + T_{approach}$	46 °C	46 °C	46 °C	46 °C
Saturated moisture content in exhaust gas (%-wet)	10.0	10.0	10.0	10.0
Recoverable water (gmol/min)	1.43	1.75	2.06	3.32
Net water produced (gmol/min)	+0.23	+0.25	+0.26	+0.32
Net water produced (ml/min)	+4.2	+4.6	+4.8	+5.8

by steam reforming and more by partial oxidation. Table 5.10 demonstrates this for an autothermal methane fuel processor [435]. When increasing the S/C ratio from 1.2 to 3.0, the O/C ratio also needs to be increased to keep the reformer self-sustaining, as far as the heat balance is concerned. Consequently, the dry concentration of hydrogen contained in the reformate drops due to dilution by nitrogen and almost 10% of fuel processor efficiency is lost. With an increasing O/C ratio (here O_2/CH_4 ratio) upwards of 10% less hydrogen is produced per mole methane.

Figure 5.45 shows a Sankey diagram of a 1.7 kW propane reformer operated by Rampe et al. [151]. A comparison of the energy flows with (italic letters) and without pre-heating of the feed is performed. In the latter case, more than 20% of the energy contained in the fuel was lost to the environment. The high heat losses shown in

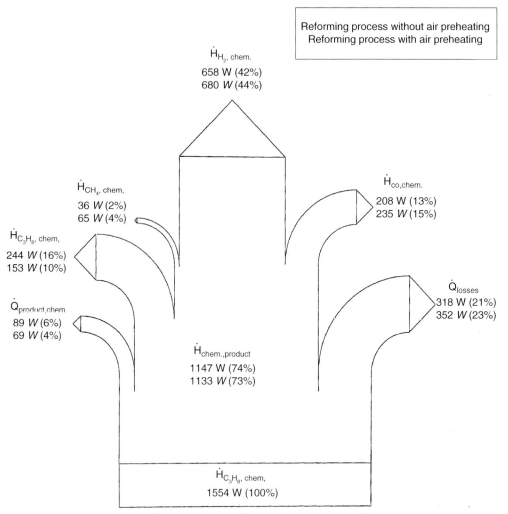

Figure 5.45 Sankey energy diagram for a 1.7-kW autothermal reformer [151].

Figure 5.45 emphasise how important efficient insulation is especially for reformers running at temperatures exceeding 700 °C (see Section 8.5). It has to be mentioned that propane conversion was incomplete for this reactor (84%), which obviously generates additional energy losses, as explained above.

Heat losses by radiation become substantial for systems with dimensions greater than the centimetre range when the temperature exceeds about 650 °C. However, for devices at the sub-millimetre to millimetre scale, the conventional correlations for heat convection are no longer valid, according to Shah et al. [436]. About three times higher convective heat losses were found by these workers for their sub-watt methanol fuel processor.

Thus Shah and Besser proposed a new correlation for the Nusselt number for convective heat transfer at the millimetre scale:

$$Nu = 1.7 Ra^{0.08} \tag{5.12}$$

In addition, Shah et al. found significant heat losses as a result of radiation at temperatures around 200 °C for their small-scale device [436]. Figure 5.46 shows the distribution of conductive and convective heat losses, radiation heat losses and heat losses through the wiring of the fuel processor of Shah and Besser.

Semelsberger and Borup compared start-up energy demand and efficiency of fuel processors fed by various fuels on an overall basis [437]. Autothermal reforming was chosen as the reforming strategy in all cases. The carbon monoxide clean-up was assumed to be performed by high temperature water–gas shift operated at 450 °C and conversion into 4 vol.% carbon monoxide followed by a low temperature water–gas shift operated at 250 °C and conversion into 0.9 vol.% carbon monoxide, while the preferential oxidation reactor was calculated with 100% selectivity towards carbon monoxide. These are very simplifying assumptions and therefore the results have to be considered with care. Table 5.11 shows the operating conditions and equilibrium

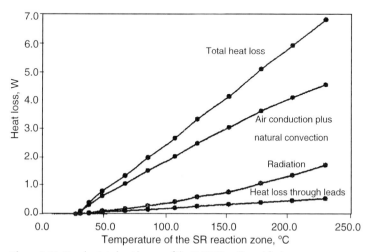

Figure 5.46 Steady-state heat losses of the 0.65-W methanol fuel processor developed by Shah and Besser [436].

Table 5.11 Operating conditions for autothermal fuel processors running on different fuels as calculated by Semelsberger and Borup.

	Total feed flow rate (L/s)	Fuel flow rate (L/s)	ATR operating conditions			Autothermal equilibrium effluent composition n_k				
			T(°C)	O/C	S/C	CO	H_2	H_2O	N_2	CO_2
Methane	55.5	12.9	827	0.85	2.33	0.09	0.37	0.26	0.23	0.06
Methanol	17.3	7.1	327	0.28	1.00	0.04	0.56	0.10	0.12	0.18
Ethanol	43.2	7.4	727	0.98	2.00	0.09	0.37	0.27	0.17	0.09
DME	25.6	4.3	427	0.36	2.00	0.04	0.48	0.20	0.12	0.16
Gasoline	39.1	2.9	827	0.73	1.54	0.17	0.37	0.14	0.26	0.06

Note: the total volumetric influent rates listed (fuel, water and air) are the flow rates needed to produce a fuel processor exit molar flow rate of 0.36 mol s^{-1} of hydrogen for a 50-kW$_{el}$ fuel processor, assuming that the carbon monoxide is shifted to hydrogen and carbon dioxide; fuel flow rate is that of the fuel only, not water or air. All flow rates are vapour phase flow rates [437].

conditions assumed. The values chosen for methane and ethanol are questionable. While the S/C ratio of 2.33 seems to be rather high for methane, the temperature chosen for ethanol is also rather high (see Sections 4.2.2 and 4.2.4). Additionally, methane formation was not considered for the reforming reactions, which falsifies the results considerably. The reactor volumes and start-up energy demand of the fuel processors are provided in Table 5.12. The start-up energy demand was calculated by simply assuming 1 kg L^{-1} reactor density, 500 J (kg K)$^{-1}$ heat capacity and heating from 20 °C to the operating temperature. A low temperature water–gas shift reactor was assumed for the methanol reformer, which might become obsolete when suitable catalyst technology is applied (see Section 4.2.1). On the other hand, the methanol reformer is the smallest of the reformer reactors, which is certainly not realistic because methanol reforming is a slower reaction compared with hydrocarbon reforming. The start-up energy demand of the methanol reformer is lowest according to the calculations, which is obvious because of the much lower operating temperature of methanol reforming.

Table 5.12 Reactor volumes and start-up energy requirement of autothermal fuel processors running on different fuel as calculated by Semelsberger et al. [437].

	Methane	Methanol	Ethanol	DME	Gasoline
ATR	7.5 L	2.3 L	5.8 L	3.4 L	5.3 L
HTS	6.9 L	0.0 L	7.2 L	0.0 L	11.0 L
LTS	25.1 L	18.9 L	24.6 L	21.9 L	21.0 L
PrOx	6.3 L	4.7 L	6.1 L	5.5 L	5.3 L
Fuel processor volume	45.8 L	25.9 L	43.7 L	30.8 L	42.5 L
Heat duty requirements	7592 kJ	2712 kJ	6632 kJ	3423 kJ	7068 kJ

Exergy analyses were performed for different methanol fuel processor concepts to identify the exergy losses in the system and provide strategies for improving the efficiency of the systems.

Delsman et al. analysed a 100-W methanol fuel processor/fuel cell system [4]. It was composed of a microstructured reformer/burner heat-exchanger reactor for methanol steam reforming and anode off-gas combustion, several heat-exchangers, a cooled microstructured heat-exchanger reactor for preferential oxidation of carbon monoxide, a heat-exchanger for feed evaporation and a low temperature fuel cell. High exergy losses were determined for the reformer, burner, evaporator and, by far highest, for the fuel cell. The efficiency was lowest for the evaporator, the preferential oxidation reactor and for the burner followed by the fuel cell. Thus, the optimisation of the burner, vaporizer and especially of the fuel cell made most sense. The reformer and fuel cell operating temperatures had little effect on exergy losses. When the efficiency of the fuel cell was increased from 40 to 60%, the exergy losses were reduced by 50%. However, this would result in a much bigger fuel cell, which was regarded as a major drawback for the small-scale portable system under investigation. When the S/C ratio of the feed was reduced from 3 to 1.5, exergy losses could be reduced by 17%. However, heat losses to the environment played a minor role according to this system analysis [4], but they are a major issue for small-scale fuel processors.

Wang et al. analysed a methanol fuel processor that consisted of an autothermal reformer operated at about 500 °C, which is too high by far for methanol reforming, two water-gas shift reactors and four preferential oxidation reactors switched in series [438]. Six heat-exchangers completed this complex system. The exergy analysis revealed optimum conditions at S/C ratios of between 1.5 and 2.0, although only the former value seems to be more realistic. The optimum O/C ratio of 0.63, which was determined, is a rather high value for autothermal reforming of methanol, especially when anode off-gas is used on top of the oxygen addition to the feed to close the fuel processor heat balance. The main exergy losses were attributed to the reformer and to the separate catalytic combustor of the fuel processor.

Hotz et al. compared the efficiency of a small PEM fuel cell system, which had a methanol fuel processor composed of a fixed-bed steam reformer and preferential oxidation reactors, with a direct methanol fuel cell system [439]. Both systems were in the power range of 2 W_{el}. The steam reformer was designed as a tubular bundle filled with copper/zinc oxide catalyst, operated at a S/C of 1.1 and heated by the combustion gases of the afterburner. Because the methanol conversion was assumed to be only 45%, most of the methanol was combusted in the afterburner, which generated sufficient energy to keep the system on temperature. However, the efficiency of such a system will be low. The exergetic efficiency (which is in fact the system efficiency as defined in Section 2.2) of the fuel processor/fuel cell system was calculated to be 30%. As discussed in Section 9.1, in reality a fuel processor/fuel cell system with a power output below 5 W_{el} will have a much lower efficiency than 20%. The size of the PEM fuel cell and of the direct methanol fuel cell was assumed to be in the same range, which, from practical experience, is rather unrealistic. The efficiency of the direct methanol fuel cell system was calculated to be 25%.

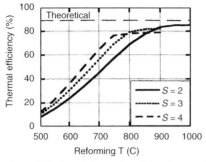

Figure 5.47 Net efficiency of methane steam reforming at various S/C ratios (given here as S) and reforming temperature; system pressure 10 bar [14].

All of the work discussed above lack a realistic consideration of heat losses, which is difficult to achieve, especially for small-scale systems, without experimental determination.

A global analysis of the overall fuel processor efficiency of a methane steam reformer coupled to a membrane separation unit was performed by Lutz et al. [14]. In the system analysed, the hydrogen was separated from the reformate by the membrane unit and the heat for steam reforming was supplied by combusting the retenate. The efficiency was calculated according to Eq. (2.4) and was applied to take the utilisation of hydrogen, carbon monoxide and unconverted methane present in the retenate into consideration. This analysis revealed a maximum efficiency for methane steam reforming at temperatures between 800 and 900 °C, depending on the S/C ratio. This was mainly related to the thermodynamic equilibrium of the methanation reaction [Eq. (3.5)], which limited the methane conversion especially at elevated pressure (see Section 3.1). Higher efficiency was determined for higher S/C ratios (see Figure 5.47) [14].

Higher S/C ratios increase the exhaust gas flow downstream of the combustion process. Because not all the heat can be regained from this off-gas flow due to the limited efficiency of the practical heat-exchangers, more low temperature heat would be lost from the system, which would reduce the overall efficiency.

However, the calculations did not take heat losses into consideration, which become dominant, especially for smaller scale systems, at temperatures exceeding 700 °C due to an increased contribution of radiation losses.

5.4.2
Interplay of the Different Fuel Process or Components

The effect of operating conditions of the reformer and the carbon monoxide clean-up reactors on their individual performance was discussed in Sections 5.1 and 5.2. However, the operating parameters of the reformer also affect the performance of the clean-up reactors. Increasing the reformer temperature increases the load on the water–gas shift reactors downstream, because the carbon monoxide content is higher due to the equilibrium of water–gas shift. A higher S/C ratio usually decreases the system efficiency, but it also pushes the equilibrium of the water–gas in a favourable direction. Thus, the water–gas shift reactors may be operated at a higher temperature, which in turn reduces their size and the time and energy demand for start-up.

Figure 5.48 shows that the system efficiency of a methanol fuel processor decreases with increasing S/C ratio [407]. The same behaviour was observed for ethanol processing [440], as shown in Figure 5.49.

This is due to the fact that an increasing amount of hydrogen needs to be recirculated from the fuel cell to the steam reformer to supply the energy for water

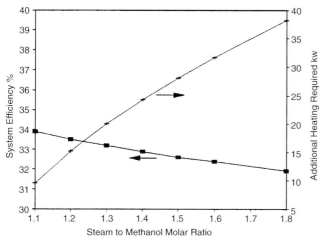

Figure 5.48 Effect of the S/C ratio on the system efficiency of a methanol fuel processor [407].

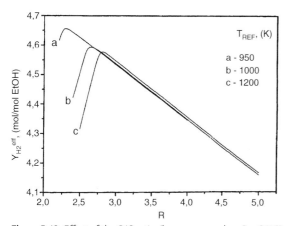

Figure 5.49 Effect of the S/C ratio (here expressed as $R = S/2C$) on the hydrogen yield of an ethanol fuel processor at different operating temperatures, T_{REF}, of the reformer; operating temperature of water–gas shift 200 °C; selectivity of preferential oxidation reaction 90% [440].

evaporation, as discussed in Section 5.4.1. The maximum of the curves in Figure 5.49 originates from the increased carbon monoxide formation at S/C ratios approaching stoichiometry. The conversion of this carbon monoxide by water–gas shift has its limitations at low S/C ratios and leads to hydrogen losses in the preferential oxidation reactor because of unselective hydrogen oxidation. The optimum S/C ratio of was calculated to be 1.15 by Ioannides [440]. This is a very low value, that originates from the very low water–gas shift operating temperature of 200 °C and the high selectivity of preferential oxidation (90%) which were assumed.

A recent publication by Francesconi *et al.* also dealt with efficiency calculations of an ethanol fuel processor working with steam reforming. The system investigated by the calculations is a fairly conventional one with catalytic carbon monoxide clean-up by high- and low temperature water–gas shift and preferential oxidation. These workers, however, consider very high S/C ratios in their calculations and identify a very high value of S/C 4 as optimum for their process (38% system efficiency and 80.5% fuel processor efficiency) and state that even a value of S/C 5 still delivers efficiency close to the optimum [441]. It is evident from many other calculations discussed in this chapter that excessive water addition to the fuel processor feed has a detrimental effect on the overall system efficiency. Another irritating result of Francesconi *et al.* was that the optimum operating temperature of the reformer was determined to be 700 °C. However, such high temperatures increase heat losses of the system, which were apparently not taken into consideration. Lower reforming temperatures of around 600–650 °C are possible for steam reforming of purified ethanol and should be chosen for practical reasons, unless the system pressure is significantly above ambient. Reforming of crude ethanol, which contains about 12 vol.% ethanol and other oxygenated hydrocarbons, might well require higher operating temperatures of the reformer due to the impurities contained in the feed.

Francesconi *et al.* calculated the efficiency of a fuel processor operating with crude ethanol [441]. This corresponds to an S/C ratio of 10. The calculations revealed an efficiency of 27% despite the high S/C ratio. This value might not be achieved in a practical system. However, the application supporting the system is interesting: when the fuel processor is integrated into a crude ethanol production plant, the distillation step could be omitted, which is required to increase the crude ethanol concentration. This of course bears considerable energy savings and makes the overall process much more attractive.

Balance-of-plant components, such as the compressor, pumps and valves, is required to run a fuel cell/fuel processor system. They consume energy. These so-called parasitic power losses are usually in the range of 15–20% of the electrical energy generation of the fuel cell. The major portion of the losses can usually be attributed to the compressor.

5.4.3
Overall Water Balance of the Fuel Processor

Another important design issue is the overall water balance of the fuel processor, because external water addition by the end-user is not favourable, especially with

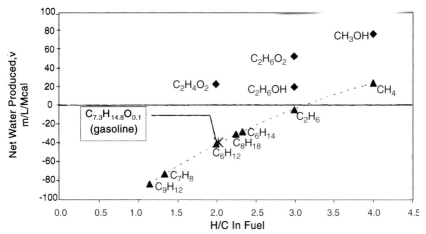

Figure 5.50 Effect of the H/C ratio of different fuels on the net water recovery of an autothermal fuel processor working at a S/C ratio of 1.5; the water recovery is normalised by the heating value of the fuel; the O/C ratio is adjusted such that thermo neutral conditions result for the overall fuel processor [435].

mobile applications. Highly purified water is required for most systems, which could create a distribution problem. Hence, recovery of water is mandatory in most instances.

The fuel determines the water balance of the system. Figure 5.50 shows the water balance calculated for autothermal fuel processors working with different fuels [435]. All systems have a S/C ratio of 1.5 in common. It is obvious that fuel processors fed with oxygenated hydrocarbons tend to have a more positive water balance. Increasing the carbon number of the hydrocarbons impairs the water balance of the fuel processor. However, higher hydrocarbons generally require a higher S/C ratio to prevent coke formation, which counter-balances this effect, as described below. Of course it is desirable to avoid water condensation downstream of the fuel processor. With low temperature PEM fuel cells the reformate must leave the fuel processor with a steam content equal to the dew point at the operating temperature of the fuel cell [15].

Ahmed et al. [435] performed calculations to highlight the effect of the various operating parameters of an autothermal methane fuel processor on the overall water balance. The system considered by Ahmed et al. included an afterburner, which combusted the anode off-gas by cathode off-gas oxygen. Water was then recovered from the burner off-gas. As shown in Table 5.10, Section 5.4.1, the water balance improved when the S/C ratio was increased.

Other factors affecting the water balance are the hydrogen utilisation in the fuel cell anode and the oxygen stoichiometry on the cathode side. Increasing hydrogen utilisation requires a surplus of cathode air and consequently cathode stoichiometry needs to be increased. This dilutes the burner off-gas, which has a detrimental effect on the water balance of the fuel cell/fuel processor system [435].

5 Fuel Processor Design Concepts

Increasing the operating pressure increases the amount of water recovered in the condenser. Higher system pressure in turn increases the load to the air supply system and increases parasitic losses. The ambient temperature determines the saturation partial pressure of water and the amount of water that is recoverable.

5.4.4
Overall Basic Engineering of the Fuel Processor

A design concept for a fuel processor based upon methane decomposition or cracking was presented by Poirier et al. [80]. In principle, the cracking process requires no carbon monoxide removal (see Section 3.4). A two-reactor concept was proposed by these workers, which is shown in Figure 5.51. The reactors had to be operated by alternating with the cracking and carbon deposition mode. Large catalytic beds were proposed with inert material at the inlet and outlet for feed pre-heating. The reactors

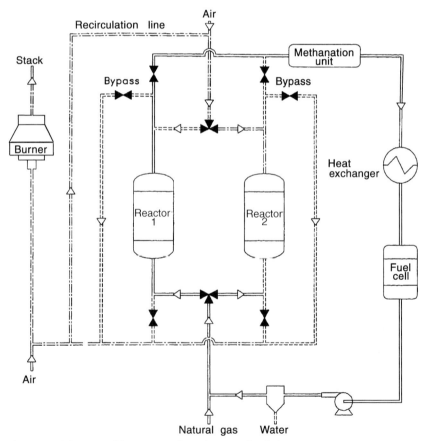

Figure 5.51 Schematic of the process of a methane cracker/fuel cell system, as proposed by Poirier et al. [80].

were operated in the non steady-state. A cracking front moved through the catalyst bed, which was cooled by the cracking reaction and had a high void fraction, taking up the carbon generated by the process. After the cracking cycle, the catalyst bed temperature had decreased from 1000 to 800 °C. Steam co-generation was proposed, which seems to be viable especially for stationary applications. Thus the calculations were performed for a 200-kW fuel cell system. The volume of the cracking reactors was estimated to be 0.5 m^3 each. The formation of 9.5 kg carbon was calculated for a single cracking cycle. The main drawback of the concept was that not only carbon but also carbon monoxide was generated from the oxides species present on the catalyst surface. Thus, a methanation stage was proposed for this cracker/PEM fuel cell system [80]. Poirier *et al.* also proposed recycling the residual hydrogen and methane from the fuel cell anode back to the cracker. The methane concentration of the anode feed was calculated to be 60 vol.%, which seems to be an extremely high value.

Specchia *et al.* compared three design concepts for a bio-diesel fuel processor by static simulations [442]. An autothermal reformer followed by catalytic carbon monoxide clean-up consisting of high and low temperature water–gas shift and water-cooled preferential oxidation was compared with an autothermal reformer followed by a medium temperature water–gas shift stage and water–cooled preferential oxidation. The third system under consideration was a cracking reactor concept, which was followed by water addition and a two-stage water–gas shift and water-cooled preferential oxidation. An anode off-gas afterburner supplied heat for steam generation and feed pre-heating in all instances. The operating conditions of the autothermal reformer were chosen to be a S/C ratio of 2.5 and an O/C ratio of 0.8. The S/C ratio of the cracking reactors, which were regenerated by carbon gasification with steam, amounted to 3.0 [442]. Calculations revealed that the lowest efficiency was achieved with the autothermal reformer coupled to the adiabatic medium temperature water–gas shift reactor. The fuel cracker was best in terms of efficiency, water management, volume and weight, but lowest in terms of controllability, cost, emissions, dynamics and operability [442].

Figure 5.52 shows the design of a complete 10-kW$_{el}$ autothermal diesel fuel processor/fuel cell system as proposed by Cutillo *et al.* [443]. Hexadecane was chosen as the model substance for diesel. The heart of the fuel processor was the autothermal reformer (ATR). A 550 °C feed inlet temperature, a S/C ratio of 2.25 and an O/C ratio of 0.76 were chosen for the calculations. The inlet temperature of the feed mixture was achieved by pre-heating the air feed (supplied by the compressor at 130 °C) in HX-A to 730 °C. The fuel was pre-heated to 365 °C in heat-exchanger HX-D. The steam feed was generated by a two-step procedure. Evaporation and superheating were performed in heat-exchanger HX-C, which utilised the heat of the purified reformate downstream of the carbon monoxide clean-up system. In a second step, the steam feed was further superheated in heat-exchanger HX-B by the afterburner (AFB) off-gas. In the autothermal reformer the feed quickly reached a much higher temperature due to the partial oxidation reaction taking place at the reactor inlet (see Section 3.3). Then steam reforming occurred downstream of the reactor inlet which cooled the reformate to the exit temperature of 744 °C, according to the calculations.

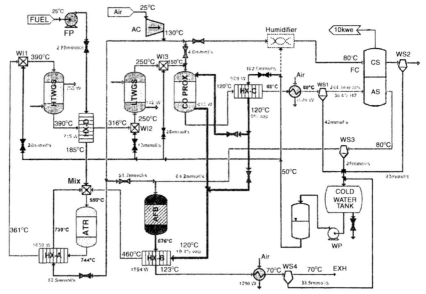

Figure 5.52 Schematic of the process of a diesel fuel processor based on autothermal reforming, as designed by Cutillo et al. [443].

Downstream of the ATR, the reformate was cooled in HX-A, further cooled and enriched with water by the water-injection system WI-1 and then entered the high temperature water–gas shift reactor (HTWGS). About 750 W of energy were removed from the reactor by internal heat-exchange. This made isothermal operation at 390 °C possible.

Downstream of the HTWGS the reformate was further cooled in HX-D and by water-injection, WI-2. Similar to the HTWGS, from the low temperature water–gas shift reactor (LTWGS) 150 W of heat were removed by heat-exchange. The reactor was operated at 250 °C to further decrease the carbon monoxide content (see Section 3.10.1). A third water injection was placed downstream of the LTWGS and air was added to the reformate upstream of the preferential oxidation reactor (CO PROX). This reactor once again carried internal heat-exchange capabilities. An O/CO ratio (λ-value) of 3 was assumed for the CO PROX reactor. Lower values could be assumed for practical systems depending on the catalyst technology applied (see Section 4.5.2). Owing to the heat formation by carbon monoxide and hydrogen oxidation in the CO PROX reactor, a substantial amount, 59 mmol s^{-1} (equivalent to about 2.5 kg h^{-1}), of water was required to cool the reactor. This water flow exceeded 50% of the water feed flow rate supplied to the ATR. Part of the water (42 mmol s^{-1}) was removed from the reformate in water separator WS1 after cooling the reformate in HX-C and in the air cooler. The last device removed 1.6 kW of latent low temperature heat from the system. Low temperature heat losses need to be minimised of course because they are the major contributor to efficiency losses. The fuel cell anode feed contained about 36.4 vol % or 150 L min^{-1} hydrogen, which corresponded to a lower heating value (LHV) of 27 kW. This value was very close to the 28.4 kW LHV of the fuel fed to the system, resulting in a 95% efficiency of the fuel processor. However, the calculations did not consider heat losses, which would reduce the efficiency further.

Hydrogen conversion of 80% was assumed for the fuel cell. The unconverted hydrogen was then fed into the AFB downstream of an additional water separation (WS2). Part of the low temperature heat contained in the AFB off-gas, which had a temperature of 125 °C downstream of HX-B was then removed by a second air cooler (about 1.8 kW of heat losses) and left the system at 70 °C. The heat formed by the oxidation reaction in the fuel cell was removed with the cathode air of the fuel cell. The cathode feed required humidification. Part of the water fed to the cathode was regained from the cathode off-gas (WS2) and fed back to the system. This improved the overall water balance.

Figure 5.53 shows the flow scheme of a 10-kW$_{el}$ diesel fuel processor/fuel cell system based on steam reforming, as designed by Cutillo et al. [443]. The energy required to run the endothermic reaction was generated in a directly coupled catalytic afterburner, which could be realised as a plate heat-exchanger, as described in Section 5.1. The S/C ratio fed to the steam reformer was substantially higher compared with the system discussed above in order to prevent coke formation. The amount of fuel fed to the steam reformer was substantially lower compared

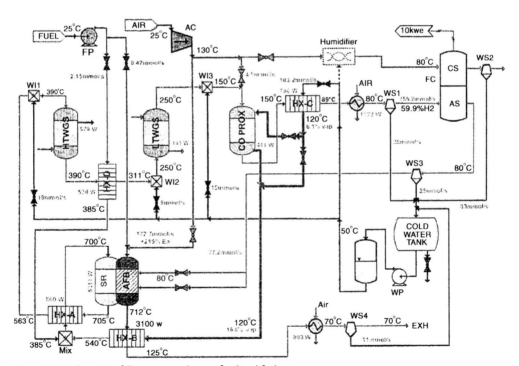

Figure 5.53 Schematic of the process scheme of a diesel fuel processor based on steam reforming, as designed by Cutillo et al. [443].

with the autothermal reformer discussed above (2.15 mmol s^{-1} compared with 2.85 mmol s^{-1}), however, 0.47 mmol s^{-1} of additional fuel had to be fed to the afterburner to achieve an equalised heat balance of the coupled reactor. Thus the steam reformer consumed approximately 8% less fuel than the autothermal reformer discussed above. The heat balance of the reactor could have been closed also by running the fuel cell at lower hydrogen utilisation. An 11% lower water addition was required for the steam reformer and the hydrogen content of the reformate was substantially higher (55.9 vol.% compared with 36.4 vol.%) due to the lack of dilution by nitrogen. Table 5.13 provides an overview of the gas compositions as calculated for the two systems being compared. Despite the initially higher water addition, because of the higher S/C ratio chosen for the steam reformer, the water injection required downstream was lower. This in turn led to a lower but still positive net water balance of the system for steam reforming (6.1% compared with 12.2% for ATR).

The efficiency of the steam reformer fuel processor of 96.6% was much higher than the autothermal reformer efficiency (88.8%), which in turn also increased the system efficiency (38.7% compared with 35.5% for the autothermal reformer) [443]. The heat removal required for the two air coolers was much lower for the steam reformer (about 2.1 kW compared with 3.4 kW for the autothermal reformer). This in turn reduced the size of these components, which was a substantial benefit because the air coolers contributed significantly to the overall system size. The volume required is a stringent factor, especially in mobile systems. All the benefits of steam reforming clearly have the drawback of a more complex reactor design, which needs to be addressed by suitable manufacturing techniques in order to become competitive in price and not just in performance (see Section 10.2).

Both systems discussed above are fairly complex, because four heat-exchangers and two air-coolers are required apart from five and four reactors, respectively. For systems with a power output of 10 kW or higher, as discussed above, this complexity may well be acceptable because it improves the efficiency. However, for portable applications, less efficiency may well be acceptable for the sake of a compact and lightweight system.

Table 5.13 Comparison of the reactor outlet composition as calculated for a diesel autothermal reformer (ATR) and a diesel steam reformer (STR) [443].

	Reformer		HTWGS		LTWGS		PrOx		FC anode inlet	
	ATR	SR	ATR	SR	ATR	SR	ATR	SR	ATR	SR
H_2	31.4	49.9	34.8	52.4	34.5	52.0	31.7	47.1	36.4	55.9
CO	8.7	10.6	1.4	2.1	0.2	0.3	–	–	–	–
H_2O	26.3	30.03	27.0	29.8	29.0	30.9	33.3	35.5	23.4	23.4
CO_2	8.7	9.2	14.3	15.7	14.8	16.8	14.0	15.7	16.0	18.6
N_2	24.9	–	22.5	–	21.5	–	21.0	1.7	24.2	2.1

Table 5.14 Comparison of product mass flow rates, reformer efficiency η_R, fuel processor efficiency η_{FP} and auxiliary power unit efficiency η_{APU} of autothermal reforming dry hydrogen molar fraction of the reformate y_{H_2}; values are determined for various fuels, feed inlet temperatures T_{in} and S/C (SCR) and O/C ratios (expressed as O_2/C ratio OCR) as calculated by Specchia et al. [371] W_B shows the water balance of the systems, which is positive when the W_B exceeds unity.

m_{fuel} (1 kg/h)	T_{in} (°C)	SCR	OCR	T_{out} (°C)	m_{H_2} (kg/h)	m_{CO} (kg/h)	m_{H_2O} (kg/h)	m_{CO_2} (kg/h)	m_{N_2} (kg/h)	η_R (%)	η_{FP} (%)	η_{APU} (%)	y_{H_2} dry	W_B
Gasoline	560	3.08	0.356	797	0.315	0	1.801	3.171	2.810	70.5	87.2	35.0	0.477	1.17
Gasoline	560	2.68	0.304	700	0.330	0	1.791	3.171	2.490	75.6	91.2	36.5	0.506	1.20
Light Diesel	550	2.97	0.418	798	0.295	0	1.860	3.189	3.317	67.5	84.4	33.8	0.436	1.17
Light Diesel	550	2.57	0.362	700	0.310	0	1.827	3.189	2.940	73.0	88.5	35.5	0.466	1.20
Biodiesel	550	2.90	0.370	800	0.263	0	1.591	2.829	2.622	68.9	86.4	34.6	0.454	1.18
Biodiesel	550	2.53	0.315	701	0.276	0	1.562	2.829	2.292	74.5	90.6	36.3	0.485	1.21

Cutillo et al. also analysed the effect of introducing a carbon monoxide tolerant fuel cell into the system, which would make the overall system less complex [443]. Because such fuel cells were expected to be less efficient, about 3% lower efficiency was assumed. Another potential simplification was the removal of one of the water–gas shift reactors. The two stage water–gas shift reactors could be replaced by a medium temperature water–gas shift reactor with higher carbon monoxide outlet concentration in combination with the high carbon monoxide tolerant fuel cell. Alternatively, a water–gas shift reactor with heat-exchange capabilities, as discussed in Section 5.2.1, could be placed into such a system and combined with preferential oxidation and low temperature PEM fuel cell technology.

A comparison of the efficiencies of the reformer, fuel processor and complete auxiliary power unit gained by autothermal reforming and steam reforming of different fuels is shown in Tables 5.14 and 5.15 [371]. It is obvious that more hydrogen is gained from gasoline than from heavier fuels. However, a higher feed temperature and lower S/C and O/C ratios are beneficial for system efficiency.

Two 50-kW fuel processors concepts were compared by Lattner and Harold [405]. Both systems had autothermal reformers combined with either catalytic carbon monoxide clean-up or membrane reactor configurations. Tetradecane was used as the model substance for diesel fuel.

To model the performance of the autothermal reformer, kinetics from the literature that had been determined for the catalytic combustion of methane over a platinum-based catalyst and for steam reforming over nickel-based catalyst were combined and fitted to the experimental data of Flytzani-Stephanopoulos et al. [153]. The water–gas shift reaction was assumed to reach thermodynamic equilibrium under all conditions in the reformer reactor, which is usually the case in reformers. Methane formation was not considered. Because catalyst pellets had been used for the determination of the kinetics, diffusion limitations were to be expected. They had been lumped into the kinetic models. The hot spot formation usually observed at

Table 5.15 Comparison of product mass flow rates, reformer efficiency η_R, fuel processor efficiency η_{FP} and auxiliary power unit efficiency η_{APU} of steam reforming; values are determined for various, feed inlet temperatures T_{in} and S/C (SCR) and O/C ratios (expressed as O_2/C ratio OCR) as calculated by Specchia et al. [371]. β_R is the fraction of the fuel which is fed to the steam reformer; $1-\beta_R$ is fed to the burner. γ_{H_2} is the dry hydrogen molar fraction of the reformate; W_B shows the water balance of the systems, which is positive when the W_B exceeds unity.

m_{fuel} (1 kg/h)	T_{in} (°C)	SCR	OCR	β_R (kg/h)	SCR_R	T_{out} (°C)	m_{H_2} (kg/h)	m_{CO} (kg/h)	m_{H_2O} (kg/h)	m_{CO_2} (kg/h)	m_{N_2} (kg/h)	η_R (%)	η_{FP} (%)	η_{APU} (%)	γ_{H_2} dry	W_B
Gasoline	650	2.88	0.24	0.839	3.43	701	0.349	0	1.329	2.661	0.186	82.1	96.6	38.7	0.722	1.17
Gasoline	700	3.2	0.29	0.802	3.99	751	0.335	0	1.266	2.545	0.131	78.9	92.2	36.9	0.728	1.14
Gasoline	700	2.6	0.21	0.857	3.03	712	0.355	0	1.363	2.718	0.244	82.1	97.6	39.0	0.716	1.19
Light Diesel	650	3.39	0.37	0.745	4.55	701	0.310	0	1.167	2.374	0.098	77.8	88.8	35.5	0.729	1.11
Light Diesel	700	3.6	0.41	0.717	5.02	743	0.299	0	1.123	2.287	0.079	74.8	85.6	34.2	0.732	1.09
Light Diesel	700	3.0	0.33	0.770	3.89	711	0.320	0	1.211	2.457	0.131	78.7	91.6	36.6	0.725	1.13
Biodiesel	650	3.08	0.29	0.794	3.88	700	0.283	0	1.083	2.246	0.115	80.5	93.4	37.4	0.720	1.15
Biodiesel	700	3.2	0.32	0.776	4.12	734	0.278	0	1.058	2.195	0.100	78.2	91.5	36.6	0.722	1.13
Biodiesel	700	2.8	0.26	0.813	3.44	713	0.290	0	1.112	2.230	0.144	80.8	95.6	38.1	0.716	1.17

the inlet of the autothermal reformer (see Section 3.3,) and also by Flytzani-Stephanopoulos et al., could be matched fairly well by the kinetic model.

The catalytic carbon monoxide clean-up worked with a two-stage water–gas shift in tubular reactors cooled by steam generation. The kinetics for a rhenium–alumina catalyst for high temperature water–gas shift and for a copper/alumina catalyst for low temperature shift had been extracted from the literature.

However, precious metal based catalysts without an oxygen carrier or additive, such as the rhenium/alumina catalyst as used here for the calculations, are known to have much lower activity compared with catalytic systems, such as platinum/ceria (see Section 4.5.1).

Heat and mass transfer limitations were not taken into consideration for the water–gas shift (see Section 5.2.1). No kinetics but instead full conversion at 50% selectivity was assumed for the preferential oxidation reactor downstream of the water–gas shift stages.

As alternatives to the catalytic carbon monoxide clean-up, two types of membrane reactors were investigated by the simulations. Namely, this was an autothermal membrane reformer carrying a 10-μm thick membrane. Permeability data from Hoelleck [444] (see also Section 5.2.4), were used to simulate hydrogen permeation. Ceramic membranes of 20-μm thickness were chosen as alternatives. Steam was used as the sweep gas for the membrane reactors (see Section 5.2.4). To achieve an optimum driving force for the hydrogen through the membrane, a co-current flow was chosen for the sweeping gas. Coke formation issues at the membranes were not taken into consideration. A methanation reactor was integrated into the membrane fuel processor concepts to remove residual carbon oxides present in the retenate as a result of leaks in the membranes, as described in Section 5.2.4. Full conversion of the carbon oxides was assumed without modelling the methanation reactor.

An 85% hydrogen utilisation was assumed for the fuel cell anode of the conventional system, while 95% utilisation was assumed for the membrane devices [405]. The electrical efficiency of the fuel cell was set to 60%. A catalytic afterburner was included in both system concepts. Residual hydrogen from the fuel cell anode was combusted therein with the cathode off-gas oxygen.

For the heat-exchangers integrated into the system design concepts, a minimum logarithmic driving force of 100 K was assumed.

The water management of the systems was also taken into consideration. The water was recovered from the afterburner exhaust. A minimum exit temperature of 48 °C was assumed for the exhaust gas, which is a rather low value, because in practical systems condensers require air cooling. In the summer time it would be difficult to cool the exhaust gases down to 48 °C. To maximise water recovery, the system pressure was set as high as possible. This would be counterbalanced by increased parasitic power losses of the compressor. The parasitic losses for compression were considered by Lattner and Harold [405], consequently a 1.2-bar fuel cell operating pressure was assumed as a compromise. Only the parasitic losses to the compressor (see also Section 8.3) were considered, while the power consumption of valves and pumps was neglected. A 75% efficiency was assumed for the compressor.

200 | 5 Fuel Processor Design Concepts

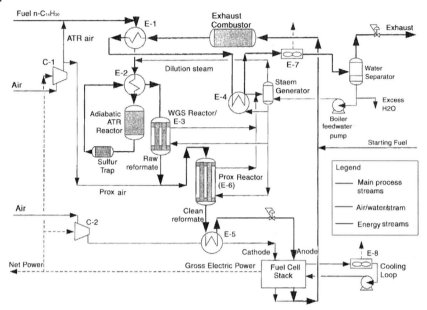

Figure 5.54 Configuration of a diesel fuel processor based upon catalytic carbon monoxide clean-up [405].

Figure 5.54 shows the system configuration of a conventional system with catalytic carbon monoxide clean-up as designed by Lattner and Harold [405]. The fuel was mixed with air from the compressor at a fairly high pressure of 3 bar and then pre-heated to 257 °C in the heat-exchanger E-1. The energy of hot combustion gases from the afterburner (here termed "exhaust combustor") were utilised for this. Downstream of heat-exchanger E-1, steam was added to the reformer feed, which was then pre-heated further by the hot reformer product in heat-exchanger E-2 to a temperature of 631 °C.

Pre-heating a fuel/air/steam mixture to such a high temperature is not feasible without the risk of initiating a homogeneous reaction. This risk was recognised by Lattner and Harold [405]. The reformer was set to an S/C ratio of 2.0 and an O/C ratio of 0.72. A hot spot of 1189 °C was calculated for the autothermal reformer. However, conventional precious metal catalysts would be damaged fairly quickly by such a high local temperature.

The reformate left the reformer with a temperature of 814 °C and entered a zinc oxide trap. However, this would be not feasible in a practical system, because zinc oxide adsorbent materials cannot tolerate temperatures exceeding 450 °C. The reformate, which was cooled to 440 °C in heat-exchanger E-2 was then passed to the water–gas shift reactor. This reactor was cooled by steam generation at 15-bar pressure and a temperature of 200 °C in a counter-current flow. The catalyst bed was assumed to be composed of a noble metal based rhenium/alumina catalyst at the inlet section followed by a copper/zinc oxide catalyst at the outlet section. Despite the fact that a water–gas shift catalyst of fairly low activity had been chosen for the calculations, the reactor volume was calculated to be 6.5 l, which is rather a low value for a 50-kW system. The product of the water–gas shift was then mixed with air from the compressor and fed into the preferential oxidation reactor. This reactor was cooled by steam generation at 15 bar and operated under isothermal conditions at a fairly high temperature of 200 °C. The purified reformate was then further cooled in heat-exchanger E-5 by the cathode air feed, which was supplied by a separate blower. This blower was used to save on the compressor energy demand, owing to the lower fuel cell operating pressure of only 1.2 bar. The reformate pressure was then reduced to the fuel cell operating pressure by a pressure control valve. However, this would translate to significant energy losses in a practical system. Valuable energy would be consumed by the compressor in order to maintain the high operating pressure of the fuel processor. Working close to ambient pressure is probably the preferred strategy for systems in the power range of tens of kilowatts and lower.

Downstream of the fuel cell, the anode off-gas was combusted in the afterburner. The afterburner off-gas was cooled in heat-exchanger E-1, as mentioned above and still further in the steam generator E-4 to 254 °C. Finally, the combustion gases were cooled by an air cooler from 254 to 48 °C. This means that this cooler would remove 35 kW of heat as waste heat out of the system, which does not seem to be an efficient solution. Instead, the heat utilisation of the afterburner off-gas should be improved for a practical system. The fuel cell was cooled by an air-cooled cooling loop (heat-exchanger E-8).

The membrane reformer concept is shown in Figure 5.55. Basically, the concept is fairly similar to the conventional concept described above. The most striking

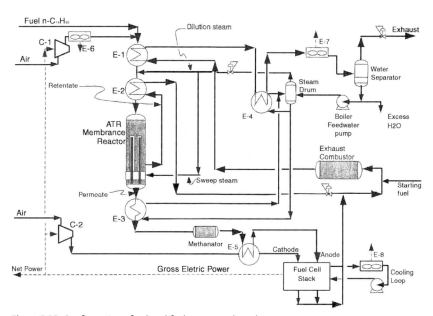

Figure 5.55 Configuration of a diesel fuel processor based upon an autothermal membrane reformer reactor [405].

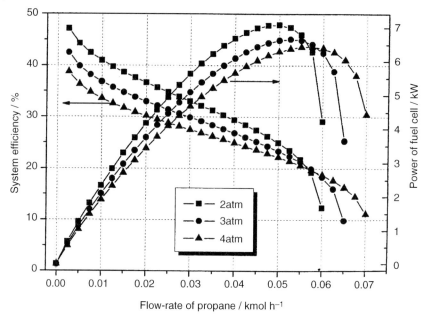

Figure 5.57 Simulated power output and system efficiency versus fuel feed rate as determined for a 5-kW propane ATR fuel processor/fuel cell system at different pressures [445].

expansion downstream of the afterburner, which is positioned at the very end of the whole chain of fuel processor components and downstream of the fuel cell itself. Besides this, expanders for flow rates as low as those present in a 5-kW fuel cell system are not readily available on the market place. 88.1% efficiency was calculated for the fuel processor at 2-bar pressure. Assuming 70% adiabatic efficiency of the compressor/expander system, a net system efficiency of 34.8% was calculated; 31.9% of the energy input was lost to the flue gases, 20% were removed by the air coolers in this design concept (Figure 5.57).

High temperature PEM fuel cells are tolerant to carbon monoxide to a large extent (see Section 2.3.1) and therefore simplify the fuel processor significantly, because carbon monoxide fine clean-up by preferential oxidation or methanation is not required. With methanol steam reforming, the reformate could even be fed to the fuel cell directly.

Ahluwalia et al. discussed the concept for a high temperature PEM fuel cell in connection with a gasoline fuel processor [446]. An autothermal reformer, and high- and low temperature water–gas shift stages with intercoolers were considered along with several other heat-exchangers. Anode off-gas combustion was performed in a catalytic afterburner. A combined compressor/gas turbine system was assumed and the process water was recovered downstream of the turbine. Thus, a system with an electrical power equivalent exceeding 10 kW$_{el}$ was obviously under consideration, which allowed the use of expander turbines. Water cooling was assumed for the fuel cell stack. An optimum S/C ratio of 0.6 was calculated for the autothermal reformer, while the optimum S/C ratio of the entire fuel processor was about 1.8, which means that water was injected downstream of the reformer. The maximum system efficiency was calculated to be 38.4%, while the fuel processor efficiency was 79.5% under these conditions. About 0.3 vol.% carbon monoxide was contained in the reformate fed to the fuel cell. The application of the expander turbine allowed recovery of energy from excess steam. This made the addition of excess water less detrimental towards the system efficiency. Similar results were found for the system pressure, which only slightly affected system efficiency in the range between 1.5 and 3.0 bar. At 2.5-bar system pressure, a comparison with a low temperature PEM fuel cell system revealed about 1.5% higher efficiency for the high temperature fuel cell. More significant were savings of catalyst weight and volume for the high temperature fuel cell. They were both in the region of 40%, mostly because no preferential oxidation and a smaller water–gas shift reactor were required. The weight of balance-of-plant components could be reduced by 50% according to the calculations.

5.4.5
Dynamic Simulation of the Fuel Processor

Few examples have been published in the open literature dealing with dynamic simulation of fuel processors or even of single components. Much of the simulation work published deals with the development of suitable start-up strategies. Of course it is desirable, in principle, to decrease the thermal mass of the components as much as possible, and to increase the start-up gas supply as much as possible to decrease the start-up time demand of the fuel processor. In general, it is not feasible to achieve a rapid start-up time with a system design that consists simply of a chain of reactors, which are supplied with heating gas at the beginning of the chain. Owing to the small dimensions of the flow ducts in most mobile systems, whether they are based upon ceramic or metallic monoliths or on plate heat-exchanger technology, the entire heat is taken out of the heating gas by the cold reactor. Not before the temperature of this reactor has approached the temperature of the heating gas, the next component downstream in the chain will increase its temperature substantially. This is a time-consuming procedure apparently. Thus, more sophisticated start-up procedures rely on *in situ* production of heat in each component of the chain, or on distributed parallel heating, which is feasible when plate heat-exchanger technology is applied.

Generally, it is preferable to utilise steam as an energy carrier instead of air, because of its higher heat capacity.

Springmann *et al.* [152] performed dynamic simulations of a combination of a monolithic autothermal reformer reactor with a tubular heat-exchanger for feed pre-heating and a high temperature water–gas shift reactor (diameter 70 mm; 1600 cells per square inch or about 630-µm channel size) [389]. Both reactors were metallic monoliths. The reformer has already been described in Section 5.1.3. During normal operation, air and steam were fed at 200 °C to the heat-exchanger and heated therein to approximately 650 °C. The reformate leaving the reformer was cooled in the

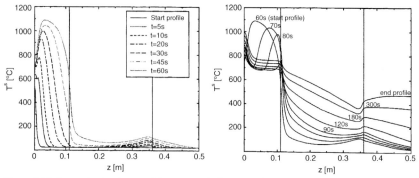

Figure 5.58 Dynamic simulation of the start-up behaviour of an autothermal gasoline reformer coupled to a heat-exchanger for feed pre-heating and a water–gas shift reactor. Left: catalytic combustion phase (5–65 s); 8-kW$_{th}$ power input; S/C = 0; O/C = 7.8. Right: autothermal reforming (65–365 s) at full load of the fuel processor (33 kW$_{th}$); S/C = 2; O/C = 0.75 [389].

heat-exchanger to 420 °C, before it entered the shift reactor. Owing to the high feed temperature and the adiabatic temperature rise of the water–gas shift, the carbon monoxide content could be reduced from about 10 to only about 5 vol.%, according to the thermodynamic equilibrium.

The analysis of the start-up procedure, as described in Section 5.1.3, was extended to the heat-exchanger and the shift reactor [152]. The reformer reactor was pre-heated by catalytic combustion with excess air, as described in Section 5.1.3, and then switched to the autothermal mode. As shown in Figure 5.58, the reformer itself could be heated to the operating temperature within 60 s, but the pre-heating of the two devices downstream of the reformer took about 6 min. An intermediate shift of the position of the maximum temperature within the reformer occurred, which was related to the introduction of cold steam into the device.

To further reduce the start-up time demand of the water–gas shift reactor, addition of air (20 l min^{-1}) was assumed upstream of the reactor after it had reached a temperature of 200 °C [389]. Thus the carbon monoxide and hydrogen present in the reformate were partially combusted in the water–gas shift reactor. In this way, the time demand for pre-heating the high temperature water–gas shift reactor could be reduced to 3 min according to the calculations.

However, it is questionable if this is really a viable option, because local over-heating of the water–gas shift catalyst is likely. This catalyst is sensitive to temperature excursions at least when highly active precious metal catalyst formulations are applied (see Section 4.5.1). Additionally, the introduction of another device for air feeding into the system increases its complexity and cost.

As a last measure, electrical pre-heating of a small and separate monolith at the inlet of the water–gas shift reactor was assumed, similar to the design of the reformer reactor (Figure 5.59) [389]. The reformer was then operated in the partial oxidation mode rather than under conditions of total oxidation. Through these means,

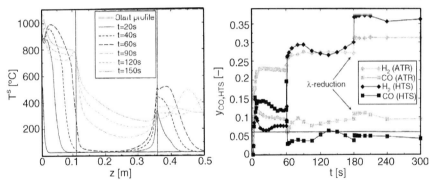

Figure 5.59 Dynamic simulation of the start-up behaviour of an autothermal gasoline reformer coupled to a heat-exchanger for feed pre-heating and a water–gas shift reactor; both reactors were pre-heated for 5 s with 2.5-kW$_{el}$ energy. Left: partial oxidation phase (5–65 s) at full load of the fuel processor (33 kW$_{th}$); S/C = 0; O/C = 0.94; 60 L min^{-1} air addition to the water–gas shift reactor. Right: gas composition downstream of the autothermal reformer (ATR) and the high temperature water–gas shift reactor (HTS) [389].

hydrogen and carbon monoxide were supplied to the water–gas shift reactor and combusted therein. The modified system provided reformate with a low carbon monoxide content even after 60 s. At this point in time steam addition to the system was assumed to begin.

Taking into consideration that another two reactors (low temperature shift and preferential oxidation reactors) are required downstream of the high temperature water–gas shift, at least in a conventional fuel processor/low temperature PEM fuel cell system, the calculations demonstrate that it is difficult to pre-heat a chain of reactors without an excessive time demand. Other measures and functionalities become possible when plate heat-exchanger technology is applied, which may help to solve this problem.

The start-up time demand as determined experimentally for breadboard and integrated fuel processors of different design, and will be discussed in Section 9.

Aoki et al. from Toyota [447] compared three possible system designs of a 60-kW$_{el}$ fuel processor/fuel cell system dedicated for the automotive drive train and fed with purified hydrocarbon fuel:

1. A conventional system, as shown in Figure 5.60. It was composed of a monolithic autothermal reformer operated at a S/C ratio of 2.0 and O/C ratio of about 0.8, high and low temperature water-gas shift reactors, a water-cooled three stage preferential oxidation reactor and a conventional low temperature PEM fuel cell cooled by refrigerant.

2. A second generation system, as shown in Figure 5.61. It was composed of a microchannel oxidative steam reformer, which was supplied with water by a humidifier. The humidifier utilised cathode off-gas for humidification. Steam reforming was performed at a S/C ratio of 1.6 and O/C ratio 0.2. The microchannel steam reformer was coupled to an integrated catalytic burner. The burner was

208 | *5 Fuel Processor Design Concepts*

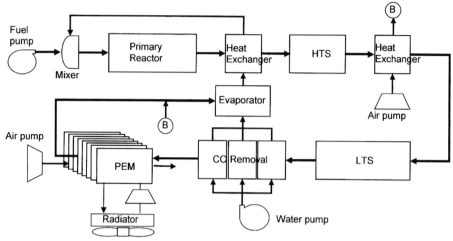

Figure 5.60 First generation fuel processor/fuel cell system as designed by Aoki et al. [447]. It was composed of a reformer, two water–gas shift stages, a water-cooled reactor for preferential oxidation of carbon monoxide and a water-cooled PEM fuel cell.

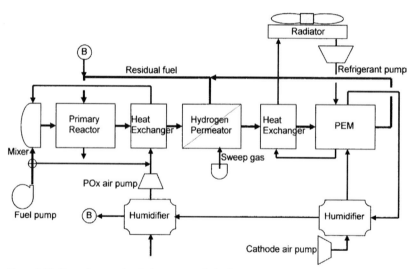

Figure 5.61 Second generation fuel processor/fuel cell system as designed by Aoki et al. [447]. It was composed of a microchannel reformer, a water–gas shift reactor (not shown), a hydrogen permeator and a water-cooled PEM fuel cell.

supplied with permeate from the membrane hydrogen separation unit and residual hydrogen from the anode off-gas. Water–gas shift coupled with membrane separation was used to purify the reformate and sweep gas humidified the permeate. The PEM fuel cell was cooled by refrigerant as in example 1.

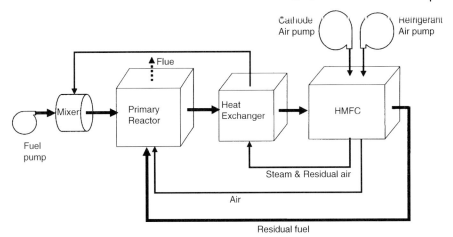

Figure 5.62 Third generation fuel processor/fuel cell system as designed by Aoki et al. [447]. It was composed of a microchannel coupled reformer/afterburner and an air-cooled medium temperature fuel cell.

3. A third generation system, as shown in Figure 5.62. It was composed of a microchannel oxidative steam reformer, which was supplied with water and air from the cathode off-gas. It was operated at a S/C ratio 1.9 and an O/C ratio of 0.15. The microchannel reformer was internally coupled to a catalytic burner, which was supplied with residual hydrogen from the fuel cell anode and cooling air from the cathode. The medium temperature fuel cell (operated between 400 and 600 °C) was cooled by air and worked with a metallic membrane. The membrane had the additional function of an anode electrode. $BaCe_{0.8}O_3$ served as the cathode electrolyte.

A comparison of the size estimated for the components of the three systems described above is shown in Table 5.16. This revealed that the second generation microchannel reactor was about 50% smaller than the monolithic reactor of the first generation. The microchannel reformer size of the third generation system amounted to only 16% of the size of the monolith. The water–gas shift reactors, contributing considerably to the overall size of the first and second generation systems, became obsolete in the third generation, as did most of the heat-exchangers.

Table 5.16 Size estimation of the components of the three systems shown in Figures 5.60–5.62 [447].

System	Primary Reactor (L)	WGS Reactor (L)	PrOx (L)	Hydrogen Permeator (L)	Heat Exchange (L)	Humidifier (L)	Fuel Cell (L)
1st Gen.	6.3	10.3 + 17.1	3	–	3.6 + 1.9	7.9	96
2nd Gen.	3	7	–	7.1	2.7 + 1.2	2.5 + 2.5	96
3rd Gen.	1	–	–	–	0.5	–	7

The medium temperature fuel cell only required 7% of the space demand of the conventional PEM fuel cell.

Next, Aoki et al. determined the start-up time demand, the power generation characteristics and the system efficiency of the three systems using dynamic simulations.

The first generation system had about 10-min start-up time demand. While the reformer was heated up within 2 min, most of the time was required to pre-heat the carbon monoxide clean-up reactors downstream. The fuel processor efficiency was calculated to be 70%, while the overall system efficiency was as low as 24%.

The start-up time demand of the second generation system was 5 min. The fuel processor efficiency was 70%, as for the first generation, while the overall system efficiency could be improved to 35% owing to the energy recovery by the coupled microchannel reformer/burner and the utilisation of cathode off-gas steam.

The start-up time demand of the third generation system could be reduced drastically to 1 min according to the calculations. The fuel processor efficiency was increased to 80% and the overall system efficiency reached 48% because of the utilisation of the cathode air in the reformer and the simplifications which became possible by application of the high temperature fuel cell.

A system rigorously designed for a low start-up time demand was presented by Ahmed et al. [448]. The target of the development work was a 50-kW gasoline fuel processor. A significant fraction of the full hydrogen flow should be delivered with a purity suitable for a PEM fuel cell within 60 s. The energy available for start-up was limited to 2 MJ, which corresponds to a heating power of 33 kW for 60-s duration.

The fuel processor concept was fairly conventional, and was composed of an autothermal reformer, water–gas shift and preferential oxidation. The reformer itself was the heat source of the system during normal operation and also during start-up. It was operated in partial oxidation mode during start-up. An O/C ratio of around 1.0 was set, which resulted in a high carbon monoxide concentration of 23%. This concept is similar to the strategy followed by Springmann et al. described above. Combustion of the partially oxidised reformate downstream of the reformer in the water–gas shift and preferential oxidation reactors was the strategy. Surprisingly for a rapid start-up system of the targeted size, a high number of clean-up reactors was chosen, namely, four water–gas shift reactors and three preferential oxidation reactors, all switched in series. Their temperature was controlled by a total of six water-cooled foam heat-exchangers. According to the simulations, the autothermal reformer could reach its operating temperature of about 800 °C within 10 s. As soon as the reformer temperature exceeded 600 °C (after 5 s), liquid water was added to the reformer feed at a S/C ratio of 0.5 for 30 s. The S/C ratio was then gradually increased up to the final value of 2.0. The hydrogen combustion in the clean-up reactors started even at ambient temperature. The temperature of the reactors was controlled by on/off valves for the air supply to prevent over-heating of the catalyst.

Figure 5.63 shows the fuel processor that was designed, based upon the results of the simulation work. It had a concentric design, with the hottest components positioned in the centre. The reformate passed through two annular paths dedicated to carbon monoxide clean-up. The reformer and the four water–gas shift reactors

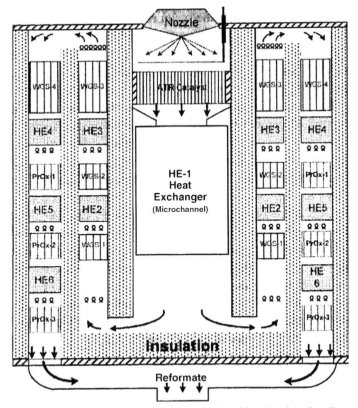

Figure 5.63 Layout of the fuel processor as developed by Ahmed *et al.* [448].

were Microlith™ devices, the heat-exchanger HE-1 a microchannel heat-exchanger, the other heat-exchangers were made of carbon foam, while the preferential oxidation reactors were made of ceramic foams. The total weight of the components amounted to 21 kg and the reformer required only 3% of this value. The heat-exchangers took up 44% of the total weight, the water–gas shift reactors 42% and the preferential oxidation reactors only 11%. However, when the system was then built by Ahmed *et al.*, and mounted, the total weight of the fuel processor amounted to 76.6 kg due to the weight of the housing.

A similar concept study was performed by Seo *et al.* [449] for natural gas reforming. However, fixed-bed reactors were used and the heat supply originated not from autothermal reforming, but from a catalytic burner in the centre of the fuel processor. Two fuel processors of this type were then built with 1- and 2-kW electrical power equivalents [450]. The smaller system was operated at a S/C ratio 3.0 and 89% methane conversion was achieved, while other hydrocarbons present in the natural gas feed were completely converted. The fuel processor efficiency was calculated as the ratio of the lower heating value of the hydrogen produced to the lower heating value of natural gas fed to the reformer and the burner. It was in the range between

76% at 75% load and 64% at 30% load, because the heat transfer efficiency from the burner to the reformer decreased at partial load. The fuel processor showed stable performance at full load for 50-h duration.

Another similar approach was built and tested by Qi et al. [451] for gasoline fuel processing (see Section 9.5).

Chen et al. [452] developed start-up strategies for a 3-kW$_{el}$ methane fuel processor through dynamic simulations. The experimental fuel processor is shown in Figure 5.64. It was composed of two heat-exchangers for feed pre-heating, an autothermal reformer, three water–gas shift stages and three stages for preferential

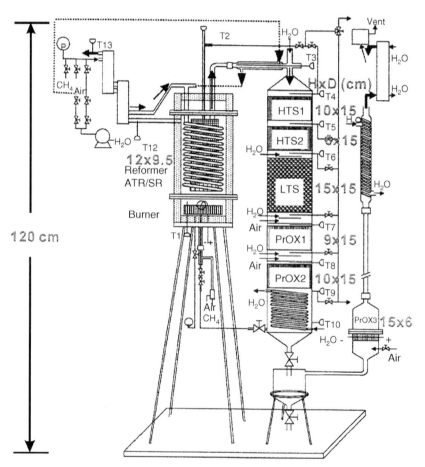

Figure 5.64 Experimental methane fuel processor as developed and analysed through numerical simulations by Chen et al. [452]. The heat-exchangers for feed pre-heating are positioned top left. The fuel processor was composed of an autothermal reformer (ATR), two high temperature water–gas shift stages (HTS1 and HTS2), one stage for low temperature water–gas shift (LTS) and three stages for preferential oxidation of carbon monoxide (PrOX1, 2 and 3).

oxidation of carbon monoxide. Water injections were used to control the temperature of each clean-up reactor separately, which resulted in a complicated system setup. Second and third stage preferential oxidation were lumped into a single reactor to simplify the design calculations. The feed was pre-heated in a coil wrapped around the autothermal reformer before it entered the reactor itself. However, heat losses were not taken into consideration for the carbon monoxide clean-up reactors. In addition, a burner was integrated into the reformer, which was capable of combusting the anode off-gas. In the steady state, the reformer was operated at a S/C ratio 1.45 and O/C ratio 0.9, the preferential oxidation was at O/CO 1.5. Astonishingly, an increasing temperature profile was calculated for the autothermal reformer over the entire length of the reactor rather than just a hot spot at the inlet. The temperature profiles determined for the shift reactors were flat and the carbon monoxide concentration decreased linearly in the preferential oxidation reactors, as shown in Figure 5.65.

All these assumptions do not reflect practical experience and results from numerous other publications. The start-up strategy was unfortunately not explained completely, it was only mentioned that fuel was combusted initially. The start-up time demand of the unit was 37.5 min. This start-up time was reduced by increasing the feed flow rate to double that of the nominal value. In fact, the strategy was to remain at a double fuel flow rate, until the temperature of the second stage preferential oxidation reactor started to rise. Then the feed flow rate was decreased gradually. By these means, the start-up time demand could be reduced to 28 min. The application of low heat capacity reactor construction material was proposed to further reduce the start-up time demand.

5.4.6
Control Strategies for Fuel Processors

Few examples of control strategies for fuel processors have been published in the open literature.

Shin and Besser presented their control concept for a small-scale methanol micro fuel processor [453]. Flow and temperature were controlled by conventional proportional/integral/differential (PID) controllers, along with a miniaturised thin film flow sensor that had been developed previously by these workers, which will not be discussed in detail here. No miniaturised valve was available and thus commercial control valves were used. The product flow rate was used as the main process variable for control as shown in Figure 5.66. It could also be used to estimate conversion. However, the determination of the reformate flow rate would not be an option in larger scale systems. The PID controllers had been embedded into LabView software. When PID controllers are used, their tuning is the most critical issue. However, self-tuning strategies do exist from commercial suppliers. Shin and Besser used a method from Ziegler-Nichols for tuning their PID controllers [453]. By adjusting their tuning parameters, allowing a certain overshoot of the controller and other means, Shin *et al.* could reduce the time demand for load changes of from 0.11 to 0.42 W to less than 15 s [453].

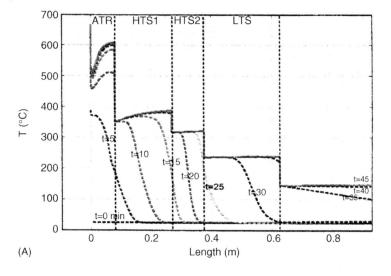

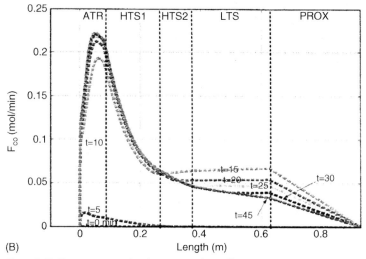

Figure 5.65 Temperature and carbon monoxide profiles versus time as determined by Chen et al. [452].

Görgün et al. presented the developments of observers for fuel processing reactors such as a partial oxidation reformer, water–gas shift and preferential oxidation clean-up reactors [454]. Simplified models were used for the observer design. The principle of an observer-based control strategy is to run a simplified reactor model in the controller software. Model parameters are then permanently adopted by measurements during the control process. For each reaction, which was considered in one of the reactors, one measurement was used to adopt the models. With catalytic partial oxidation and preferential oxidation the reactor temperature and one species

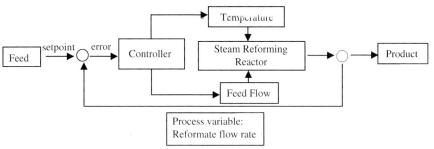

Figure 5.66 Feedback control strategy for a small-scale methanol fuel processor; the reformate flow rate is measured and used for system control [453].

concentration were required to control the reactors, because two reactions were taken into consideration by the models. For water–gas shift, only one reaction was included into the model and thus only the temperature was required to control the reactor.

A very simple control strategy was proposed by Capobianco *et al.* for a methanol fuel processor composed of a methanol steam reforming membrane reactor and a hydrogen buffer tank [455]. The strategy was to run the reformer at constant load and increase the pressure in the buffer tank when the fuel cell required no hydrogen. Consequently, the pressure on the permeate side of the membrane would increase until all of the reformate would have passed into the permeate tail gas with no hydrogen separation taking place any longer. This seems to be an inefficient approach.

5.5
Comparison with Conventional Energy Supply Systems

A few selected examples of smaller scale conventional electrical power generators based upon internal combustion engine technology will be discussed below.

One is the DACHS combined heat and power system developed by the German company SenerTec. The device has an electrical power output of 5.5 kW$_{el}$, a thermal power output of 12.5 kW$_{th}$, a high electrical efficiency of 25%, while the thermal efficiency is about 65%. The overall efficiency is therefore high at 90%. It has a size of 1.06-m length, 0.72-m width and 1-m height, a weight of 520 kg and noise emissions of less than 56 dB. Possible fuels are natural gas, liquefied petroleum gas, bio-diesel and diesel.

Honda, amongst others, offers an electricity generator, EU1000i [456]. The system has dimensions of 450-mm length, 240-mm width and 380-mm height and a weight of 13 kg. A 50-cm^3 gasoline engine generates a rated electrical power of 900 W with an efficiency that was calculated to be 14.2%. This corresponds to a fuel consumption of 0.6 L h^{-1} at rated power. The device has noise emissions of 59 dB at rated power and a turn down ratio of 1:4. The efficiency was calculated to decrease to 7.7% at 25% load (225-W electrical power generation).

6
Types of Fuel Processing Reactors

While in many instances conventional fixed catalyst beds, such as steam reforming, suffer from low utilisation of the catalyst pellet, structured reactors, such as monoliths and heat-exchangers, which are coated with catalyst, are limited by their capacity for taking up the catalyst. Consequently, more active catalysts are required for structured reactors [57]. This may frequently lead to a switch in the catalyst technology from low cost materials to precious metal formulations. The higher price of these materials needs to be counterbalanced by the lower catalyst mass required.

6.1
Fixed-Bed Reactors

Adiabatic fixed-bed reactors on a smaller scale have a relatively simple design. An insulated stainless steel vessel or tube with a mesh at the reactor inlet and outlet to maintain the catalyst within the reactor is sufficient. If heat needs to be removed from or introduced into the catalyst bed, smaller tubes are usually chosen, which are then cooled or heated from outside by heat carriers or, in the high temperature range, by flames or hot combustion gases of homogeneous burners. Thus the reactor then has the design of a shell and tube heat-exchanger.

However, monolithic reactors and plate heat-exchangers are more suitable than fixed-beds for the rapid start-up and transient operation requirements of fuel processors on the smaller scale [57].

6.2
Monolithic Reactors

The most prominent applications of monolithic reactors are in the environmental field, especially automotive exhaust gas treatment. However, they are also widely used as reactors in the field of fuel processing, which will be described in Section 7.

Figure 6.1 View of ceramic monoliths as produced by Corning (photograph: © (2005) Corning incorporated. All rights reserved. Reprinted by permission).

Ceramic monoliths (see Figure 6.1) are produced mainly by extrusion techniques [130]. The cross-section of the monoliths is usually elliptic or square. Up to 1600 cells per square inch (cpsi) may be achieved by this technique [130], which corresponds to a channel width of about 500 μm. The shape of the channels is usually rectangular or hexagonal. More than 75% porosity may be achieved in ceramic monoliths [136]. The geometric surface area is 2.8 m^2 L^{-1} for 400 cells per square inch ceramic monoliths [136]. Commercially available ceramic monoliths have a thermal shock resistance of 800 °C and higher. Some of the leading producers of ceramic monoliths are Corning, Cormetech and Engelhard [136]. Further details of the production techniques of ceramic monoliths will be discussed in Section 10.2.1.

Once the low surface area monolith body has been produced, a catalyst carrier needs to be deposited onto the monolith, which may be achieved in most instances by washcoating (see Section 4.1.3). The materials of ceramic monoliths are very compatible with catalyst coatings. They do not migrate into the catalyst coatings and neither do the active species of the precious metal catalysts migrate into the monolith body [57].

The development of metallic monoliths started in the 1960s in the chemical industry, while consideration of their use for automotive exhaust gas treatment began in the 1980s [136]. Metallic monoliths have several advantages over ceramic honeycombs, namely higher mechanical stability, lower wall thickness and higher heat conductivity of the wall material. Cell densities up to 1600 cells per square inch may be realised for metallic monoliths [130]. The geometric surface area of metallic monoliths is higher, 3.7 m^2 L^{-1} are achieved for a 400-cells per square inch monolith

Figure 6.2 Metallic monolith with electrical heating developed by Emitec (photograph: courtesy of Emitec).

[136]. Owing to the lower wall thickness of metallic monoliths their porosity is higher. More than 90% may be achieved and thus the pressure drop is lower compared with ceramic monoliths of similar cell density [57].

Start-up time demand is a critical issue for automotive exhaust gas treatment and also for fuel processing (see Section 5.4.5). To decrease the start-up time demand of exhaust systems, electrically heated metallic monoliths have been developed (see Figure 6.2).

Its is obvious that ceramic and metallic monoliths have been developed for automotive exhaust gas treatment, where isothermal conditions are not the performance target [457]. Both the low void fraction of the solid wall material and the low heat conductivity of the Fecralloy of metallic monoliths and the even lower heat conductivity of ceramic monoliths favour the formation of hot spots within the monolith for exothermic reactions. An obvious way out of this dilemma is to increase the heat conductivity of the wall material, which suggests materials such as copper and aluminium. The latter has about ten times higher heat conductivity than Fecralloy. These materials, however, have operating temperature limitations. Increasing the fraction of wall material increases not only the weight and size of the reactor, but also its thermal mass, which raises the energy demand during start-up. Groppi *et al.* [457] have calculated the effective heat conductivity in the axial ($k_{e,ax}$) and radial ($k_{e,r}$) directions for cylindrical monoliths. The effective conductivities were normalised by the heat conductivity of the wall material k_s. As shown in Figure 6.3, decreasing the void fraction of the monolith from the usual value of 0.9 in commercial monoliths to 0.8, increases the heat conductivity in both dimensions by a factor of two.

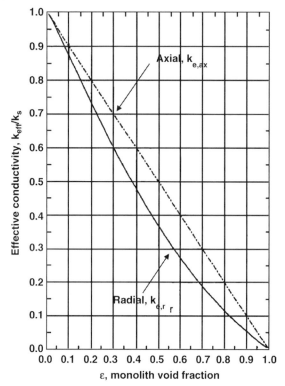

Figure 6.3 Normalised axial and radial heat conductivity $k_{e,ax}$ and $k_{e,r}$ of a cylindrical monolith versus monolith void fraction [457].

A patented metallic gauze technology known as Microlith, from Precision Combustion Inc., was presented by Castaldi et al. [458] (see Figure 6.4). These gauzes generate very high cell per square inch values of around 2500. Because the length of the gauze beds is very low (ratio of length to diameter is in the range of from 0.1 to 0.5), no

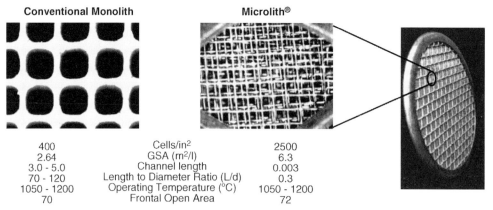

Conventional Monolith		Microlith®
400	Cells/in²	2500
2.64	GSA (m²/l)	6.3
3.0 - 5.0	Channel length	0.003
70 - 120	Length to Diameter Ratio (L/d)	0.3
1050 - 1200	Operating Temperature (°C)	1050 - 1200
70	Frontal Open Area	72

Figure 6.4 Microlith packed metal gauzes developed by Precision Combustion Inc. [458].

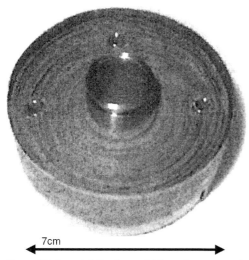

Figure 6.5 Ceramic jell-roll monolithic catalyst tapes as developed by Bae *et al.* for autothermal reforming [459].

laminar flow regime is formed in the gauzes and thus higher mass transfer coefficients may be achieved compared with conventional metallic or ceramic monoliths.

Another type of monolithic microstructured catalyst was presented by Bae *et al.* [459]. It was produced by ceramic tape-casting. Gadolinium-doped ceria containing 0.5 wt.% platinum was used as the catalyst material, which was dispersed with organic binders, dispersion agents and solvents, xylenes and alcohols [459]. This slurry was then mixed with organic binder resins. The slurry was cast at a thickness of 50–200 µm and dried in air. The resulting tapes could be cut and rolled to a jell-roll shape, as shown in Figure 6.5. The resulting monoliths showed activity for isooctane autothermal reforming.

Ceramic monolithic structures made from inverted opal silicon carbonitride (see Figure 6.6) were presented by Mitchell *et al.* [460]. The inverse opal structure was achieved using polystyrene templates. They prepared monoliths of 74% porosity, typically 350-µm wide, 100-µm high and 3-mm long. Propane steam reforming was then successfully performed in the reactor (see Section 7.1.2).

Ceramic foams (see also Section 10.2.1) are excellent catalyst supports [461]. They have a high porosity of between 40 and 85% formed from pores of 40–1500-µm diameter. Owing to the laminar flow regime, the pressure drop is low in ceramic foams. The first application of ceramic foams was catalytic combustion.

6.3
Plate Heat-Exchanger Reactors

The application of plate heat-exchanger reactors carries advantages for all types of fuel processing reactors, namely for the reformer, the water–gas shift reactor and the

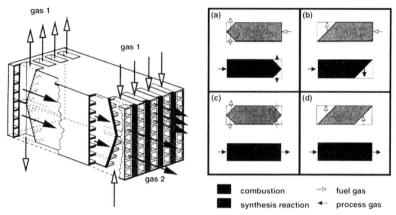

Figure 6.8 Preparation of a ceramic monolith to obtain a co- or counter-current heat-exchanger [463].

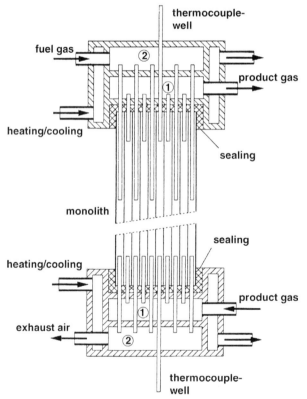

Figure 6.9 Reactor heads for the upgrading of ceramic monoliths to heat-exchangers as developed by Frauhammer et al. [463].

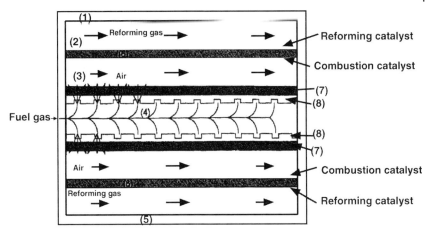

Key:
(1) upper holder
(2) reforming chamber
(3) combustion chamber
(4) fuel supply chamber
(5) lower holder
(6) heat transfer plate
(7) porous plate
(8) fuel dispersion plate

Figure 6.10 Plate reformer concept for stationary applications [5].

temperature of 1200 °C the temperature of the reactor heads did not exceed 200 °C and thus silicon was used as sealant. A 500-mm long monolith with a cross-sectional area of 26×26 mm^2 and 100 cells per square inch (cpsi, which corresponds to a channel diameter of approximately 2.5 mm) was modified with the reactor head described above. This reactor was then applied for a coupled methane combustion and steam reforming in separate flow paths [375], which will be discussed in Section 7.1.3.

Plate heat-exchangers for steam reforming on a small scale stationary size have been developed by companies such as IHI and British Gas, amongst others [5]. Figure 6.10 shows a drawing of such a plate heat-exchanger with alternating combustion and reforming channels operated in the co-current mode. The air/fuel mixture was fed stepwise along the combustion chamber to distribute the heat generation over the reactor length. Either fixed catalyst beds or catalyst coatings were used in these reactors [5].

6.3.2
Microstructured Plate Heat-Exchanger Reactors

Ramshaw was one of the first to propose coating catalysts onto the surface of microreactors [466].

Nowadays, catalytic microreactors for gas phase reactions have become established as tools for laboratory and pilot scale investigations. The first examples of microreactors becoming established in chemical production have been published, but many other applications are being kept proprietary. Microreactors are heading for widespread introduction into the distributed and into the large scale production of chemicals and pharmaceuticals. Reviews by Kolb and Hessel [467] and Kiwi-Minsker and Renken [468] have dealt with the advantages of microtechnology for gas phase reactions and their multiple applications. Books on micro reaction technology have been published by Hessel *et al.* [469–471].

7
Application of Fuel Processing Reactors

This chapter describes experimental work published in the open literature that is dedicated to specific fuel processing reactors, while entire fuel processors are presented in Chapter 9.

7.1
Reforming Reactors

7.1.1
Reforming in Fixed-Bed Reactors

The famous HotSpot fuel processor developed by Johnson Matthey was actually a fixed-bed reactor. Platinum and chromium catalyst were applied for the reactor, which was able to start from ambient conditions when methanol was used as the feedstock [5,472]. Through the initial methanol combustion, the reactor was pre-heated and then able to produce hydrogen containing reformate under autothermal conditions.

For methanol steam reforming over a copper/zinc oxide catalyst, Peters et al. used tubular reactors designed for a 50-kW fuel processor, which were heated externally by superheated steam at 65 bar to achieve the desired reaction temperature of between 260 and 280 °C [473]. A temperature drop over the catalyst bed was observed, which approached 50 K for the fresh catalyst. It originated from the limited heat transfer through the catalyst bed. Owing to the deactivation of the catalyst during 1000 h of time on stream, the temperature gradient decreased gradually. The hydrogen productivity of the first 10% of the catalyst bed also decreased considerably from 10 to 5.8 m^3 (h g$_{cat}$)$^{-1}$ after 700 h of time on stream at a S/C ratio 1.5 and 3.8 bar pressure [474]. Emonts reported the carbon monoxide concentration determined for the reactor [18]. At full conversion, S/C ratio 1.5 and 3.8 bar pressure, between 1.6 and 2.1 vol.% carbon monoxide were detected in the dry reformate at 260 and 280 °C reactor temperature, respectively.

Lindström et al. developed a fixed-bed autothermal methanol reformer designed for a 5-kW$_{el}$ fuel cell [168]. The system was started, without pre-heating, from ambient

Fuel Processing for Fuel Cells. Gunther Kolb
Copyright © 2008 WILEY-VCH Verlag GmbH & Co. KGaA, Weinheim
ISBN: 978-3-527-31581-9

temperature by methanol combustion in a start-up burner. The start-up burner was operated at a six-fold air surplus during start-up to avoid excessive temperature excursions. It took only 300 s to achieve the final operating temperature of 450 °C in the start-up burner, while the reformer reached stable operation after 400 s. During normal operation, fuel cell anode off-gas was utilised as a partial heat source for the reformer. Significant selectivity towards carbon monoxide was observed for the autothermal reforming process and therefore a water–gas shift stage became mandatory [168]. Copper/zinc oxide catalyst was used for reforming and water–gas shift, in the former case this was also doped with zirconia. The reformer was operated at a S/C ratio of 1.3 and O/C ratio of 0.3 at 260 °C temperature. More than 90% methanol conversion was achieved, while the carbon monoxide concentration downstream of the water–gas shift reactor reached 0.5 vol.% under these conditions. The reformer showed a stable performance for a period of 12 h.

Lattner and Harold performed autothermal reforming of methanol in a relatively large fixed-bed reactor carrying 380 g BASF alumina supported copper/zinc oxide catalyst, which was then modified with zirconia [174]. The catalyst particles had a cylindrical shape of 1.5-mm diameter and length. The air feed was introduced into the fixed bed through 12 porous ceramic membrane tubes of 4-mm diameter and 11.7-cm length. The methanol/steam mixture was introduced at a temperature of 220 °C, while the air was pre-heated to 300 °C. The O/C ratio was set to 0.22 while the S/C ratio varied from 0.8 to 1.5. The axial temperature profile of the reactor, which had a length of 50 cm, was rather flat, the hot spot temperature did not exceed 280 °C, which was achieved by the air distribution system. More than 95% conversion was achieved. When the air was accidentally introduced at a single point of the reactor, the temperature of the catalyst bed rose to more than 550 °C at this position. At a S/C ratio of 1.5 and O/C ratio of 0.22, very low carbon monoxide formation was observed, only 0.4 vol.% were found in the reformate. The feed temperature was 200 °C under these conditions. When the S/C ratio was decreased to 0.8, while the feed temperature increased to 220 °C, the carbon monoxide content increased to 1.7 vol.%. However, the weight hourly space velocity calculated from the data of Lattner and Harold revealed a low value of only 6 L (h g_{cat})$^{-1}$ for the highest gas hourly space velocity of 10 000 h^{-1} reported.

Mitchell *et al.* described early work by the company Arthur D. Little, with a breadboard ethanol reformer that worked with partial oxidation [475]. Ethanol containing 6 vol.% water was used as feedstock. The reformer itself worked with coaxial air and fuel injection. The air feed was pre-heated to 800 °C by an electrical pre-heater. Ethanol was also vaporised and superheated to 150 °C by electrical energy [475]. Steam was added to the feed up to a S/C ratio of 0.8. A nickel catalyst was used in the reformer. Not more than 80% conversion could be achieved in the reformer despite the high feed temperature. A 50-kW prototype of the reformer was then scaled up, which was 87 kg in weight and 72 L in volume.

An example of a larger scale fixed-bed reactor is the heat-exchange reformer developed by Haldor Topsoe [476]. It was used for fuel cell applications in the power range of from 50 to 250 kW. As shown in Figure 7.1, it is composed of a central homogeneous burner for natural gas fuel and anode off-gas and two annular

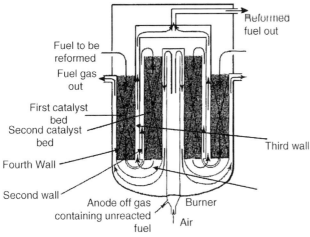

Figure 7.1 Schematic diagram of the Haldor Topsoe methane steam reformer/burner reactor [476].

concentric catalyst beds for steam reforming. The reforming catalyst is heated by the central tube and the annular flue gas mantle.

A fixed-bed reformer for catalytic partial oxidation of methane was developed by Recupero et al. [477]. It was designed for an electrical power equivalent of 5 kW$_{el}$. The argument for choosing partial oxidation was the lower system complexity compared with steam reforming. The particle size of fixed-bed catalyst was 3 mm. The reactor was operated at a low system pressure of between 1 and 3 bar. The O/C ratio ranged between 1.0 and 1.1, while the reaction temperature ranged between 800 and 900 °C. The tubular reactor had a volume of 637 cm^3, while the catalyst had to be diluted in a ratio of 1:3 with inert material to suppress hot spot formation. The actual volume of the catalyst bed was only 85 cm^3 [477]. This was commercially (CRG-F from British Gas) operated at 80 000 h^{-1} gas hourly space velocity [477]. However, the weight hourly space velocity amounted to about 125 L (h g$_{cat}$)$^{-1}$, which is a rather low value compared with catalyst coatings, which achieve up to ten times higher space velocities (see Sections 4.2.4 and 4.2.5). This probably originated from the poor utilisation of the relatively large catalyst particles due to mass transfer limitations. Around 97% methane conversion and 99% selectivity towards hydrogen and carbon monoxide could be achieved for hundreds of hours of test duration. The selectivity towards carbon deposits exceeded the theoretical yields and was attributed to the Boudouart and methane cracking reactions (see Section 3.2) [477]. A hot spot of 200 K was observed in the reactor and reached almost a 930 °C peak temperature. Recupero et al. also calculated the efficiency of a fuel processor working with partial oxidation. They drew the conclusion that it was more efficient compared with a steam reformer. However, more detailed analysis of various systems, carried out by other workers, came to the contrary conclusion (see Section 5.4.4).

Wang et al. performed autothermal reforming of liquefied petroleum gas in a fixed-bed reactor over different nickel-based catalysts [106]. The fixed-bed had a length of

45.7 cm and a diameter of 1.9 cm. The hydrocarbon feed was either pure propane or liquefied petroleum gas from a fresh (first time filled) or, alternatively, from a re-used bottle. The commercial feedstock contained propane, butane, propylene, ethane and 40 ppm ethyl mercaptane, apart from some nitrogen. Autothermal reforming was carried out at 800 °C, a S/C ratio of between 1.75 and 1.80 and O/C ratio of between 0.9 and 1.1. Almost complete conversion of the feed was achieved. However, the catalyst deactivated due to the presence of sulfur in the feedstock, which was indicated by an increasing content of methane and ethylene in the reformate. The catalyst degradation was faster in the case of the re-filled tank, because the sulfur accumulated in the tank (see Section 3.9).

Moon et al. [67] performed steam supported partial oxidation of isooctane and reformulated naphtha (boiling range from 153 to 208 °C) over a commercial naphtha reforming catalyst from ICI. The reaction was performed between 550 and 700 °C. A S/C ratio of 3.0 was chosen for most experiments and the O/C ratio was set to 1.0 according to the stoichiometry of partial oxidation. Gas hourly space velocity was about 8800 h^{-1}. When increasing the S/C ratio from 0.5 to 3.0 at a constant O/C ratio of 1.0, the methane content in the reformate could be reduced from about 10 vol.% (dry basis) at S/C 0.5 to less than 1% at an S/C 3.0 due to the thermodynamic equilibrium of methanation. The reactor showed a stable performance for more than 24 h when fed with sulfur free (<5 ppm) isooctane under standard conditions. It degraded significantly within a few hours when 100 ppm sulfur were added to the isooctane feed. The deactivation continued with increasing methane and carbon dioxide formation, while the carbon monoxide formation decreased [67]. This probably originated from the increasing selectivity towards combustion instead of partial oxidation.

Lee et al., from Samsung, reported on the development of a natural gas reformer coupled to a water–gas shift reactor with an electrical power equivalent of 1 kW [478]. The steam reformer fixed bed was placed in the centre of the system, while the annular water–gas shift fixed bed surrounded the reformer, separated by a layer of insulation. A commercial ruthenium catalyst served for steam reforming, and a copper-based catalyst for the water–gas shift. A natural gas burner supplied the steam reformer with energy. The reformer was operated between 850 and 930 °C, while the shift reactor worked between 480 and 530 °C. The carbon monoxide content of the reformate was reduced to 0.7 vol.% despite the high operating temperature of the water–gas shift stage, because the reformer was operated at a high S/C ratio of between 3 and 5. At full load the efficiency of this sub-system was 78%, it decreased to 72% at 25% load.

7.1.2
Reforming in Monolithic Reactors

Horng described the start-up behaviour of their monolithic autothermal methanol reformer [479]. The ceramic monolith was coated with a mixed platinum and copper/zinc oxide catalyst. The monolith had 117-mm diameter and 50-mm length, while the whole reactor was more than 510-mm long. Glow plugs were used for the start-up.

The reformer could be heated to 200 °C at the outlet within 220 s by pre-heating 70 L min^{-1} of air with 960-W heating power and adding 14 mL min^{-1} of liquid methanol, which corresponded to an O/C ratio of 3.75. Thus, an air excess compared with the O/C ratio 3.0 required for total methanol combustion was used for start-up. A portion of the methanol reacted immediately at the glow plugs [480]. After 220 s, the monolith had a temperature of more than 500 °C at the inlet, while the outlet was still rather cold at 80 °C. Then the feed composition was switched to autothermal conditions. The hydrogen content of the reformate reached 35 vol.% within 100 s, while it took another 100 s to reach stable conditions. The carbon monoxide content of the reformate was in the region of 5 vol.% under stationary conditions. Decreasing the heating power to 480 W under otherwise same conditions did not change the start-up time demand significantly. These experiments demonstrated that it is feasible to pre-heat a methanol reformer by catalytic methanol combustion from ambient temperature by simply injecting methanol in a slightly pre-heated air flow. The latter is necessary to avoid excessive cooling of the air feed through evaporation of the methanol.

Lyubovsky and Roychoudhury reported the development of an autothermal reformer for methanol based upon the Microlith® technology of Precision Combustion Inc. [190]. Two types of reactors were tested. The first reactor consisted of cylindrical metallic gauzes of 41-mm diameter, wash-coated with a precious metal catalyst and piled up to a total height of 6 mm. The reactor thus had a low volume of 8 cm^3 and was operated at very high flow rates (gas hourly space velocity up to 400 000 h^{-1}), which corresponded to a residence time of 3.5 ms [190]. The O/C ratio was varied from 0.05 to 0.58 and the S/C ratio between 0 and 2. Figure 7.2 shows the

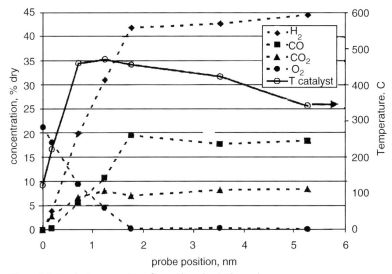

Figure 7.2 Product composition for methanol autothermal reforming as determined in a Microlith® reactor prepared by Lyubovski et al. [190]: O/C ratio 0.58; S/C ratio 2; feed inlet temperature 100 °C.

product composition obtained for this reactor. Oxygen was completely converted close to the reactor inlet, while large amounts of carbon monoxide, namely about 18 vol.% on a dry basis were formed; 100% methanol conversion was achieved under these conditions [190]. A hot spot of 470 °C was observed 1 mm downstream of the reactor inlet. Decreasing the O/C ratio to 0.44 decreased the conversion considerably to 70%.

The second reactor was prepared by winding the Microlith® gauzes onto a cylindrical coil. The feed was added to the centre of the reactor, thus generating a radial flow throughout the device, as shown in Figure 7.3. The reactor had a 5-cm diameter and length and about 100-cm³ volume. It was tested under the same feed flow rates as the first reactor, consequently the gas hourly space velocity was reduced to $30\,000\,h^{-1}$. A platinum/alumina catalyst was incorporated into this reactor. Because the methanol/air/steam feed was pre-heated to a temperature of between 100 and 200 °C, light-off occurred rapidly (see Section 4.6). The concentration of carbon monoxide decreased because the water–gas shift reaction took place at the higher residence time. However, 5 vol.% carbon monoxide was still formed at a reaction temperature of about 500 °C. The catalyst also produced methane above 450 °C reaction temperature. When the feed rate was further decreased to $3000\,h^{-1}$, the carbon monoxide concentration could be reduced to below 2 vol.% on a dry basis. Under the low residence time conditions at a S/C ratio 2 and O/C ratio 0.6, the reactor showed stable performance for 50 h [190].

These investigations document the problem of hot spot formation in monolithic reactors. The temperature of methanol autothermal reforming should not exceed 350 °C to minimise carbon monoxide formation.

Strings of brass wires with a diameter of 200–400 μm were applied by Horny et al. [481]. The strings formed a quasi-monolithic reactor, which was used for steam reforming and autothermal reforming of methanol. This design is an interesting alternative to conventional monoliths for low temperature applications. The resulting voids between the strings had a hydraulic diameter of around 100 μm. Because methanol reforming is generally carried out at relatively low reaction temperatures,

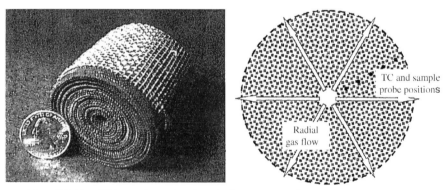

Figure 7.3 Monolithic autothermal methanol reformer prepared by Lyubovski et al. [190]; the gases are fed into the centre of the reactor and pass through it in a radial direction.

well below 350 °C, the advantages of brass, namely its high thermal conductivity [120 W (m K)$^{-1}$] and its composition (here 20–37% copper, 63–80% zinc), which is similar to the composition of a copper/zinc oxide/alumina catalyst commonly applied for methanol steam reforming, could be exploited. The catalyst was prepared by forming an aluminium/metal alloy on the outer surface of the brass. Through acid or basic leaching of the aluminium, a specific surface area in the range between 30 and 55 m^2 g^{-1} could be achieved. The samples prepared by acid leaching showed superior performance in methanol steam reforming. An alumina free sample showed no activity, which emphasized the need for alumina as the active phase. At 290 °C reaction temperature and a feed composition of 9 vol.% methanol and 11% water, which corresponded to S/C 1.2, 65% methanol conversion could be achieved at 99% hydrogen selectivity over CuZn$_{37}$ samples treated by acid leaching for 20 min. Under autothermal conditions, more than 25% methanol conversion was achieved at S/C 1.2 and O/C 0.3, while the oxygen was fully converted. Later, Horny et al. improved their catalyst by doping with chromium [482]. At S/C 1.0 and O/C 0.25, axial temperature profiles were determined over the reactor to determine the hot spot formation. The "hot spot" did not exceed 3 K due to the high heat conductivity of the brass. A fixed catalyst bed showed a hot spot of about 20 K under comparable conditions.

Catillon et al. reported the application of a copper/zinc oxide catalyst coated onto copper foams for methanol steam reforming [483]. Significant improvement of the heat transfer and consequently higher catalyst activity was achieved compared with fixed catalyst beds.

Jung et al. investigated the partial oxidation of methane on noble metal coated metallic monoliths heated by electricity [484]. Over a palladium catalyst supported by alumina, auto-ignition of the reaction took place at 270 °C and an O/C ratio of 1.1. However a high O/C ratio of 1.4 was required to achieve 98.9% methane conversion. This originated from the low temperature of the reactor at low O/C ratios, which did not exceed 700 °C at the exit under these conditions. The electrically heated monolith ignited within 90 s.

Fichtner et al. used a microchannel honeycomb reactor for partial oxidation of methane [485]. These workers presented various reasons for the application of microchannel technology for partial oxidation, namely the safe operation of the explosive mixture, a higher surface to channel volume compared with conventional ceramic monoliths and, finally, a smaller pressure drop compared with packed beds or porous solid foams. The reaction was carried out at a temperature of 1000 °C, 25-bar pressure and residence times in the order of few milliseconds. The adiabatic hot spot formation was calculated to be at 2320 °C. This excessive hot spot was expected to be reduced in the metallic honeycomb by axial heat transfer from the oxidation to the steam reforming reaction zones. Astonishingly, pure rhodium was chosen as the construction material for the reactor and it served as the active catalyst species at the same time. Rhodium has a high thermal conductivity of 120 W (m K)$^{-1}$. Channels 120-µm wide, about 130-µm deep and 5-mm long were introduced into rhodium foils of 220-µm thickness. Then 23 foils, each carrying 28 channels, were sealed by electron beam and laser welding. The stack of foils formed a honeycomb, which was pressure resistant up to 30 bar. The maximum operating temperature of the reactor was 1200 °C. The feed was mixed in a micromixer upstream of the reactor, pre-heated to 300 °C and then fed to the reactor. The experiments were carried out between ambient pressure and 25 bar at an O/C ratio of 1.0. The activity of the rhodium monolith increased within the first 20 h of the experiments and then remained constant for 200 h. At a 650 °C reaction temperature, mainly water and carbon dioxide were produced and only 10% conversion was achieved. After ignition of the reaction between 550 and 700 °C, a 1000 °C reaction temperature was then achieved within 1 min and mainly carbon monoxide and hydrogen were formed. Only 62% conversion of methane but 98% conversion of oxygen were finally achieved at 1190 °C. Selectivity towards hydrogen was 78%, while the carbon monoxide selectivity amounted to 92%. However, neither hydrogen nor carbon monoxide selectivity reached the thermodynamic equilibrium, which was attributed to the low residence times and operation in the kinetic regime. Methane conversion increased to 96% when the O/C ratio was increased to 1.3, while the carbon monoxide selectivity remained almost unchanged at 90%, but hydrogen selectivity increased slightly from 75 to 83%. When the gas hourly space velocity was increased from $2 \times 10^5 \, h^{-1}$ to $1.2 \times 10^6 \, h^{-1}$, methane conversion dropped by 10% and inferior hydrogen and carbon monoxide yields were observed. This was assumed to originate from mass transport limitations. However, this assumption could not be confirmed by experiments with several monoliths carrying channels of different widths [485]. The performance of the reactor deteriorated when the system pressure was increased. By-product and even soot formation then occurred downstream of the reactor.

Aartun et al. investigated the partial oxidation and autothermal reforming of propane in microstructured monoliths [486]. The rhodium monolith previously used by Fichtner et al., as discussed above, and a Fecralloy monolith composed of plane foils of the metal were used for the experiments. After bonding, the Fecralloy monolith was oxidised in air at 1000 °C for 4 h to form a 10-μm thick α-alumina layer (see Sections 6.2 and 10.2.1) The alumina layer was impregnated with rhodium chloride and alternatively with a nickel salt solution. The catalyst loading with nickel (30 mg) was much higher as compared with rhodium (1 mg). The content of rhodium on the monolith channel surface was determined to be 3% by XPS. Partial oxidation was carried out at an O/C ratio of around 1.25 and a 12.6-ms residence time. The Fecralloy monolith impregnated with rhodium showed the highest hydrogen yields at temperatures above 600 °C. Below this temperature, total oxidation dominated. At 12.6-ms residence time, full conversion of propane could not be achieved below 1000 °C reaction temperature, while hydrogen selectivity was still as low as 58% under these conditions. Under conditions of oxidative steam reforming, again the Fecralloy monolith impregnated with rhodium performed best. The feed composition was set to a S/C ratio of 1.0 and O/C ratio of 1.0. Full conversion could not be achieved below 1000 °C, similar to the experiments under conditions of partial oxidation. Owing to the presence of water in the feed, the hydrogen selectivity increased to 87% at this temperature. The reactor impregnated with nickel showed inferior performance. Deactivation was observed, which was assumed to originate from coking, sintering, oxidation of the nickel or even losses of volatile nickel species. With increasing temperature, enhanced formation of by-products, namely methane and ethane, was observed in the reformate both under partial oxidation conditions and in the autothermal mode, which was attributed to thermal cracking. Aartun *et al.* later reported improved performance of the reactor under conditions of partial oxidation and oxidative steam reforming [487]. Full propane conversion could be achieved for a 12.6-ms residence time and lower reaction temperature of 800 °C. However, light hydrocarbons such as ethane, ethylene, acetylene and propylene were observed as by-products, in addition to methane. Axial temperature measurements upstream of the reactor revealed the ignition of homogeneous reactions as soon as the temperature of the reactor itself exceeded 750 °C [487]. Aartun *et al.* then impregnated alumina foams directly with rhodium [488]. However, the propane conversion could be improved under conditions of partial oxidation and of oxidative steam reforming, but by-product formation could not be suppressed completely, even when the residence time was increased to 120 ms. Surprisingly, homogeneous reactions were not observed upstream of the alumina foams.

These experiments confirmed once again that pure noble metal surfaces are not suitable catalysts, even at reaction temperatures exceeding 1000 °C. Even though low surface area carrier materials such as α-alumina probably provide insufficient surface area to achieve satisfactory catalyst selectivity, fine dispersion of noble metals on carrier materials of substantial surface area improves the catalytic performance of noble metals considerably. In Section 4.2.5 results of catalyst optimisation for partial oxidation of propane performed by Pennemann *et al.* [59] were discussed, which always revealed full propane conversion over rhodium catalysts above a reaction temperature of 750 °C and short contact time, because catalysts of sufficient surface area were applied.

Rampe *et al.* developed a monolithic autothermal propane reformer with feed gas pre-heating functions (see Figure 7.4). The reactor had a thermal power output of up to 1.7 kW while the volume of the monolith was 283 cm^3 [151]. The reactor was operated at a high O/C ratio of 1.33, and S/C ratio of 1.0.

Thus, the conditions of the reformer were preferably steam supported partial oxidation operated with an air surplus (see Section 3.2). The steam prevented coke formation rather than serving as a feedstock for steam reforming. Consequently, only 14% of the steam feed was converted, however, not by steam reforming, but rather by the water–gas shift (assuming thermodynamic equilibrium of the water–gas shift) under these conditions. The hydrogen content of the reformate was only 29 vol.% on a dry basis due to the operating conditions chosen. A platinum catalyst was used, which is known to be a good oxidation catalyst (see Section 4.2.5). Consequently, even under an oxygen deficient atmosphere (O/C < 1) no substantial steam reforming activity was observed in the reactor. Up to 75% efficiency (based upon hydrogen and carbon monoxide formation) could be achieved with the reactor. The air pre-heating improved the efficiency by about 4%. Propane conversion did not exceed 90%.

Residual propane would damage catalytic carbon monoxide clean-up catalysts downstream of the reformer in a practical application.

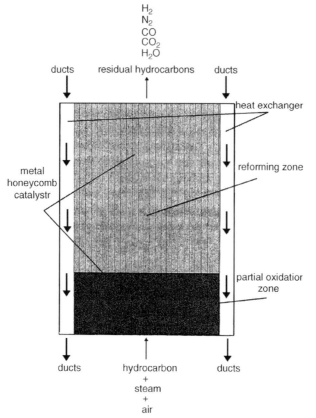

Figure 7.4 Monolithic autothermal propane reformer developed by Rampe et al. [151].

The ceramic monolithic structures made from inverted opal silicon carbonitride (see Section 6.2), which had been developed by Mitchell et al. [460], were applied for propane steam reforming. A 5 wt.% ruthenium catalyst was coated onto the monoliths. Full conversion was achieved at fairly high reactor temperatures of 900 °C. Despite the low S/C ratio of 1.33, stable operation of the catalyst and reactor at 800 °C and 60% conversion was achieved through several hours of operation and more than 15 thermal cycles without apparent coke formation.

Docter et al. [489] presented an autothermal reformer for gasoline with an electrical power equivalent of 10 kW$_{el}$. It consisted of a mixing zone with fuel, air and water injection and a metallic monolith of 0.5-L volume coated with catalyst. The monolith was electrically heated at the inlet section. The reactor was operated at a very high O/C ratio of 1, which once more corresponds to partial oxidation stoichiometry. Steam was added to the feed at a S/C ratio of 1.5. From these operating conditions a low hydrogen content of about 27 vol.% resulted, which was determined for the reformate. The reactor could be turned down by a ratio of 1:10 within 2 s while operating temperatures decreased from 800 to about 660 °C. The efficiency of the reactor was in the range of 80% at more than 2-kW power output.

A microstructured monolith for autothermal reforming of isooctane was fabricated by Kolb et al. from stainless steel metal foils, which were sealed to a monolithic stack of plates by laser welding [73]. A rhodium catalyst developed for this specific application was coated by a sol–gel technique onto the metal foils prior to the sealing procedure. The reactor carried a perforated plate in the inlet section to ensure flow equi-partition. At a weight hourly space velocity of 316 L $(h\,g_{cat})^{-1}$, S/C 3.3 and O/C 0.52 ratios, more than 99% conversion of the fuel was achieved. The temperature profile in the reactor was relatively flat. It decreased from 730 °C at the inlet section to 680 °C at the outlet. This was attributed to the higher wall thickness of the plate monolith compared with conventional metallic monolith technology. The reactor was later incorporated into a breadboard fuel processor (see Section 9.5).

Microlith packed metal gauze technology (see also Section 6.2) was applied by Castaldi et al. and Roychoudhury et al. for gasoline reforming [458]. The autothermal reformer was composed of a stack of four Microlith screens coated with either platinum catalyst supported by lanthanum-stabilised alumina carrier [458] or rhodium on a ceria/zirconia carrier [490]. Ceramic seals were put between the screens. The diameter of the screens was 40 mm, while the length of the entire stack including seals and features for sampling and temperature measurement was only between 12 and 40 mm, depending on the number of gauze elements integrated [490]. The isooctane feed rate supplied to this tiny reactor was very high with a thermal equivalent of 3.4 kW [458], which corresponded to gas hourly space velocities of between 30 000 and 120 000 h^{-1}. Isooctane was used as a surrogate for gasoline. The S/C ratio ranged between 0.5 and 2.1, while the O/C ratio was set to between 0.65 and 1.1. With a very low pre-heating temperature of 200 °C of the reactor and of the steam, air and fuel feed, light-off of the platinum catalyst was achieved at a S/C of 1.5 and O/C of 0.62. The reactor front (first screen) reached the operating temperature of 800 °C within 10 s. The reactor temperature dropped to 650 °C towards the outlet, while only 49% conversion of the fuel was achieved. When increasing the O/C ratio to 0.98 and the S/C ratio to 1.5, 94% conversion could be achieved over the platinum catalyst. Owing to the very low residence times, only 0.5 vol.% of methane was formed in the reactor under these conditions, because the methanation reaction could not reach thermodynamic equilibrium. No evidence for coke formation was observed over the reactor for a 5-h duration. When increasing the bed length to 38 mm, the fuel conversion could be increased to 97% at S/C 1.24 and O/C 0.89 for the platinum catalyst. The peak temperatures of the catalyst bed were somewhat lower for the rhodium catalyst and as much as 98% conversion could be achieved at S/C 1.53 and O/C 1.07, while only 0.2 vol.% of methane was found on a dry basis [490]. This was attributed to the better steam reforming capabilities of the rhodium catalyst compared with platinum according to these workers [490]. The O/C ratio was then increased at constant S/C ratio of 1.5 for the rhodium catalyst for a bed length of 38 mm [490]. As shown in Figure 7.5, the conversion increased with increasing O/C ratio [490]. This demonstrates that almost no steam reforming took place in the reactor due to the low residence times applied, but, rather, that partial oxidation was accompanied by some water–gas shift.

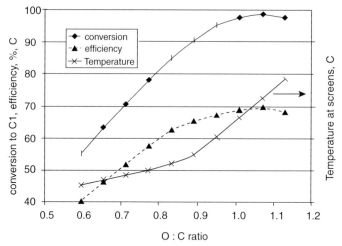

Figure 7.5 Conversion efficiency and peak temperature in the gasoline reformer developed by Roychoudhury et al.: S/C ratio 1.5; GHSV, 55 000 h^{-1} [490].

Liu et al. described the performance of an autothermal diesel reformer operated with dodecane and hexadecane [491]. It consisted of four ceramic monoliths switched in series, with 600 cells per square inch, which each had a diameter of 3.8 cm and a length of 2.54 cm. Typically, less than 0.5 vol.% of methane was detected in the reformate, while higher hydrocarbons only appeared in concentrations of a few hundred ppm. The thermal power equivalent of fuel fed to the reformer was in the range up to 5 kW. Moderate space velocities of up to 30 000 h^{-1} were set for the experiments. At a S/C ratio 2.0 and O/C ratio 0.74, the temperature in the four monoliths decreased from 800 °C (determined downstream of the first monolith) to 700 °C after the fourth monolith. The radial temperature profile did not exceed 30 K [491]. The initiation of homogenous reactions upstream of the first monolith was assumed by these workers. The reformate composition was determined downstream of the four monolith. Interestingly, the hydrogen and carbon oxide concentration increased over all the reactors except for the last, where methane formation occurred, which then consumed some of the hydrogen produced in the three other monoliths upstream. When changing the O/C ratio from 0.9 to 0.34, the carbon monoxide concentration decreased, while the carbon dioxide concentration remained close to the thermodynamic equilibrium, as shown in Figure 7.6. The hydrogen concentration in the reformate deviated from the thermodynamic equilibrium due to the increased formation of light hydrocarbons, which are not shown in Figure 7.6. The highest efficiency of the reformer was determined at a S/C ratio of 2.0 and an O/C ratio of around 0.84 [491]. Changing the S/C ratio from 1.1 to 2.8 at a constant O/C ratio of 0.68 decreased the temperature of the first reactor by not more than 40 K. However, according to the equilibrium of the water–gas shift, the carbon monoxide concentration decreased with the steam excess. Coke formation was mainly observed downstream of the reactors in the zones of lower temperature [491]. About 1% of the feed was

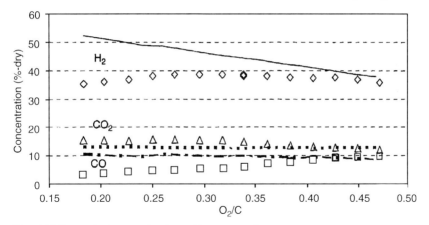

Figure 7.6 Dry gas hydrogen and carbon oxide concentrations as determined experimentally for different O/C ratios (here shown as the O_2/C ratio) for an autothermal diesel reformer; thermodynamic equilibrium compositions are provided as lines: solid line, hydrogen; dotted line, carbon dioxide; dotted–dashed line, carbon monoxide [491].

converted into coke, according to an estimation given in this work [491]. The coke formation probably originated from the Boudouart reaction (see Section 3.2).

Aicher et al. [72] developed an autothermal reformer for diesel fuel dedicated to supplying a molten carbonate fuel cell system from Ansaldo Fuel Cells S.p.A., Italy. The diesel fuel (which contained less than 10 ppm sulfur for the pilot plant application) was injected into the steam and air flows, which were pre-heated by a diesel burner to 350 °C. The reactor itself was operated at 4 bar, a S/C ratio of 1.5 and high O/C ratio of 0.98, which makes the reactor into a steam supported partial oxidation device. Consequently, the dry hydrogen content of the reformate was rather low with less than 35 vol.%. The operating temperature of the honeycomb had to be kept well above 800 °C to prevent coke formation and the presence of light hydrocarbons such as ethylene and propylene in the reformate. The reactor was operated for 300 h, which led to a slight deterioration in the catalyst performance.

Lenz et al. described the development of a 3-kW monolithic steam supported partial oxidation reactor for jet fuel, which was developed to supply a solid oxide fuel cell [492]. The prototype reactor was composed of a ceramic honeycomb monolith (400 cells per square inch) operated at temperatures between 950 °C at the reactor inlet and 700 °C at the reactor outlet [10]. The radial temperature gradient amounted to 50 K, which was attributed to inhomogeneous mixing at the reactor inlet. The feed composition corresponded to a 1.75 S/C ratio and an 1.0 O/C ratio at a 50 000 h^{-1} gas hourly space velocity. Under these conditions, about 12 vol.% of both carbon monoxide and carbon dioxide were detected in the reformate, while methane was below the detection limit. Later, Lenz and Aicher described a combination of three monolithic reactors coated with platinum/rhodium catalyst switched in series for jet fuel autothermal reforming [10]. An optimum S/C ratio of 1.5 and an optimum O/C ratio of 0.83 were determined. Under these conditions 78.5% efficiency at a 50 000 h^{-1} gas hourly space velocity were achieved. The conversion did not exceed 92.5%. In the product these workers detected about 0.5 vol.% methane, while the concentration of all higher hydrocarbons up to C_6 did not exceed 0.2 vol.% [10]. Increasing the space velocity did increase the hydrocarbons content in the reformate and decreased conversion. Interestingly, the steam formation exceeded thermodynamic equilibrium at the reactor inlet of the first reactor but approached the thermodynamic equilibrium downstream.

This indicated that total oxidation was predominant at the reactor inlet but then steam reforming consumed the water that had been produced. Heat losses of about 1.14 kW were calculated for the reactor, which underlined the need for sufficient insulation, especially at temperatures exceeding 600 °C. When the feed was switched from sulfur-free jet fuel to jet fuel containing 300 ppm sulfur, conversion dropped quickly and then deteriorated slowly with the time on stream.

7.1.3
Reforming in Plate Heat-Exchanger Reactors

An early proposal to apply coated catalyst systems for reforming applications, which is not yet heading for mobile fuel cell systems, was made by Ramshaw more than 20 years ago [466].

De Wild and Verhaak described a plate heat-exchanger for combined methanol steam reforming and methanol or fuel cell anode off-gas combustion [493], which was of the plate and fin type. It contained catalyst coated corrugated aluminium foils. The performance of the heat-exchanger was compared with aluminium foams (10 cm length, 5 cm diameter) coated with catalyst and a fixed catalyst bed filled with pellets of 3-mm diameter. A commercial copper-based catalyst was used for the steam reforming part of the reactor, while a platinum catalyst served to catalyse the combustion reaction. The plate heat-exchanger showed higher activity of the catalyst and worked almost isothermally, while along the fixed catalyst bed a temperature drop of up to 80 K was observed. However, catalyst deactivation was also higher in the heat-exchanger reactor, which was attributed to the higher space velocity that was applied [493].

Pan and Wang developed a plate–fin reformer for methanol steam reforming [382]. The system, which was also investigated by simulations (see also Section 5.1.1), was an integrated device. It contained functionalities for fuel and water evaporation, steam reforming and catalytic combustion. The overall dimensions amounted to 150 mm × 125 mm × 40 mm. Fixed catalyst beds, namely 160 g of a copper/zinc oxide catalyst on an alumina carrier and 45 g of a platinum/alumina catalyst, were used for steam reforming and catalytic combustion, respectively. The catalyst particles were ground and sieved to a size fraction of between 0.44 and 0.68 mm [382].

A significant pressure drop is to be expected for such a system in practical use compared with catalyst coatings. A cross-flow arrangement was applied, which is also questionable, because temperature mal-distributions are inevitable.

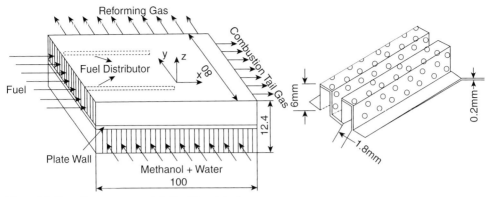

Figure 7.7 Cross-flow arrangement (left) of the plate–fin methanol steam reformer coupled with a catalytic combustor as developed by Pan and Wang [382]; the diagram on the right side shows details of the gas distribution system, which was introduced into the combustor fixed-bed.

The system was arranged on a breadboard level and pre-heated with hydrogen and air from gas cylinders. The reformer catalyst required reduction in hydrogen prior to operation. During steady operation, about 40% of the reformate was fed to the combustion chamber to simulate fuel-cell anode off-gas [494]. This would correspond to a 60% hydrogen utilisation of the fuel cell, which is a low value. To avoid the formation of significant hot spots, perforated fins were introduced into the combustor fixed-beds to distribute the hydrogen addition over the length of the catalyst bed, as shown in Figure 7.7. Nevertheless, a temperature difference of almost 50 K was observed over the length of the reformer reactor [382], which certainly prevented optimum performance of the reformer catalyst. Additionally, the poor heat transfer in the fixed catalyst beds generated a temperature difference of up to 40 K between the combustion zone and the steam reforming zone. However, the reformer reactor part of the device was operated at a low S/C ratio of 1.2. The reaction temperature never exceeded 250 °C. Owing to the extremely high residence times applied by Pan and Wang, full conversion of methanol could be achieved. Low carbon monoxide formation, which did not exceed 1 vol.%, was observed, because the reverse water–gas shift reaction was suppressed at the low reaction temperatures. The fuel processor was then scaled-up by a factor of 14 [494]; 1000 h of operation were performed with this bigger device. The methanol conversion decreased from 100% to about 93% within this time while the carbon monoxide concentration remained below 2 vol.%. The reformate recycled to the combustor part could be reduced to 30%, because heat losses played a less dominant role in the larger scale device. Pan and Wang then placed a carbon monoxide clean-up system downstream of the reformer, which will be discussed in Section 7.3.1.

Two different concepts for natural gas pre-reformers for solid oxide fuel cells were developed by Meusinger *et al.* [39]. Steam reforming was applied and thus a homogeneous burner was used to supply heat to the reforming reaction. The first concept was a stainless steel heat-exchanger coated with nickel carbonyl as the catalyst for the steam reforming reaction. However, owing to the low catalyst surface area, its activity was low and consequently high reaction temperatures of around 900 °C were required for the process. A second heat-exchanger reactor carried commercial Raschig-ring steam reforming catalyst (size 6–12 mm) between its plates. At a high S/C ratio of 3 and a 700 °C reaction temperature, up to 41% methane conversion and 73% ethane conversion could be achieved [39], which was regarded as sufficient for feeding this reformate into the solid oxide fuel cell stack [495].

Subsequently, Peters et al. continued these investigations. Their natural gas feed contained, besides methane, ethane, propane and butane, also 0.11 vol.% pentane and 0.1 vol.% higher hydrocarbons. Feed pre-heating, water evaporation and super-heating to a temperature between 350 and 400 °C was supplied with energy from the hot off-gas of a catalytic burner. The pre-reformer was then operated at temperatures between 536 and 785 °C. Around 20% methane conversion was achieved in the reactor, while the ethane conversion ranged between 40 and 50%.

Another early plate heat-exchanger prototype reformer for methane steam reforming was presented by Polman et al., from the Gastec company [496]. It was composed of three layers of corrugated stainless steel foils, one steam reforming layer in the centre and two combustion layers attached on top and at the bottom. The device had a cell density of 250 cells per square inch (cpsi), which corresponds to a channel size of about 1.6 mm. The foils were wash-coated with catalyst for steam reforming and catalytic combustion. The device was operated as a co-current heat-exchanger. The steam reformer feed was pre-heated to 380 °C and the combustor feed to 530 °C. A 99.98% conversion was achieved in the combustion section, while 97% conversion was observed for steam reforming. The reactor was operated in the temperature range between 650 and 700 °C [496].

The feasibility of the folded wall reactor concept described in Section 6.3.1 was investigated in a single combustion layer combined with two neighbouring reforming layers, by Gritsch et al. [377]. Combined methane steam reforming and methane combustion were applied as test reactions. The test reactor was fabricated of structured metal foils and the channel dimensions were at the millimetre scale. The reactor had a length of 250 mm, was 50-mm wide and the combustion and reforming channels were 2.3-mm and 1.1-mm wide, respectively. The methane burner was operated at lambda values between of 1.4 and 1.8, which corresponded to residence times of between 9 and 14 ms. The burner feed was pre-heated to 650 °C before it entered the reactor. More than 99% conversion was achieved when the methane concentration in the burner feed exceeded 5%. The distributed hot spot of the reactor was limited to 150 K, but the combustion reaction was already completed after the first 30 mm of the reactor length. Homogeneous combustion was observed for feed containing both hydrogen and methane. To achieve distributed energy generation, the feed had to be dosed at various positions into the reactor as soon as hydrogen was contained in the feedstock. Alternatively, carbon dioxide was added as an inhibitor to eliminate the homogeneous reactions.

These problems could have been removed by flow-channels in the micro-scale, because they act as flame arrestors (see Section 6.3.2).

The whole reactor length was required, however, to achieve complete conversion of methane steam reforming [377].

Numerous microreactors were developed for methanol steam reforming. Some of these reactors were still heated by electricity and carried exchangeable plates for catalyst screening [497–503]. However, these devices are bulky because space is required to carry gaskets and screws, and they are still far from any practical applications [504].

Cominos et al. developed a stack-like microreactor for testing catalysts under methanol steam reforming conditions for PEM fuel cells [498]. The stainless steel device had outer dimensions of $75 \times 45 \times 110$ mm^3 and required a stack of from 5 to 15 plates carrying microchannels of 500-μm width, 350-μm depth and of 50-mm length. The feed was distributed between the channels and plates from a common inlet region. Through numerical simulations the flow distribution was proven to deviate by less than 2% for both of the channels of one plate and for the individual plates. Laboratory-made copper/zinc oxide catalysts were prepared by introducing γ-alumina wash-coats of 10-μm average thickness as the carrier material into the microchannels. The BET surface area of the catalysts was determined to be 72 m^2 g^{-1} and the average pore diameter was 45 nm. The active components, Cu and Zn, were introduced by wet impregnation onto the catalyst carrier. Experiments were carried out at residence times of between 100 and 600 ms and total gas flow of between 500 and 900 mL min^{-1} using five coated plates. At reaction temperatures exceeding 250 °C carbon monoxide formation started. When 15 plates were introduced into the reactor, 80% methanol conversion was achieved at a 290 °C reaction temperature, a residence time of 600 ms and an S/C ratio of 2, resulting in a product containing more than 50 vol.% hydrogen and 0.25 vol.% carbon monoxide.

A combined evaporator and methanol reformer was developed by Park et al. [500] to power a 15-W fuel cell. The device was heated by electrical heating cartridges. Both the evaporator and the reformer channels, which were identical in size, were prepared 200-μm thick metal sheets by wet chemical etching. The channels were 33-mm long, 500-μm wide and 200-μm deep. Therefore, the channels were completely etched through the sheets and the channel depth could be varied by introducing several of these sheets into the reactor. The evaporator was fed by a mixture of methanol and water and operated at 120 °C. Prior to coating the channels with a commercial copper/zinc oxide catalyst supported by alumina (Synetix 33-5 from ICI), an alumina sol was coated as the interface onto the channel surface. The catalyst was reduced by 10 vol.% hydrogen in nitrogen at 280 °C prior to the experiments. Methanol conversion increased at S/C 1.1 from 55 to 90% when the reaction temperature was raised from 200 to 260 °C at 6 mL h^{-1} liquid feed flow rate. At a 260 °C reaction temperature, the carbon monoxide concentration observed was less than 2 vol.% [500].

Thus a water–gas shift reactor would have been necessary in addition to preferential oxidation to purify the reformate for a low temperature PEM fuel cell.

Pfeifer et al. used a stack-like reactor heated by cartridges for methanol steam reforming [180]. The aim of this work was a microstructured reformer supplying a 200-W fuel cell for small scale mobile applications. To highlight the dynamic behaviour of their microstructured testing reactors, three electrically heated reformers, two of them made from stainless steel and one copper reactor were fabricated by Pfeifer and coworkers [180]. Up to 62 aluminium foils, which carried 80 channels, each 64-mm long and 100-µm wide and deep, were introduced into the stainless steel reactors, and 46 into the copper reactor. The characteristics of the devices are summarised in Table 7.1. For two of the reactors, fins were introduced over the whole length of the foils to improve the heat transfer throughout the reactor perpendicular to the flow direction. A palladium/zinc oxide catalyst was introduced by wash-coating before the reactors were sealed by electron beam welding. Calcination and reduction of the catalyst was carried out after the welding procedure. The total heating power of the six heating cartridges of each reactor amounted to 1.5 kW. The feed inlet temperature was set to 140 °C. At 310 °C reaction temperature, 1.25-bar system pressure and a S/C ratio 1.9, 80% methanol conversion was achieved in the stainless steel reactors compared with more than 90% for the copper reactor. The carbon monoxide concentration in the reformate was 0.5 vol.%. Lower temperature gradients were found for the copper reactor (7 K) compared with the stainless steel devices (18 K). Dynamic measurements were performed. It took 60–70 s for the stainless steel reactors to reach 90% of the degree of conversion achieved under stationary conditions when switching from bypass to a reactor on stream.

However, much faster response times can be achieved when minimising the reactor volume further. The electrical heating made the reactors discussed above fairly bulky compared with their hydrogen power output equivalent.

Microstructured plate heat-exchangers, which require no electrical heating, or only for start-up, are more likely to be future applications in fuel processors. Some of these devices will be presented below.

Several microstructured, but no longer microscaled, reactors were developed for methanol steam reforming in the power range of a few hundred watts. Owing to the relatively low activity of the copper/zinc oxide catalyst applied, the size of the reactors was considerable.

Table 7.1 Characteristic data of three electrically heated reactors for methanol steam reforming [183].

	Type RS1	Type RS2	Type RC
Foil Material	AlMg$_3$	AlMg$_3$	Copper
Housing Material	Steel 1.4301	Steel 1.4301	Copper
Channel width/depth [µm]	100/100	100/100	100/150
Additional heat transfer fins	3 (2 mm width)	none	3 (2 mm width)
Number of Foils	62	62	46
Foil length/width [mm]	100/50	100/50	100/50
Reaction volume [cm^3]	17.3	19.8	19.3
Catalyst mass [g]	13.5	13.8	13.2
Total reactor weight [g]	1,900	1,907	2,136

Cremers et al. [505] presented a stainless steel methanol steam reforming reactor, which was supplied with heat by heating oil in a co-current mode. The reactor was designed to supply a fuel cell with an electrical power output of 500 W. Over a commercial copper/zinc oxide catalyst from Süd-Chemie, complete methanol conversion was achieved at 275 °C reaction temperature but at a very low weight hourly space velocity of 7.8 L/(h g_{cat})$^{-1}$. Because catalyst activity was too low compared with the design criterion, a micro fixed-bed reactor was built with integrated heat exchanging capabilities. The steam reforming passages of the reactor formed a fixed reactor bed to incorporate 15.9 g of catalyst; 62 plates served for the heating, while 60 plates carried the steam reforming catalyst. Surprisingly, the heat supply of the reactor could not be achieved by methanol combustion within the reactor, because the catalyst required an even higher operating temperature than the steam reforming process [505]. Therefore, a separate methanol burner was built and an evaporator was switched between the two devices. The power supply of the steam reformer was transferred by a heat transfer fluid (oil) to the evaporator and the reformer. During the first 4 h of operation, 15% of the initial activity of the catalyst was lost, but then the activity remained stable for another 4 h. The microstructured reactor had a start-up time demand of 18 s after being heated to the operating temperature, which was considered as an improvement to conventional fixed-bed technology. Load changes were performed in ratios up to 1:5 without significant changes to the product composition.

Reuse et al. were among the first who successfully combined endothermic methanol steam reforming with the exothermic methanol combustion [499]. Their reactor consisted of a stack of 40 foils, 20 dedicated to each of the two reactions (see Figure 7.8). The foils were 78-mm long and 200-μm thick. The foils carried 34 "S"-shaped channels, each being 30-mm long, 100-μm deep and 310-μm wide. The plates were made of Fecralloy and an α-alumina film of 5-μm thickness was generated on their surface by temperature treatment at 1000 °C for 5 h to improve the adherence of the catalyst coatings. Then 206 mg of commercial copper/zinc oxide catalyst from Süd-Chemie (G-66MR) was coated onto the channels at a 5-μm thickness for steam reforming; 434 mg of cobalt oxide catalyst were impregnated onto the α-alumina for methanol combustion in the second set of channels. The reactor was operated in the co-current mode. Steam reforming was performed at a S/C ratio of 1.2. Under stationary conditions, with a reaction temperature of between 250 and 260 °C, more than 95% conversion and more than 95% carbon dioxide selectivity was achieved at 25 L (h g_{cat})$^{-1}$ weight hourly space velocity, which is within the usual range for copper/zinc oxide catalysts.

Figure 7.9 shows a methanol steam reformer based on plate heat-exchanger technology designed for a 100-W_{el} fuel cell, which was developed at IMM (Institute of Microtechnology Mainz) with project partners, within the scope of the European project Mirth-e.

Park et al. [507] increased the size of their methanol steam reformer described above to an electric power equivalent of 28 W and combined steam reforming with catalytic combustion. The reactor was sealed by brazing. While the same steam reforming catalyst as described above was coated onto etched channels of 200-μm depth and 300-μm width, the catalytic combustor was a small scale fixed-bed of platinum/alumina catalyst spheres

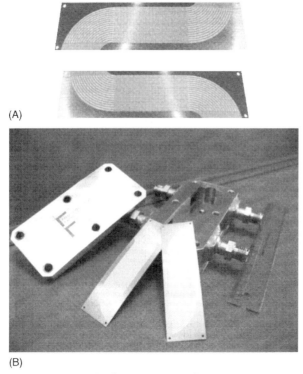

Figure 7.8 Integrated reformer/combustor for methanol steam reforming (A) "S" shaped metal foils; (B) reactor housing [499].

[507]. The reactor had dimensions of 70 mm × 40 mm × 20 mm [508]. More than 99% conversion could be achieved at 240 °C reaction temperature.

Early work from Tonkovich *et al.* dealt with a heat-exchanger/reactor containing catalyst powder for the partial oxidation of methane [509]. The intention was to run the

Figure 7.9 100-W_{el} methanol steam reformer/catalytic afterburner [506].

reaction without the usual hot spot formation known from adiabatic reactors in a microstructured reactor because of the short residence times and integrated cooling by air; 5 wt.% rhodium catalyst on mesoporous silica was used in a fixed catalyst bed put into the microchannels. The reactor had a total size of 70 mm × 38 mm × 4 mm and consisted of 9 plates each carrying 37 channels, 35-mm long, 254-μm wide and 1.5-mm deep. The significant channel depth was required to take up the catalyst powder. Methane (6 L h^{-1}) and oxygen (3 L h^{-1}) were fed into the reactor at a pressure drop of less than 2.5 mbar. The residence time was 50 ms. No more than 60% conversion was achieved at a 700 °C reaction temperature and no carbon formation was observed.

Cremers *et al.* [510] developed a combined microstructured methane steam reformer/anode off-gas burner (see Figure 7.10). NiCroFer 3220HT® Fecralloy was applied as the construction material. The reformer was designed to power a 500-W$_{el}$ fuel cell. The reactor was composed of 200 foils, which were 300-μm thick and carried channels 320-μm wide and only 110-μm deep. The foils were 79-mm long and 23-mm wide [511]. Steam reforming was performed over a nickel/nickel spinel catalyst described in Section 4.2.4. The catalyst was coated onto the foils before the foil stack was sealed by diffusion bonding. The platinum combustion catalyst was introduced into the reactor after diffusion bonding [511], because it could not withstand the elevated temperature of the bonding procedure (see Section 10.2.2). The combustion catalyst was impregnated onto the α-alumina layer on the Fecralloy surface (see Section 10.2.1). The catalysts were reduced prior to testing, which does not seem very viable for a later practical application. The anode off-gas was mixed upstream of the reactor by laboratory equipment and then further mixed with air in a separate micro-mixer. The steam reforming side of the reactor was operated at an

Figure 7.10 Combined methane reformer/combustor designed for 500-W electrical power output [512].

extremely high S/C ratio of 6.0 in most instances, which seems to be too high for a practical application. When the reactor was operated in a counter-current mode, methane conversion did not exceed 70%, while 90% conversion were achieved with co-current operation, which is in line with results from simulation work of numerous workers, as discussed in Section 5.1.1. At 90% methane conversion the hydrogen product stream was sufficient to power a 50-W fuel cell. When the S/C ratio was decreased to 2.5 in the short term, the dry gas composition of the reformate did not change significantly.

Ryi et al. presented another combined microchannel methane steam reformer/hydrogen combustor [513]. This was different to the reactor presented by Cremers et al., which was discussed above, as it had a separate inlet for hydrogen (which acted as a surrogate for future anode off-gas) and the mixing was performed internally by adding the hydrogen through a small hole. The reactor, which had no electrical heating, was heated from ambient to the operating temperature of 700 °C by hydrogen combustion. However, it took 2.5 h to reach this temperature. While a platinum/tin catalyst was used for hydrogen oxidation, a rhodium/alumina catalyst stabilised by magnesia served for the methane steam reforming, which was performed at a S/C of 3.0. At 700 °C reforming temperature, about 95% methane conversion was achieved. The thermal power output of the hydrogen product was approximately 67 W [513]. However, the hydrogen feed required to keep the small reactor on temperature by the combustion reaction had a much higher power equivalent of about 170 W. This emphasises the need for efficient insulation of such small scale devices and the problem of heat losses of small scale reactors in general, especially at the high operating temperature of hydrocarbon reforming. Next, Ryi et al. made some modifications to their reactor [514]. They added microstructured heat-exchangers for pre-heating the air fed to the combustor and evaporating the water and pre-heating the steam and methane feed of the reformer part of the reactor. Clearly a mixing problem had existed previously in the combustion side of the first reactor, which was probably due to the local addition of hydrogen through the holes. The laminar flow regime in the microchannels prevented mixing within the channel system. This was possibly the reason why Ryi et al. inserted Fecralloy mesh into the outlet conduit of the combustor side of the reactor. The mesh was coated with platinum/zirconia catalyst. The combustion reaction increased the gas outlet temperature downstream of the mesh to 800 °C within 1 min from ambient [514]. Unfortunately, the reactor core temperature was not provided by Ryi et al. These modifications made it possible to increase the thermal power output of the hydrogen present in the reformate to 220 W. However, the hydrogen feed, which was required to keep the reactor in operation was still higher at about 310 W.

An integrated microstructured 5-kW combined methanol steam reformer/catalytic combustor was fabricated and the results were presented by Hermann et al., from GM/OPEL [515]. The specifications for the reactor and a future 50-kW fuel processor system were fairly ambitious. Amongst others:

- volumetric power density >5 kW L^{-1}
- gravimetric power density <2.5 kW kg^{-1}
- transient response to load changes from 10 to 90% in milliseconds.

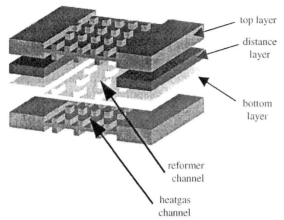

Figure 7.11 Layer assembly of a 5-kW micro-structured methanol reformer developed by GM/Opel [515].

The reactor was composed of three types of plates forming a stack. Instead of microchannels, a fin-like structure was chosen (see Figure 7.11). The fins served as a mechanical support and improved the heat transfer. A total of 225 plates formed the reactor. Plates coated with steam reforming catalyst alternated with plates for heating, because the prototype device was heated externally using hot air. Distance plates were added to increase the height of the steam reforming channels. The coating density of the reaction walls was 250 $g_{cat}\,m^{-2}$ which corresponded to a very large thickness of 200 µm. The theoretical power output of the device was 4.97 kW with a corresponding power density of 5 kW L^{-1}. The reactor was designed for 4-bar maximum operating pressure and a 350 °C maximum reaction temperature. At a 30 $m^3\,h^{-1}$ nitrogen flow rate the pressure drop of the reactor was 600 mbar. Results were determined for partial load of the device (1–2 kW for the LHV of the hydrogen produced). At 350 °C burner off-gas (heating gas) inlet temperature, a S/C ratio of 1.5- and 3-bar pressure full conversion of the methanol was achieved and 0.9 $m^3\,h^{-1}$ hydrogen was produced.

Subsequently, a meso-scaled combined reformer/catalytic combustor with 10-kW power output was realised by GM/OPEL, which was not presented in detail. For this bigger reactor, the carbon monoxide content of the reformate increased, as expected, with increasing reformer outlet temperature from 0.5% at 250 °C to 2% at 300 °C. Increasing the residence time increased the carbon monoxide concentration of the reformate due to the reverse water–gas shift reaction. Increasing the S/C ratio from 1.2 to 1.8 at a 300 °C reaction temperature increased the hydrogen concentration in the reformate slightly from 72 to 73% and decreased the carbon monoxide content from 1.5 to 1.0%, which originated from the beneficial effect of steam addition on the equilibrium of the water–gas shift reaction.

Whyatt *et al.* [516] presented early results generated with a combined system of independent and microstructured evaporators, heat-exchangers and reformers for isooctane steam reforming for a fuel cell of 13.7-kW electrical power output. Four integrated reformers/cross-flow heat-exchangers switched in series were fed by four independent evaporators. The reformer and evaporators were supplied with energy by anode-off gas combustion in an independent burner. Water pre-heating, fuel evaporation and superheating was performed in a second group of four units, which were heated by the hot reformate. The micro-structured components were fabricated by photochemical etching and diffusion bonding. The reformers achieved up to 98.6% conversion at a 750 °C reaction temperature and a S/C ratio of 3. The product composition was 70.6 vol.% hydrogen, 14.6 vol.% carbon monoxide, 13.7 vol.% carbon dioxide and 0.9% methane. Modulation to doubled feed rate was feasible within 20 s.

Then Whyatt et al. [517] designed two reformer reactors with total volumes of 107 and 68 cm^3, made from Inconel, for reforming of methane, propane, butane, ethanol, methanol, isooctane and a mixture of hydrocarbons, the last as a surrogate of sulfur-free gasoline. The bigger reactor is shown in Figure 7.12. The two reactors were tested in a test-bench in combination with microstructured heat-exchangers for feed pre-heating by hot reformate and evaporators supplied by hot process gases. The experiments were carried out at ambient pressure. All hydrocarbons were tested at a S/C ratio of 3 and 725 °C reaction temperature.

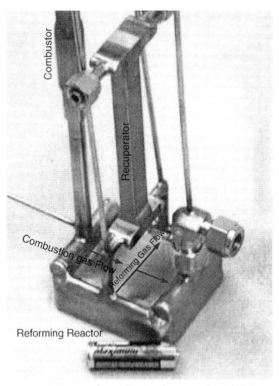

Figure 7.12 Steam reformer/heat-exchanger with catalytic burner and additional heat-exchanger as developed by Whyatt et al. [517].

Table 7.2 Conversion, power density, reaction temperature and reformate composition for various fuels generated with a microstructured steam reformer, by Whyatt et al. [517].

Fuel	Methane	Propane	Butane	Methanol	Ethanol	Iso-octane	Sulphur-free Gasoline
C_1 Feed-rate [10^{-4} mol/s]	8.13	8.34	11.62	8.07	10.4	7.51	10.8
Conversion [%] to C_1	95	99.8	99	100	98.9	100	99.6
Power density [kW/dm³]	2.26	4.06	5.98	1.69	2.25	1.70	2.27
Reaction Temperature [°C]	725	775	751	725	724	725	725
Dry Reformate Composition							
H_2 [%]	75.7	72.3	71.6	70.4	70.0	70.9	70.1
CO [%]	12.7	14.8	14.6	14.1	13.7	14.7	14.9
CO_2 [%]	10.4	12.0	12.9	14.9	15.2	13.3	14.1
CH_4 [%]	2.0	0.7	0.8	0.6	0.9	1.1	0.9

Table 7.2 summarises some of the results generated for various fuels. Load changes of the liquid feed flow rate from 100 to 10% changed the reformate flow rate within 5 to 10 s.

To decrease the start-up time and the electrical power demand required for the air supply system, Whyatt et al. [518] re-designed their system completely. The target was to meet the US Department of Energy ambient temperature start-up time demand targets, which amounted to <1 min by 2005 and <30 s by 2010. Figure 7.13 shows the flow schematics of the device, the prototype is shown in Figure 7.14. The energy for start-up and continuous operation was provided by homogeneous fuel combustion

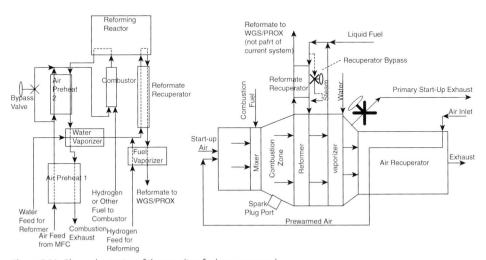

Figure 7.13 Flow schematics of the gasoline fuel processor sub-system consisting of steam reformer, evaporator and additional heat-exchangers, as developed by Whyatt et al. [518].

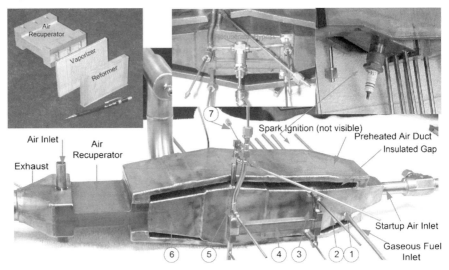

Figure 7.14 Prototype microstructured gasoline fuel processor sub-system, as developed by Whyatt et al. [518].

rather than catalytic combustion, in order to decrease the time demand of start-up. The combustion was ignited by a spark-plug. Hydrogen was used as the energy source with the option of combusting atomised gasoline later on. The combustion gases heated both the reformer and the evaporator in cross-flow, placed behind the reformer in a duct. The pressure drop of the heating gas passing through the reformer and evaporator was less than 3 and 0.2 mbar, respectively. The combustion gases finally pre-heated the combustion air in a low-temperature heat-exchanger, which was placed further downstream. It was bypassed during start-up to further minimise the combustion air supply pressure drop. To compensate for the lack of air pre-heating during start-up, it was expected to burn an increased amount of fuel. As soon as the reformer and evaporator had reached an adequate temperature, water was evaporated by the heat of the combustion gases. Fuel evaporation was achieved by injection of liquid fuel into the superheated steam, which reduced the start-up time demand further. The fuel/steam mixture was then super-heated by hot reformate in a small separate heat-exchanger and finally fed into the reformer. The steam reformer was designed to operate at an outlet temperature of 650 °C and a S/C ratio of 3. It was composed of reforming reaction channels of low height, which were as wide as the whole device. The combustion gases passed around these single channels in a cross-flow, as described above. The overall pressure drop on the reformate side over the reformer and evaporator amounted to 56 mbar. It was expected to produce a higher amount of steam (S/C ratio between 18 and 24) during start-up, to heat the gas purification units downstream. However, they had not been connected to the system in the work presented. The very high S/C ratio during start-up would have released the load from the water–gas shift reactor, which would have been below its design operating temperature at that point in time.

Rapid cold-start testing of the device was carried out. The duration of the whole test was as short as 60 s. After 5 s of heating and dosing water, the fuel pump was switched on and after 15 s the excess water flow was reduced to the level of stationary operation. At this point in time, the temperature of the reformer feed amounted to about 110 °C and the reformate left the reactor at a temperature of 220 °C. The full reformate flow was achieved after 30 s and the temperature of the reformate was then 500 °C. After 60 s the steam left the evaporator at a temperature of 400 °C and the reformate outlet temperature was already 700 °C. The total mass of a future 50 kW fuel processor was estimated to be 55 kg, which corresponded to a total energy demand of 7 MJ for start-up heating. From this value, the power demand of an air blower, which had to provide some 22 m^3 min^{-1} air to the system during start-up, was calculated to about 1 kW. As this power would have been required only during the rapid system start-up (60 s) a normal battery would have been sufficient as the power supply. An efficiency of 78% was calculated for the entire future fuel processor.

Figure 7.15 shows a microstructured coupled diesel steam reformer/catalytic afterburner developed by Kolb et al. [312], which carried dedicated inlets for both anode off-gas and cathode air for burning the residual hydrogen from the fuel cell anode at temperatures exceeding 800 °C. The reactor, into which catalyst from Johnson Matthey Fuel Cells was coated, had two inlets for anode off-gas, two inlets for the air supply to the burner, one inlet for the diesel/steam supply and two outlets for the combustion gases, apart from the outlet for the reformate. Through this arrangement, the feed gases were mixed after entering the microstructured device, which prevented them from uncontrolled reaction upstream of the reactor. Full conversion of the diesel fuel was achieved for a total operation time of 40 h with this reactor, which had a power equivalent of 2 kW of thermal energy of the hydrogen produced.

Figure 7.15 Combined diesel steam reformer/anode off-gas burner designed for 2000-W electrical power output [312].

7.1.4
Reforming in Membrane Reactors

Membrane reactors combine chemical conversion with a membrane separation step (see Section 5.2.4). Within the field of fuel processing membranes, reactors are usually either combined with the reforming step or with a water–gas shift, as described in Section 7.2.3.

In particular, with methane steam reforming the application of membranes allows the equilibrium of the reaction (see Section 3.1) to shift in favourable directions. Thus, the membrane reformer could be operated at a lower temperature.

Early work by Uemyia *et al.* [519] dealt with methane steam reforming in an annular nickel catalyst bed placed around a porous glass support, which had been coated with palladium by an electroless plating technique (see Section 10.2.5). As alternative membrane material, porous Vycor glass, was used which separated the gases by diffusion limitations according to the Knudsen law. The membranes were 20-µm thick. The reactor was operated at a reaction temperature of between 350 and 500 °C at a S/C ratio of 3.0 and 3-bar pressure on the reaction side. Methane conversion of 90% was achieved at 500 °C reaction temperature with the palladium membrane. With the porous glass membrane, conversion did not exceed the thermodynamic equilibrium conversion of 45%. The porous membrane had a much lower permeation rate for hydrogen, which explained its inferior performance. Increasing the flow rate of the argon sweep gas increased the hydrogen partial pressure difference from the reaction to the permeate side and hence increased methane conversion. The methane conversion increased with increasing reaction pressure, a phenomenon that is discussed in detail for methane steam reforming membrane reactors in Section 5.2.4.

Joergensen *et al.* described the steam reforming of methane over a nickel/magnesia catalyst in a membrane reactor at a S/C of 2.9, temperatures between 350 and 500 °C and pressures between 6 and 10 bar [410]. The weight hourly space velocity was very low at 0.5 L $(h\, g_{cat})^{-1}$. The catalyst was put into a stainless steel tube. A palladium/silver tube of 100-µm wall thickness was positioned in the centre of this catalyst bed. Nitrogen was used as the sweep gas for the permeate. Because the product hydrogen was removed through the membrane over the length of the catalyst bed, the equilibrium conversion of methane steam reforming could be exceeded. At 500 °C and 10-bar pressure, 61% methane conversion could be achieved, which corresponded to the equilibrium conversion calculated for 707 °C.

A similar experimental arrangement to that applied by Joergensen *et al.* was used by Kikuchi *et al.* to compare the performance of palladium membranes prepared by electroless plating with platinum, palladium and ruthenium membranes prepared by chemical vapour deposition [520]. The membranes were deposited onto commercially available alumina tubes. While the palladium membranes created by the plating technique appeared as an 8-µm thick dense layer on the alumina surface, the noble metals were deposited preferably inside the carrier pores by chemical vapour deposition. While the hydrogen/nitrogen selectivity was infinity for the plated membrane, it was about 280 for the platinum membrane created by chemical vapour deposition. Much lower values were determined for the ruthenium and palladium membranes. The permeance was similar for all membranes and in the range between 3.0×10^{-6} and 3.5×10^{-6} mol $(m^2 \, s \, Pa)^{-1}$. Kikuchi et al proposed that permeation through membranes prepared by chemical vapour deposition is not based on solution and diffusion, as it is for dense membranes, but rather on surface diffusion. The membranes degraded through sintering mechanisms.

A commercial nickel catalyst was used for methane steam reforming performed at a 500 °C reaction temperature, a S/C ratio of 3.0 and atmospheric pressure, while the permeate side was evacuated. The performance of the vapour deposited platinum membrane was similar to the plated dense palladium membrane. In the permeate of the deposited ruthenium and palladium membranes, small amounts of carbon oxides and also methane were observed. While it was expected that all these species had passed through the membranes by diffusion, in addition some methane was converted into carbon dioxide over the noble metals of the membranes. Kikuchi et al. demonstrated by simulations that conversion and hydrogen permeation in a membrane reactor is higher, where the first portion of the catalyst bed is not coupled to the membrane. Such an arrangement as shown in Figure 7.16 would clearly save expensive membrane surface area. Experimental work by Itoh et al. performed for methanol steam reforming [521] confirmed the assumptions of Kikuchi et al.

Tong and Matsumura highlighted the important effect of catalyst activity on the performance of membrane reactors for methane steam reforming [522]. They

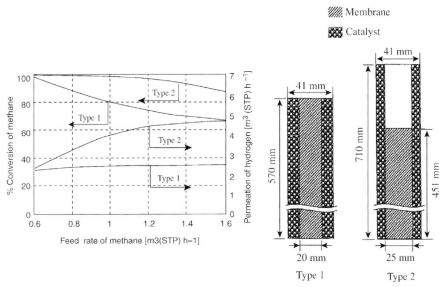

Figure 7.16 Methane conversion and hydrogen permeation through two membrane reactor set-ups. Type 1: catalyst bed packed in the same length as the membrane. Type 2: catalyst bed in front of and along the membrane [520].

prepared an 11-μm thick membrane by electroless plating onto a porous stainless steel tube. Nickel catalysts were used for methane steam reforming, one was highly active and contained 47 wt.% nickel and the other less active containing only 10 wt.% nickel. Methane steam reforming was performed at 723 °C and a S/C ratio 3.0. The more active catalyst generated higher methane conversion and consequently higher hydrogen flux, regardless of the space velocity and reaction pressure, while the carbon monoxide selectivity was comparable. Increasing the sweep gas flow for the catalyst with high activity had less effect on the hydrogen flux than increasing the feed pressure.

Barbieri et al. [523] prepared palladium membranes by solvated metal atom deposition, a deposition method that created a 0.1-μm thick film of palladium on an alumina tube. Another membrane was prepared from a palladium/silver alloy (21 wt.% silver), which had a thickness of 10 μm. Both membranes suffered from pinholes and cracks and thus did not show infinite hydrogen selectivity. However, conversion exceeding the thermodynamic equilibrium could still be achieved for methane steam reforming at temperatures exceeding 400 °C.

Kikuchi described a natural gas membrane reactor, which had been developed and operated on a larger scale by Tokyo Gas and Mitsubishi Heavy Industries supplying PEM fuel cells with hydrogen [524]. It was composed of a central burner surrounded by a catalyst bed filled with a commercial nickel catalyst. Into the catalyst bed 24 supported palladium membrane tubes were inserted. The membranes had been prepared by electroless plating and were 20-μm thick. Steam was used as a sweep gas for the permeate. The reactor carried 14.5 kg of catalyst. It was operated at 6.2-bar pressure, S/C ratio 2.4 and a 550 °C reaction temperature. The conversion of the natural gas was close to 100%, while the equilibrium conversion was only 30% under the operating conditions used. The retenate composition was 6 vol.% hydrogen, 1 vol.% carbon monoxide, 91 vol.% carbon dioxide and 2 vol.% methane.

Besides methane, methanol is the fuel most frequently investigated for membrane reformers.

Wieland et al. combined different membranes with a fixed catalyst bed for methanol steam reforming in a plate-type reactor [403]. Figure 7.17 shows the conversion determined in the membrane reactor with different membranes. It is compared with the thermodynamic equilibrium conversion and the conversion achieved in a fixed catalyst bed without a membrane. The conversion was higher for the membrane reactors and exceeded the thermodynamic equilibrium of the feed at more than 20-bar pressure. The space velocity, at which the fixed catalyst bed was operated was fairly low. Wieland et al. performed numerical calculations to highlight the effect of lower membrane thickness on the performance of their reactor. Figure 7.18 shows that a reduction in the membrane thickness increases the range of full methanol conversion to higher space velocities.

Basile et al. performed steam reforming and autothermal reforming of methanol in a fixed-bed of copper/zinc oxide catalyst, which was positioned below a palladium/silver membrane containing 20 wt.% silver which had a thickness of 60 μm [525]. The pressure on the reaction side was 1.2 bar, while the permeate side was under

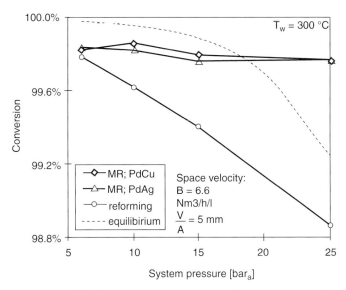

Figure 7.17 Methanol conversion versus system pressure for methanol steam reforming in membrane reactors (MR) with two different palladium membranes: PdCu was a 25-μm thick palladium membrane containing 40 wt.% copper; PdAg was a 25-μm thick palladium membrane containing 25 wt.% silver; "reforming" corresponds to the conversion achieved with a fixed copper/zinc oxide catalyst bed without membrane; the dashed line is the equilibrium conversion; space velocity 6.6 N m^3 (h L$_{cat}$)$^{-1}$; operating temperature 300 °C [403].

atmospheric pressure. However, the reactions were performed at a very low temperature of 200 °C and consequently only 50% conversion could be achieved at S/C 1.3 despite the extremely low weight hourly space velocity of 1.2 L (h g$_{cat}$)$^{-1}$.

Lin *et al.* found much higher methanol conversion over a copper-based commercial catalyst for methanol steam reforming in a membrane reactor compared with a conventional fixed-bed [416]. However, at least 25 vol.% hydrogen were still present in the retenate. Then Lin *et al.* designed a tubular membrane reactor for methanol steam reforming with a central palladium membrane tube, an annular catalyst bed for methanol steam reforming carrying copper-based catalyst that surrounded the membrane and a second concentric annular bed containing palladium/alumina catalyst for combustion of the retenate. This arrangement was similar to the fuel processor design described by Edlund *et al.* from Idatec, which was discussed in Section 5.2.4. Up to 85% of the hydrogen could be recovered from the reformate. An heat balance was calculated by Lin *et al.* It revealed that 92% hydrogen recovery by the membrane separation was possible leaving sufficient hydrogen in the retenate to supply the steam reforming reaction with energy by combustion of this residual hydrogen. This value was reduced to 74% when the energy required for evaporation and pre-heating of the feedstock was included in the calculations. However, neither heat-exchangers were included in these calculations nor heat losses to the environment.

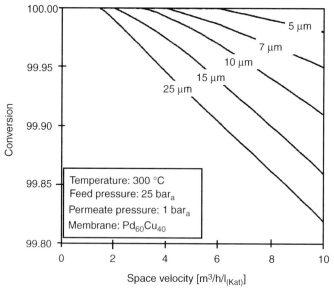

Figure 7.18 Calculated methanol conversion versus space velocity for methanol steam reforming in a membrane reactor with a palladium membrane containing 40 wt.% copper of different thickness; feed pressure 25 bar, operating temperature 300 °C [403].

A membrane separation device was prepared by Wilhite et al. by micro-electromechanical techniques [526]. The palladium/silver membrane deposited onto a silicon oxide support was only 20-nm thick, which was possibly the lowest membrane thickness ever reported for hydrogen separative purposes. A lanthanum/nickel/cobalt oxide catalyst ($LaNi_{0.95}Co_{0.05}O_3$) catalyst for partial oxidation of methanol was deposited onto the membrane. At a O/C ratio of 0.86 and 475 °C reaction temperature, up to 64% methanol conversion and more than 90% hydrogen selectivity could be achieved. These workers claimed a deviation from Sievert's law (see Section 5.2.4) for their membrane. Namely, the hydrogen flux did not depend by a power of 0.5 of the retenate and permeate pressure but rather by a power of 0.97, which they attributed to the absence of internal solid-state diffusion limitations in their ultra-thin membrane.

Ferreira-Aparicio et al. reported the development of a laboratory-scale membrane reactor for the partial dehydrogenation of methylcyclohexane into toluene in a membrane reactor [527]. A platinum/alumina catalyst containing 0.83 wt.% platinum was put into a porous stainless steel tube, which had been prior coated with a palladium membrane by electroless plating. At 350 °C reaction temperature and a pressure of 1.4 bar at the reaction side, 99% of the hydrogen product could be separated through the membrane, which had a thickness of 11 μm. However, the sweep stream required on the permeate side was more than 20 times higher than the hydrogen permeate flow rate that could be achieved.

Ceramic membranes are low-cost alternatives to palladium membranes. Typically the catalyst applied for the reforming reaction is then impregnated onto the membrane.

Some examples of the application of ceramic membranes for fuel processing applications will be discussed below.

Kurungot *et al.* developed a silica membrane and incorporated it into a catalytic membrane reactor for the partial oxidation of methane [528]. Rhodium catalyst on a γ-alumina carrier containing 1 wt.% rhodium was coated with a 9-μm thickness onto an α-alumina tube. Then the silica membrane, which was only 1.5-μm thick, was coated onto the catalyst. Hydrogen and methane permeation through the membrane were measured revealing increased permeability for hydrogen with increasing temperature, while methane permeability remained at a lower level. Between 100 and 525 °C, the separation factor (see Section 5.2.4) increased from 7.5 to 31. The hydrothermal stability of the membrane was verified at 525 °C for 8 h with feed composed of 18 vol.% hydrogen, 18 vol.% methane and 74 vol.% steam. The separation factor decreased from 31 to 26 within the course of the experiment. Then partial oxidation of methane was carried out at an O/C ratio of 1.0 and various S/C ratios. The reaction was performed under atmospheric pressure and nitrogen sweep gas flows on the permeate side. The results obtained with the membrane were compared with the performance without catalytic membrane in the temperature range between 400 and 575 °C. At 525 °C reaction temperature, S/C 3.5 and a weight hourly space velocity of 240 L (h g_{cat})$^{-1}$, equilibrium conversion could be exceeded by 37% due to the hydrogen removal from the reformate. Lowering the S/C ratio from 3.5 to 2.5 under these conditions decreased the conversion considerably from 58.7 to 34.2%. At a higher weight hourly space velocity, carbon formation was suspected, because the membrane colour changed from yellow to grey.

A different type of membrane reactor for the partial oxidation of methane was presented by Ikeguchi *et al.* [529]. While air was fed to the membrane on the retenate side, methane was fed to the permeate side. In this way oxygen ions permeated through the membrane and reacted with methane on the permeate side over a 1 wt.% rhodium catalyst supported on magnesia. The membrane was about 800-μm thick. The oxygen flux through the membrane was about 5.5 L oxygen through 100 cm^2 membrane area per hour in the absence of any methane feed on the permeate side. This value increased to 5.2 L h^{-1} when 5 vol.% of methane was fed to the permeate side. At 900 °C reaction temperature, 30–40% of the methane feed could be converted with high selectivity towards carbon monoxide and hydrogen. The catalyst activity suffered from carbon formation.

Kusakabe *et al.* prepared yttria-stabilised zirconia membranes on α-alumina support tubes by a sol–gel procedure [530]. This material was investigated as an alternative to silica, which is also highly selective for hydrogen permeation, but degraded in the presence of steam. The membrane was then impregnated with platinum and rhodium as the catalyst for methane steam reforming. The membrane was 1-μm thick, while the platinum and rhodium loadings were 12 and 9 wt.%, respectively. Figure 7.19 shows the permeance of the membrane for hydrogen, nitrogen and carbon dioxide with increasing temperature. At 500 °C the permeance for hydrogen was as high as 10^{-6} mol (m^2 s Pa)$^{-1}$. Higher conversion could be achieved in the membrane reactor compared wit a fixed-bed reactor, but the permeate also contained methane and carbon oxides.

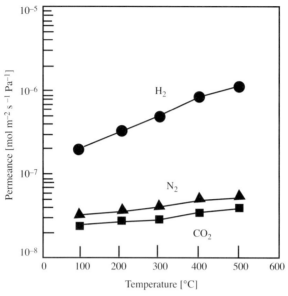

Figure 7.19 Permeance of hydrogen, nitrogen and carbon dioxide versus temperature for a platinum loaded membrane made of yttria stabilised zirconia [530].

7.1.5
Reforming in Chip-Like Microreactors

For the low power supply in the range of few watts, chip-like reactors were developed by several groups. Frequently, production techniques usually applied for fabrication of Micro Electro Mechanical Systems (MEMS) were adapted to fabricate these small reactors. The devices are actually "micro", not just as far as their channel dimensions are concerned, but also due to their outer dimensions.

Target applications are the replacement of batteries for mobile phones and, in larger sizes, power supplies for electronic equipment such as laptops and camcorders.

Pattekar et al. [531] fabricated a microreactor for methanol steam reforming from a silicon-chip (see Figure 7.20). The reactor housing was still made of stainless steel, electrically heated and sealed with graphite. The micro-channel was a long serpentine, of 1000-µm width and 230-µm depth fabricated by photolithography and KOH etching. Copper catalyst was sputtered with a thickness of 33 nm onto the chip. Preliminary simulations revealed non-uniform temperature distribution in the reactor housing, which was caused by heat losses. This emphasized the need for efficient insulation especially with low power systems. Then Pattekar et al. presented a silicon reactor fabricated by Deep Reactive Ion Etching (DRIE). It carried 7 parallel microchannels of 400-µm depth and 1000-µm width filled with commercial copper/zinc oxide catalyst particles, from Süd-Chemie. The catalyst was trapped by a 20-µm

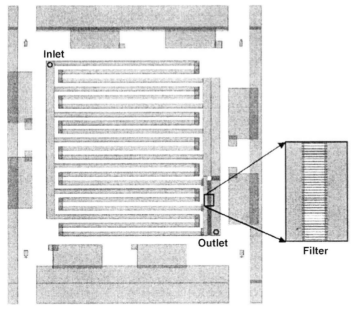

Figure 7.20 Mask layout of the chip-like methanol steam reformer developed by Pattekar and Kothare [531].

filter within the reactor, which was also fabricated by DRIE. The reactor was covered by a Pyrex wafer via anodic bonding. For details of the fabrication process see Section 10.2.4. The catalyst was introduced as a suspension into the ready-made reactor. Resistance devices for temperature sensing and a meander for heating were made of platinum and incorporated into the reactor, which had overall dimensions of 40 mm × 40 mm × 3 mm. A pressure drop of approximately 1.6 bar was calculated over the steam reforming channels at the design point, which is a fairly high value. Numerical simulations revealed deviations from the flow equi-partition over the individual channels of less than 10%. An 88% methanol conversion was achieved at 200 °C reaction temperature, S/C ratio 1.5, and 72 cm^3 min^{-1} hydrogen were produced under these conditions.

As a next step, Pattekar and Kothare presented a radial shaped microreactor heated by electricity, which had a similar principle set-up as the reactor described above, but lower pressure drop due to the radial flow path (see Figure 7.21). An evaporation region was placed in the centre of the reactor, which produced 145 cm^3 min^{-1} hydrogen.

Kundu *et al.* from Samsung [533] reported on the development of a microreactor for methanol steam reforming in the power range from 5 to 10 W, which was fairly similar to the second generation device developed by Pattekar *et al.* described above. Metallic connectors were used instead of the Teflon connectors used by Pattekar. The microreformer was 30-mm wide and long. It consisted of evaporation and steam reforming zones. Parallel and serpentine channels for steam reforming were tested

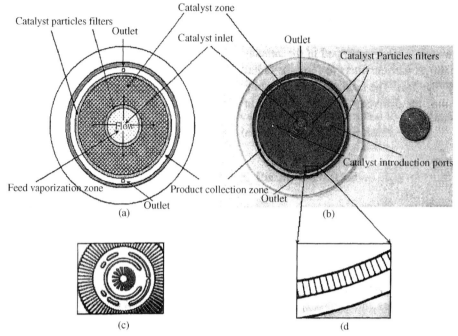

Figure 7.21 Radial flow chip-like methanol steam reformer developed by Pattekar et al. [532]: (a) schematic; (b) fabricated device; (c) evaporation region; and (d) catalytic particle filter.

as alternatives, and the latter showed superior results [533]. Then 140 mg of commercial copper/zinc oxide catalyst from Johnson Matthey were introduced as a slurry into the channels and held back by filters similar to those described by Pattekar above. A 0.01–0.02 mL min^{-1} methanol–water mixture was fed to the reactor at a S/C ratio of 1.2, which corresponded to a gas hourly space velocity of between 6000 and 13 000 h^{-1}. The catalyst was activated by the water–methanol mixture itself. It required about 30 min to achieve maximum activity by these means. 75 vol.% hydrogen, 24 vol.% carbon dioxide and 1.5 vol.% carbon monoxide was contained in the reformate. Up to 90% methanol conversion could be achieved at 260 °C reaction temperature. The system suffered from catalyst durability issues, which were attributed to sintering of copper oxide particles [503].

Kim and Kwon described methanol conversion of greater than 80% in their electrically heated microreactor, which carried a copper/zinc oxide catalyst [143]. About 4 mL min^{-1} hydrogen were produced by the reactor. At 300 °C reaction temperature and S/C 1.1, full conversion of the methanol was achieved. However, the weight hourly space velocity of about 13 L (h g$_{cat}$)$^{-1}$ was rather low, as is typical for copper/zinc oxide catalysts.

Yoshida et al. presented a small fuel processor sub-system for methanol steam reforming consisting of an evaporator, a steam reformer and a methanol burner for power supply [534]. The dimensions of the whole arrangement amounted to

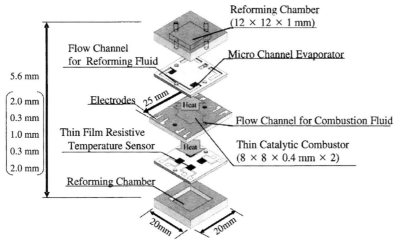

Figure 7.22 Small scale methanol reformer/evaporator/burner system with 4-W electrical power equivalent [534].

20 mm × 20 mm × 5.6 mm. It was composed of silicon and glass substrates, the latter structured by sand blasting. The burner was positioned in the centre of the device, while two reforming sections and an evaporator were placed on its top and bottom, as shown in Figure 7.22. The evaporation channels had a narrow width of 50 μm and various depths of between 30 and 80 μm were tested. Commercial copper/zinc oxide catalyst was used for steam reforming, while a platinum/titania catalyst served as the combustion catalyst. Combustion power of 11 W was required to generate hydrogen with a thermal energy of 5.9 W, which transfers to a reformer efficiency of 36%, an electrical power output of 2.25 W and an efficiency of 14% of a future fuel cell/fuel processor system, which is a good value considering the size. At S/C 1.3 the carbon monoxide concentration of the reformate was in the range of 1 vol.%. As shown in Figure 7.23, heat losses become a dominant issue for such small scale systems.

Park et al. presented a chip-like methanol fuel processor subsystem [535]. It was composed of a combustion chamber for hydrogen oxidation filled with a platinum catalyst supported on carbon nanotubes and a combined evaporator/steam reformer, which was merged in a single device by two intertwined, spiral-wound channels for evaporation and steam reforming, as shown in Figure 7.24. The burner was capable of heating the device up to the operating temperature within about 5 min. However, the system suffered from excessive temperature gradients and the hydrogen flow rate required for the combustion reaction exceeded the hydrogen flow rate produced by the reformer, mostly because of incomplete methanol conversion, which did not exceed 25%.

Other micro methanol steam reformers were developed, heated either by electricity [536] or coupled with a catalytic burner [537].

A chip-like glass reactor was developed by Kim et al., which was supported with energy from the decomposition of hydrogen peroxide over a platinum/alumina/silica catalyst containing 5 wt.% platinum [538]. More than 90% methanol conversion

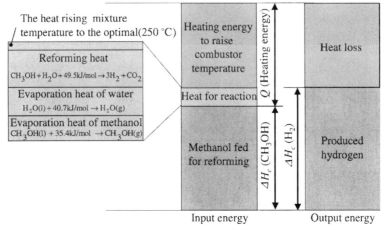

Figure 7.23 Energy diagram of a small scale methanol reformer/evaporator/burner system with 4-W electrical power equivalent [534].

could be achieved over the copper/zinc oxide/alumina/silica catalyst at a 270 °C temperature and low weight hourly space velocity of 2.4 L (h g_{cat})$^{-1}$. However, hydrogen peroxide conversion was incomplete. The electrical power equivalent of the reformate corresponded to 1.5 W_{el}.

7.1.6
Plasmatron Reformers

Plasmatron reformers are suitable for the conversion of all types of fuel, including heavy feedstock such as biomass and diesel. Their lower size is limited to an electrical power equivalent of about 1 kW_{el}.

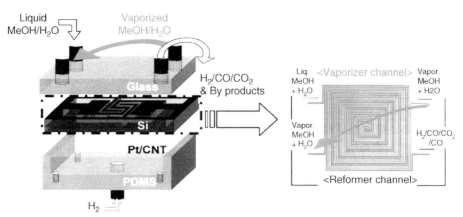

Figure 7.24 Combined evaporator, methanol steam reformer and hydrogen burner [535].

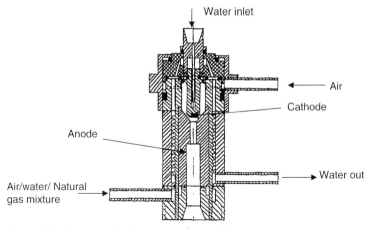

Figure 7.25 Schematic of a plasmatron reformer [87].

Figure 7.25 shows the schematic of a gliding arc plasmatron [87]. An arc was generated between the cathode with a hafnium tip and the tubular anode. The arc was then moved by the feed gas into a chamber downstream. For smaller scale applications, the cathode could also be designed as a copper rod with a zirconium tip [86]. Both electrodes were cooled by water. Their lifetime was usually in the range of 2000 h and their exchange simple. Westinghouse developed a plasmatron in which the arc was rotated at speeds of up to 1000 rev s^{-1} by interaction of the arc current with a magnetic field [86]. Reforming catalyst sections could be combined with the plasmatron, reducing its electrical power demand [87].

Bromberg *et al.* performed methane reforming in their plasmatron reformer [539], and 80% methane conversion could be achieved. The energy input of more than 100 MJ kg^{-1} hydrogen was rather high. It would have consumed 80% of the thermal energy of the hydrogen produced. When a nickel catalyst bed was positioned downstream of the reformer, the overall conversion could be increased and the energy input into the plasmatron could be decreased because the conversion in the plasmatron was reduced to 40%.

Another type of gliding-arc plasmatron consists of two diverging electrodes, which create an arc. This arc is pushed forwards by the gas flow, moves along the electrodes and extinguishes at their end as shown in Figure 7.26. Several arcs exist in parallel. Such a plasma reformer was developed by Paulmier and Fulcheri [92] for the autothermal reforming of gasoline. It worked at a pressure up to 3 bar, a feed inlet temperature of 500 °C and up to 40-kW thermal energy of the gasoline feed. A cross-sectional view is shown in Figure 7.27. The vessel had a volume of 6.3 L. Fuel vapour and steam were injected via a stainless steel tube of 4-mm inner diameter. Hot air was fed through six holes of 3-mm diameter concentric to the feed injection. The feed flow velocity was high at up to 80 m s^{-1}, and thus the residence time in the plasma and the reaction chamber was in the range of from 3 to 12 ms. The S/C was varied from 1.0 to 3.0 for autothermal reforming and even up to 5.0 for steam reforming, while the O/C ratio ranged between 0.3 and 1.25 for autothermal reforming. Two consecutive

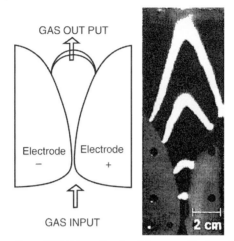

Figure 7.26 Schematic of a gliding arc with diverging electrodes [92].

sets of tungsten electrodes were used, each 45-mm long, 20-mm high and 2-mm wide [92]. Electrical connections of the electrodes were composed of Inconel tubes surrounded by an alumina/silica tube for electric insulation. The power supply to the electrodes worked with ac or dc and at a voltage of 5000 V for the first electrode and 10 000 V for the second, while the current was in the range of from 400 to 450 mA. An ac power supply was the preferred option, because the electrode erosion was non-uniform (higher at the anode) for dc operation. The gap between the electrodes was 2-mm wide. For autothermal reforming, the hydrogen content in the product gas did not exceed 7 vol.% on a dry basis, which was achieved at the lowest O/C ratio of 0.3 and S/C 2.5. Gasoline conversion was 10% under these conditions. When the O/C ratio was raised to 1.25, the conversion reached 50%, however, the hydrogen content decreased to 1 vol.% on a dry basis. The results for steam reforming were better. Up to 30 vol.% hydrogen content was determined on a dry basis, while the gasoline conversion was still low with 57% at S/C 3.0 [92].

Sobacchi et al. presented a corona plasma reformer coupled to a catalytic reformer downstream [540]. The corona plasmatron consisted of a stainless steel tube as the outer electrode, which was 1.2-m long and had 22.2-mm diameter, while the inner electrode was an Inconel wire put into the centre of the tube. The reactor was heated externally by a furnace to 400 °C. The power to the plasma source was in the range of from 1 to 20 W at a 20-kV voltage and 100-ns pulse duration at a frequency of 0.2–2.0 kHz. The catalytic reformer was a conventional fixed-bed, also heated externally. The reaction was performed at an O/C ratio of 0.9. In fact, the corona plasmatron generated no hydrogen at all when operated without the catalytic reformer. However, the pre-treatment of the gasoline in the plasmatron increased the hydrogen yield in the catalytic reactor by a factor of two at a temperature of 640 °C. To achieve full conversion in the catalytic reactor, its temperature had to be substantially higher than 630 °C. Under these conditions the beneficial effect of plasma reforming was negligible.

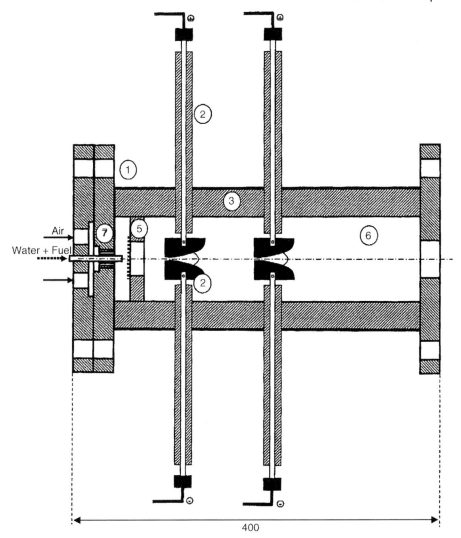

Figure 7.27 Cross-sectional view of the gliding arc plasma reactor developed by Paulmier and Fulcheri [92]: 1, stainless steel vessel; 2, electrodes, 3, electrode supports; 4, inner thermal insulation; 5, flame guide; 6, post-discharge chamber; 7, injection system.

Another gliding arc reformer was presented by Czernichowski *et al.* [541]. It was operated with natural gas, cyclohexane, heptane, toluene, gasoline, JP-8 and diesel fuel. Sulfur compounds were converted into hydrogen sulfide in the system and up to 4 wt.% of sulfur did no affect the operability of the device. Only 2% of the fuel cell energy output was required for the plasma generation. Figure 7.28 shows a GlidArc reformer, which had a power equivalent of 7 kW and volume of 0.6 L. The plasma was

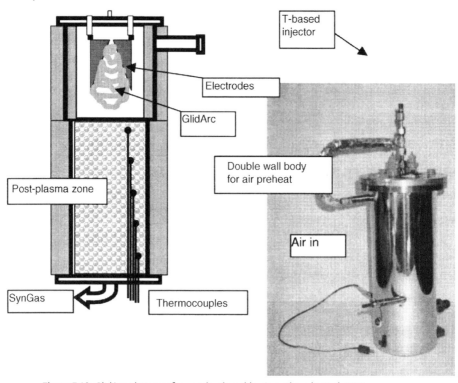

Figure 7.28 GlidArc plasma reformer developed by Czernikowski et al. [541].

generated with a voltage of 10 kV but a current of only 25 mA, and thus the time-averaged power consumption was only 100 W. JP-8 jet fuel was converted in the reformer at an O/C ratio of 1.4 and the reformate contained between 10 and 16 vol.% hydrogen on a dry basis along with up to 20 vol.% carbon monoxide, 2.8 vol.% methane and 2.8 vol.% ethylene. The thermal efficiency of the reformer was reported to be in the range of 75%. Thus the device is suitable as a pre-reformer and also for solid oxide fuel cell applications.

Figure 7.29 shows the design of a spray-pulse type catalytic reactor system for isooctane developed by Biniwale et al. [93]. The main plasmatron reactor was a vertical cylinder carrying an atomiser at the top, which was supplied with liquid feed at a pressure of 120 bar. Air was introduced through separate inlets positioned at the top of the reactor. The catalyst was positioned at the reactor bottom, coated onto an alumina mesh and heated by a tungsten wire coil.

The isooctane feed was pulsed at frequencies of from 0.001 to 0.5 Hz at a pulse duration of 2 ms corresponding to a gaseous isooctane feed of 4.2 cm^3 per pulse. At the maximum pulse frequency of 0.5 Hz, 480 cm^3 min^{-1} isooctane were therefore fed to the reactor along with a nitrogen flow at 200 cm^3 min^{-1}. The S/C ratio was set to 1 and the O/C ratio to 0.64, corresponding to autothermal conditions. Up to 98% conversion could be achieved over a rhodium/ceria catalyst. Nickel/tungsten and

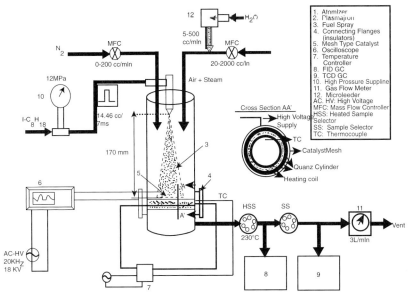

Figure 7.29 Spray-pulse type catalytic reactor system [93].

nickel/manganese catalysts showed up to two orders of magnitude lower hydrogen production rates. Figure 7.30 shows the isooctane conversion achieved in the testing reactor for a duration of more than 2 h. Not only the activity, but also the stability of the precious metal catalyst was superior to its nickel-based counterparts. Light hydrocarbons, mostly methane, but also C_2–C_6 species were detected in the product, but their concentration was not further specified unfortunately. The weight hourly space velocity of the catalyst was 525 L $(h\,g_{cat})^{-1}$, which is a rather high value. This method of feed injection is thus a viable option for pre-reforming hydrocarbon fuels. However, the drawback of the system presented here is the low amount of catalyst in a fairly bulky system, which would require optimisation before introduction into a practical system.

7.2
Water–Gas Shift Reactors

7.2.1
Water–Gas Shift in Monolithic Reactors

Dokupil et al. [542] described monolithic water–gas shift reactors operated at a gas hourly space velocity of 20 000 h^{-1} for medium temperature shift and 10 000 h^{-1} for low temperature shift. The adiabatic temperature rise in the medium temperature reactor operated at 300 °C amounted to 68 K, in the low temperature reactor operated at 280 °C a temperature rise of only 15 K was determined.

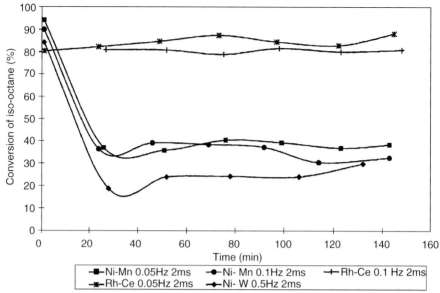

Figure 7.30 Spray-pulse reforming of isooctane over different catalysts, pulse frequency and pulse duration [93].

Microlith packed metal gauze technology (see also Section 6.2) was applied by Castaldi et al. and Roychoudhury et al. for a water–gas shift [458,490]. A single water–gas shift stage was tested separately at very high gas hourly space velocities up to $50\,000\,h^{-1}$. Reformate surrogate containing 3.6 vol.% carbon monoxide was fed to the reactor, which was operated at a reaction temperature of between 220 and 300 °C. However, conversion was very low at the high space velocities. The space velocity had to be reduced to values around $5000\,h^{-1}$ to reach 50% conversion, which then corresponded to the thermodynamic equilibrium.

7.2.2
Water–Gas Shift in Plate Heat-Exchanger Reactors

The implementation of a declining temperature gradient into a water–gas shift reactor by counter-flow cooling has been discussed extensively in Section 5.2.1.

An early stage of plate a heat-exchanger for water–gas shift in the kW size range was described by Kolb et al. [543]. The reactor still had a three stage cross-flow design for the sake of easier fabrication. Platinum/ceria catalyst was wash-coated onto the metal plates, which were sealed by laser welding. The reactor was tested separately and showed equilibrium conversion under the experimental conditions. It was subsequently incorporated into a breadboard fuel processor (see Section 9.5).

Figure 7.31 shows a microstructured plate heat-exchanger reactor developed by Kolb et al. [312], which was designed for a power equivalent of 2 kW of the corresponding fuel cell. The temperature profile, which was determined experimentally in

Figure 7.31 Integrated water–gas shift reactor/heat-exchanger designed for a 2-kW fuel cell system [312].

the reactor, is shown in Figure 7.32. After a slight temperature rise at the reactor inlet, which was required to achieve a high initial rate of reaction, the reactor temperature gradually decreased downstream. The content of carbon monoxide in the reformate surrogate could be reduced from 10.6 to 1.05 vol.%, which corresponded to 91%

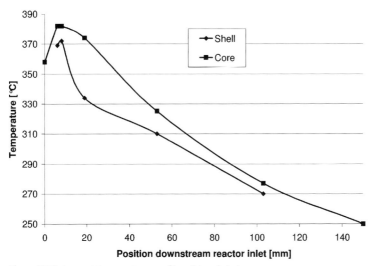

Figure 7.32 Internal temperature profile achieved during operation of the reactor shown in Fig. 7.31 [312].

conversion. Somewhat lower temperatures were observed at the shell of the reactor, especially in the inlet section, which was attributed to heat losses to the environment.

7.2.3
Water–Gas Shift in Membrane Reactors

Barbieri et al. developed a membrane reactor for water–gas shift [544]. A palladium/silver film containing 23 wt.% silver, which was between 1- and 1.5-μm thick was produced by sputtering. This film was coated onto a porous stainless steel support. This patented production method allowed a much higher ratio of pore size to film thickness compared with conventional methods. Tubular membranes of 13-mm outer diameter, 10–20-mm length and 1.1–1.5-μm thickness, respectively, were prepared. A commercial Cu based catalyst supplied by Haldor-Topsoe was used for the water–gas shift reaction. At 210 °C a permeating flux of 4.5 L $(m^2 s)^{-1}$ was determined for pure hydrogen at 0.2-bar pressure drop. At a reaction temperature of 260–300 °C, and 2085 h^{-1} gas hourly space velocity, the thermodynamic equilibrium conversion could be exceeded by 5–10% with this new technology.

7.3
Catalytic Carbon Monoxide Fine Clean-Up

7.3.1
Carbon Monoxide Fine Clean-Up in Fixed-Bed Reactors

Lee et al. described the development of tubular two stage preferential oxidation reactors for a 10-kW_{el} fuel cell system [545]. Platinum/ruthenium/alumina catalyst was used, which converted 1 vol.% carbon monoxide in the feed to values below 100 ppm for temperatures between 100 and 140 °C at an O/CO ratio of 4.0 and a weight hourly space velocity of 40 L $(h\ g_{cat})^{-1}$. Then a single stage reactor with 1-kW_{el} power equivalent was tested. The carbon monoxide concentration could be reduced to less than 200 ppm at an O/CO 3.2, however, the reactor exit temperature increased to 240 °C. By using dual stage reactors, the carbon monoxide could be reduced below 10 ppm at an O/CO 2.0 and 4.0 in the first and second stage, respectively. The addition of air to the second stage was low, because only 1000 ppm carbon monoxide had to be converted.

Figure 7.33 shows the dynamic response of the dual stage reactors for a load change from 100 to 35% and back to 100%, which proved that the carbon monoxide in the reformate effluent could be maintained below 100 ppm.

To gain a power equivalent of 10 kW_{el}, 128 parallel tubes with 1.27 mm inner diameter were switched in series (see Figure 7.34) and a cooling stage was introduced between the reactors. The dimensions of the reactor were rather low with 200-mm diameter and 200-mm length. When this reactor was tested, no steam was added to the feed, because the effect of steam was regarded as negligible. However, as discussed in Section 4.5.2, steam has a beneficial effect on the performance at least

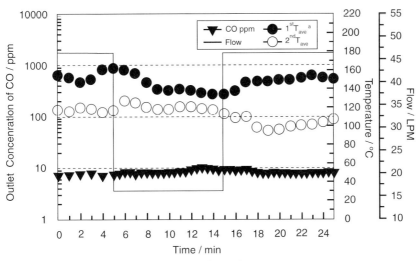

Figure 7.33 Response of dual stage tubular preferential reactors (1-kW$_{el}$ power equivalent) to load changes from 100 to 35% and back to 100%: ●, the average temperature of the first stage reactor; ○, the average temperature of the second stage reactor [545].

of precious metal preferential oxidation catalysts. Load changes from 100 to 65% and back were demonstrated for the 10-kW$_{el}$ reactors, proving that carbon monoxide could be reduced to values below 100 ppm. Start up of the reactors was performed within 3 min by heating the reactor tubes with reformate at a temperature of 150 °C and adding oxygen at O/CO 3.0. Through these means the carbon monoxide could be reduced to below 20 ppm within 3 min.

Pan and Wang switched four adiabatic preferential oxidation reactors in series downstream of the reformer/evaporator described in Section 7.1.3 [546]. Heat-exchangers were installed after each reactor. The four reactors were operated at the same inlet temperature of 150 °C, and the O/CO ratio in the feed increased from 1.6 to 3, to minimise the heat formation in the first reactors. Despite these measures, the temperatures rise in the first reactor was as high as 121 K, in the second reactor it was still at 82 K and decreased to 28 K in the third and 8 K in the last reactor. While only 50% carbon monoxide conversion could be achieved in the first reactor, conversion was complete after the last stage. The combined steam reformer/clean-up system was operated for 24 h. The carbon monoxide content of the reformate could be maintained below 40 ppm.

7.3.2
Carbon Monoxide Fine Clean-Up in Monolithic Reactors

Dokupil *et al.* [542] described a monolithic preferential oxidation reactor operated at 100 °C and 15 000 h^{-1} gas hourly space velocity. It carried an integrated heat-exchanger to improve its thermal management.

Figure 7.34 Dual stage preferential oxidation tubular reactor with 10 kW$_{el}$ power equivalent as developed by Lee et al. [545].

Zhou et al. reported the operation of a monolithic preferential oxidation reactor system in a 5-kW methanol reformer for a duration of 14 h [322]. Carbon monoxide was reduced to less than 50 ppm, however, four stages of oxygen addition were required to achieve this result.

Chin et al. performed the reaction over metal foams of 2.54-cm diameter and 5.1-cm length with different pore densities [547]. The catalyst was composed of 5 wt.% platinum and 0.5 wt.% iron on alumina. The foams had either 40 or 20 pores per inch and contained either 4 or 12 wt.% of the catalyst. Higher conversion was achieved for the smaller pores and lower catalyst loading on the foam. However, conversion decreased when the reaction temperature was increased starting from 100 °C and full conversion could not be achieved. This was attributed to hot spot formation despite the low O/CO ratio between 1.0 and 2.0 that was applied. Comparable results were achieved with ceramic monoliths of similar size.

Ahluwalia et al. reported on experiments with cordierite monoliths applied for the preferential oxidation of carbon monoxide [548]. The monoliths had 7.62-cm diameter and 12.7-cm length. A commercial catalyst from Engelhard was used for the experiments, which lit-off even at room temperature. Reformate surrogate was fed to the reactors, which consisted of 48 vol.% hydrogen, 31 vol.% nitrogen, and varying amounts of carbon oxides and steam. The feed temperature was varied between 85 and 100 °C. For 1.3 vol.% carbon monoxide feed concentration and the required minimum O/CO ratio of 1.0, 90 K adiabatic temperature rise was observed in the monolith, while it exceeded 170 K for O/CO 2.0. The carbon monoxide selectivity was almost independent of the carbon monoxide concentration in the feed but rather depended on the O/CO ratio for a certain gas hourly space velocity. Surprisingly, steam addition had little effect on the carbon monoxide selectivity, which contradicts numerous other results (see Section 4.5.2) and might have been an inherent property of the catalyst under investigation, which clearly not did promote reverse water–gas shift. Modelling of the reaction with the help of experimental results revealed that a single stage monolith was not able to reduce the carbon monoxide concentration to values below 10 ppm, where the carbon monoxide concentration in the reformate feed exceeded 1.05 vol.%. A concept for a two stage preferential oxidation was developed. The proposal was to convert about 85% of the carbon monoxide in the first reactor, where the carbon monoxide concentration in the feed was 1 vol.%. The optimum O/CO ratio was in the range of 1.2 for the first reactor, while the second reactor required a higher value of 2.4. By addition of a third stage, the performance of the chain of monoliths could be further improved and up to 3.5 vol.% carbon monoxide or more in the feed could be converted completely according to these calculations. The hydrogen loss was low at 1 vol.%. However, the attractive performance of this technical solution is counterbalanced by a rather complex set-up.

Microlith packed metal gauze technology (see also Section 6.2) was applied by Castaldi et al. and Roychoudhury et al. for preferential oxidation [458,490]. A single stage reactor was tested at a high gas hourly space velocity of 150 000 h^{-1} with a surrogate of water–gas shift product containing 5000 ppm carbon monoxide. The O/CO ratio of the feed was set to 2.4. The adiabatic temperature rise of the reformate was calculated to be 45 °C under these conditions. The preferential oxidation reactor was operated at a reaction temperature of about 250 °C for 500 h. About 93% conversion was achieved, which corresponded to an outlet concentration of carbon monoxide in the range of 350 ppm. Thus a second stage preferential oxidation reactor was required, which was operated at a higher O/CO ratio of 8.0 and a much higher gas hourly space velocity of 220 000 h^{-1}. It converted the remaining carbon monoxide to a concentration below 20 ppm even under conditions of load changes.

7.3.3
Carbon Monoxide Fine Clean-Up in Plate Heat-Exchanger Reactors

Dudfield et al. performed investigations of compact aluminium fin heat-exchanger reactors for the preferential oxidation of carbon monoxide [549]. The reactors had the dimensions of 46-mm height, 56-mm width and 170-mm length, which corresponded to a 0.44-L volume and 590-g weight. They contained 2 g of catalyst each [328]. Platinum/ruthenium catalyst formulations of various composition were incorporated into different reactors and tested. Reformate surrogate with a composition equivalent to methanol reformer product was applied for the tests. The feed contained, on a wet basis, 62.9 vol.% hydrogen, 20.2 vol.% carbon dioxide, 0.5 vol.% carbon monoxide, 0.36 vol.% methanol, 6.3 vol.% steam, 2.0 vol.% oxygen and 7.5 vol.% nitrogen. The reactors were operated at 2920 h^{-1} gas hourly space velocity, which corresponded to a feed rate of 24.3 L min^{-1}, a high weight hourly space velocity of 720 L (h g$_{cat}$)$^{-1}$ and an O/CO ratio of 8.0 [549]. The reactors were cooled by a cooling oil cycle, which was externally cooled by an additional external heat-exchanger and a fan (see Figure 7.35). This cooling oil was also used for pre-heating the feed in a heat-exchanger switched upstream of the reactor.

At 160 °C reaction temperature, the carbon monoxide concentration found in the reformate was 11 ppm. It became obvious that not only the carbon monoxide, but also the methanol contained in the feed was completely oxidised under the reaction conditions. The conversion decreased from 99.8 to 85%, when the feed flow rate was increased from 24.3 to 150 L min^{-1} [549]. The latter corresponds to a very high weight hourly space velocity of 4500 L (h g$_{cat}$)$^{-1}$. At a flow rate of 100 L min^{-1} and 160 °C reaction temperature, the exit concentration of carbon monoxide was maintained at

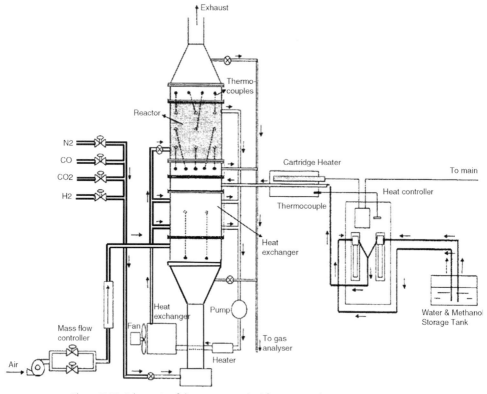

Figure 7.35 Schematic of the test rig applied for testing of preferential oxidation plate-heat-exchangers by Dudfield et al. [549].

200 ppm almost independent of the feed concentration, which was varied from 500 to 10 000 ppm. When the O/CO ratio was changed from 1.0 to 6.0, the conversion increased from 50 to 100% at the lowest feed flow rate of 24.3 L min^{-1}. To maintain conversion at a high level, a second reactor carrying a slightly modified platinum/ruthenium catalyst formulation was switched in series [549]. Carbon monoxide could be reduced to below 20 ppm by this dual-stage configuration. Interestingly, these workers observed a reduction in carbon monoxide to below the detection limit of 5 ppm, when steam addition was stopped. This contradicts several investigations performed with precious metal based preferential oxidations catalysts as described in Section 4.5.2. When the carbon monoxide content of the feed was increased from 1000 to 7000 ppm, the carbon monoxide content of the product could be maintained below 20 ppm with this two-stage configuration. Figure 7.36 shows a comparison of the performance of the single and combined reactors under slightly different operating conditions as described above [328]. The optimum distribution of the air fed to both reactors switched in series was determined to be 70/30 [328]. Stable performance of the reactors for 40 h could be demonstrated [550].

To meet the requirements of the 20-kW fuel processor, which was the overall aim of the investigations, the reactor design described above was scaled up to a dual stage design of total volume 4 L. Each reactor was now 108-mm high, 108-mm wide and

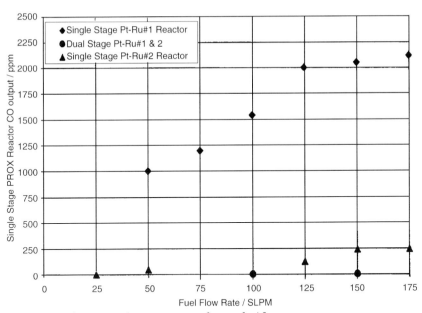

Figure 7.36 Carbon monoxide output versus reformate feed flow rate for two different compact aluminium heat-exchanger/reactors tested separately and switched in series [328], test conditions: 25–175 L min^{-1} reformate; 2.5–17.5 L min^{-1} air; O/CO ratio 3.5; synthetic reformate composition, 68.9 vol.% H_2, 0.6 vol.% CO, 22.4 vol.% CO_2, 6.9 vol.% H_2O and 0.4 vol.% CH_3OH.

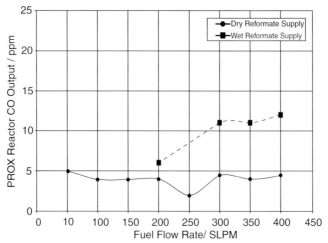

Figure 7.37 Carbon monoxide output versus time for various reformate feed flow rates of two compact aluminium heat-exchanger/reactors switched in series [550], test conditions: 300–400 L min^{-1} synthetic reformate; O/CO ratio 6.0; feed composition, 62.4 vol.% H_2, 0.6 vol.% CO, 20.2 vol.% CO_2, 6.7 vol.% H_2O; 1.7 vol.% O_2, 8.3 vol.% N_2 and 0.3 vol.% CH_3OH.

171-mm long, had a volume of 1.85 L and weight 2.5 kg, and each carried 8.5 g catalyst [328]. The air feed was split in a ratio of 2/1 between the first and second reactor. For reformate flow rates up to 400 L min^{-1}, which corresponds to a weight hourly space velocity of about 2800 L (h g$_{cat}$)$^{-1}$ the carbon monoxide concentration in the reformate could be maintained below 12 ppm for dry synthetic reformate and below 20 ppm for wet synthetic reformate, as shown in Figure 7.37. Some fluctuations in the carbon monoxide concentration occurred during the load changes, which had not been observed for the dry reformate. Lower operating temperatures of between 130 and 150 °C were applied for these larger prototypes [550]. In general, very high O/CO ratios between 6 and 8 were set for all tests described above, which obviously generated excessive losses of valuable hydrogen. Then Dudfield *et al.* changed the composition of the synthetic reformate by decreasing the hydrogen content and increasing the steam content from 7 to 13 vol.%. In parallel, the content of residual methanol was decreased to 0.2 vol.%. The reactors were operated at a higher temperature of 160 °C and a lower feed rate of 200 L min^{-1}. The carbon monoxide content could be decreased from as much as 2.0 vol.% to values below 15 ppm with a much lower O/CO ratio of 5.0 [550]. This originated from the beneficial effects of additional steam, higher reactor temperature, lower space velocity and from the lower concentration of methanol.

The reactors were operated at full load (20 kW equivalent power output) for approximately 100 h without deactivation [328].

In connection with the 20-kW reformer, the CO output of the two final reactors was less than 10 ppm for more than 2 h at a feed concentration of 1.6% carbon monoxide.

Because the reformer was a combination of steam reformer and catalytic burner in plate and fin design, this may be regarded as an impressive demonstration of the capabilities of the integrated heat-exchanger design for fuel processors in the kilowatt range.

The performance of one of the plate and fin heat-exchanger (0.25-L reactor volume) of the first generation, wash-coated with 2.0 g catalyst, was assessed in comparison with similar alternative technologies, by Dudfield et al. [328]. The first was a shell and tube heat-exchanger filled with 4.66 g of catalyst microspheres (0.25-L reactor volume). The second was a heat-exchanger into which steel granules had been sintered to generate a porous structure. The sintered metal was wash-coated with 3.4 g catalyst (0.25-L reactor volume). The plate and fin design was produced as a sandwich and made of aluminium, the other devices were made of stainless steel. All reactors were cooled by oil. The pressure drop was lowest for the plate and fin reactor. It was 30-times higher for the sintered structure and 8-times higher for the shell and tube heat-exchanger. The performance of the reactors was compared at 150 °C, 10–175 L min^{-1} flow rate of reformate surrogate and 1–17.5 L min^{-1} air addition. The reformate surrogate was composed of 69.7 vol.% hydrogen, 0.6 vol.% carbon monoxide, 22.4 vol.% carbon dioxide, 6.9 vol.% steam and 0.4 vol.% methanol. The thermal management of the sintered porous structure was worst, leading to hot-spot formation up to 150 K despite the integrated heat-exchange capabilities. The two remaining reactors showed comparable performance, slightly in favour of the plate and fin design. Hot spots were limited to 20 K. For the plate and fin heat-exchanger reactor a maximum of 250 ppm carbon monoxide were detected in the product at 175 L min^{-1} feed rate (see Figure 7.38).

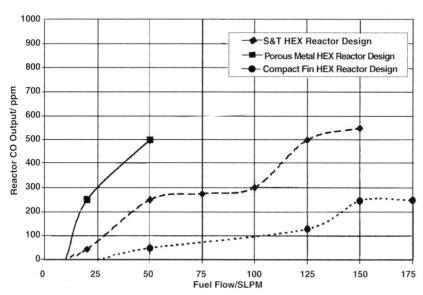

Figure 7.38 Carbon monoxide content in the product versus reformate flow rate for three different reactor heat-exchanger (HEX) designs [328].

A microstructured plate heat-exchanger for preferential oxidation in the kW-size range was developed by Kolb et al. [543]. The reactor had a three stage cross-flow design for sake of easier fabrication. Platinum catalyst supported by alumina and zeolite A was coated onto the metal plates, which were sealed by laser welding. The reactor was tested separately with a feed flow rate of 185 L min^{-1}, which contained 0.4 vol.% carbon monoxide. The reactor converted more than 90% of the carbon monoxide at a 206 °C reaction temperature and O/CO ratio 4.1 at a weight hourly space velocity of between 48 and 98 L (h g$_{cat}$)$^{-1}$. It was subsequently incorporated into a breadboard fuel processor (see Section 9.5).

Cominos et al. developed a microstructured testing reactor with integrated heat exchanging capabilities for the preferential oxidation reaction [324]. The reactor had outer dimensions of 52 × 53 × 66 mm^3. It was split into three parts, namely, two heat-exchangers composed of six plates and the reactor itself, which were thermally decoupled by insulation material. The reactor/heat-exchanger arrangement was designed for a fuel processor/fuel cell system with 100-W electrical power output, which was presented by Delsman et al. [332]. The heat-exchanger was designed to cool the reformate coming from the reformer with a temperature of 280 °C to a temperature of 175 °C before it was fed to the preferential oxidation reactor, operated at 160 °C. The reformate should be cooled within the reactor. The purified reformate was then further cooled to the operating temperature of the low temperature fuel cell as shown in Figure 7.39. The fuel cell anode off-gas served as coolant for each of the three devices.

Four thermocouples were integrated into the reactor to determine the temperature at the outlet of each component. The reactor itself contained 19 plates each with 82 microchannels coated with catalyst and sealed by graphite gaskets. The dimensions of the channels were 250-µm width, 190-µm depth and 30-mm length. The 10 plates through which the cooling flow had to pass had 75 microchannels with dimensions of

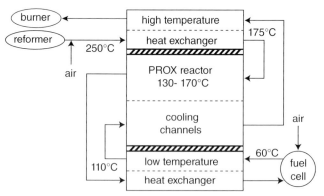

Figure 7.39 System design for a 100-W preferential oxidation reactor coupled to two heat-exchangers and cooled by anode off-gas [332].

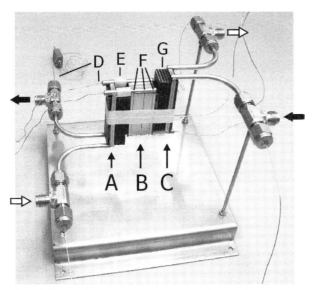

Figure 7.40 Coupled low temperature heat exchanger (A), preferential carbon monoxide oxidation reactor (B) and a high temperature heat exchanger (C); D, thermocouples; E, tube connection; F, thermocouple channels; G, insulating material; the devices were sealed by laser welding [332].

250-µm width, 125-µm depth and 30-mm length. Catalyst development was performed in the reactor, which is discussed in Section 4.5.2.

Later, the reactor concept of Cominos *et al.* discussed above was optimised to a design that was less bulky and closer to a practical application, which was presented by Delsman *et al.* [332]. They performed numerical simulations for the reactor, which showed that a two-dimensional model was required to describe the reactor performance accurately. A high length/width ratio of 3 was chosen for the heat-exchangers and the reactor in order to minimise axial heat transfer, which had a detrimental effect on the heat-exchanger performance otherwise. To minimise the volume of the device the reactor/heat-exchanger assembly was sealed by laser-welding (see Figure 7.40). The subsystem was considerably smaller ($45 \times 17 \times 50\,mm^3$) than the first generation.

The heat recovery efficiency of the assembly was determined experimentally from the outlet temperature of the coolant gas [332]. The overall heat recovery varied between 73 and 95%, with the higher numbers corresponding to higher flow rates and higher oxygen excess.

Ouyang *et al.* studied the preferential oxidation of carbon monoxide in a silicon reactor [117] fabricated by photolithography and deep reactive ion etching (see also Section 10.2.4). Each reactor carried two gas inlets, a pre-mixer, a single reaction channel and an outlet zone where the product flow was cooled. The single channel was 500-µm wide, 470-µm deep and 45-mm long. Platinum/alumina catalyst containing 2 wt.% platinum was deposited onto the channel walls with 2–5-µm thickness of the coating. Then the channel was sealed by anodic bonding with a Pyrex glass plate.

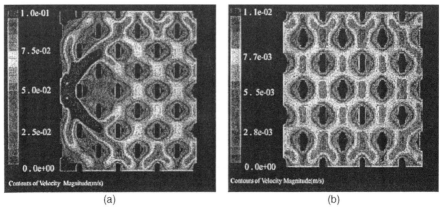

Figure 7.41 CFD simulation of the flow profile in the silicon microreactor developed by Srinivas et al. [551]; left, fluid inlet region; right, interior region.

The surface area of the catalyst was very high (400 m² g⁻¹). It was reduced in undiluted hydrogen at 400 °C for 4 h before the tests. Simulated reformate at 5 mL min⁻¹ containing 1.7 vol.% carbon monoxide, 21 vol.% carbon dioxide, 68 vol.% hydrogen and 9 vol.% nitrogen and 0.5 mL min⁻¹ air were fed to the reactor, which corresponded to an O/CO ratio of 2.5. The experiments were performed applying a temperature ramp program from ambient to 300 °C. Full conversion of carbon monoxide was achieved at a 170 °C reaction temperature. For up to 300 °C reaction temperature, full conversion could be maintained, which is usually not the situation for fixed-bed operation. This was attributed to the improved heat transfer in the microchannel, which avoided hot spots leading to reverse water–gas shift.

Srinivas et al. designed a silicon microreactor for preferential oxidation, which was 6 cm × 6 cm wide and long, but the flow path was only 400-μm high [551]. Instead of microchannels, pillars were chosen for the flow distribution in the reactor. As shown in Figure 7.41, numerical calculations using FLUENT software revealed that uniform flow distribution was achieved after 2% of the reactor length. The reactor was coated with 2 wt.% platinum catalyst supported by alumina to a thickness of 10 μm. Tests were performed at an O/CO ratio 2.0 and high weight hourly space velocity of 120 L (h g_{cat})⁻¹. Not more than 90% conversion of carbon monoxide could be achieved in the reactor at 210 °C and similar results were obtained for a small fixed catalyst bed.

7.3.4
Carbon Monoxide Fine Clean-Up in Membrane Reactors

Sotowa et al. prepared zeolite Y membranes ion-exchanged with platinum and different silica-coated/rhodium loaded γ-alumina membranes for the preferential oxidation of carbon monoxide [552]. The membranes were coated onto α-alumina tubes. While the defect-free thickness of the zeolite membranes was around 3 μm, the combined silica/γ-alumina membrane was of 200- and 700-nm thickness, respectively. The hydrogen permeance through the zeolite membrane was 6.4×10^{-7} mol $(m^2 \, s \, Pa)^{-1}$ at 250 °C, while the carbon monoxide permeance was one order of magnitude lower. When feeding a mixture of hydrogen, carbon monoxide and carbon dioxide to the reactor, the carbon monoxide flux through the membrane was approximately three orders of magnitude lower than the hydrogen flux. When oxygen was added to the feed, the carbon monoxide concentration could be reduced below the detection limit on the permeate side. The silica/alumina membrane showed up to 10 times higher hydrogen/carbon monoxide permeation selectivity at 250 °C. The hydrogen permeance through this membrane was 2×10^{-7} mol $(m^2 \, s \, Pa)^{-1}$. Thus the silica membrane reduced the carbon monoxide flux drastically from 50 000 ppm in the feed to 500 ppm, while this residual low amount of carbon monoxide, which had passed the silica, was then oxidised over the rhodium catalyst on the γ-alumina membrane.

Kusakabe et al. prepared yttria-stabilised zirconia membranes on α-alumina tubes and impregnated them with platinum and rhodium for the preferential oxidation of carbon monoxide [530]. The carbon monoxide concentration could be decreased from 10 000 ppm in the feed to 30 ppm in the permeate, at 150 °C reaction temperature by the platinum containing membrane. However, the ceramic membrane also showed permeance for carbon dioxide and methane. The latter was oxidised to a certain extent over the platinum catalyst of the membrane. The rhodium catalyst showed inferior performance and even some activity towards methanation at a 250 °C reaction temperature in the absence of oxygen in the feed.

7.4
Membrane Separation Devices

Membrane separation (see also Section 5.2.4) as a unit operation is available for many applications on an industrial scale. However, for smaller scale mobile applications the separation unit is, in most instances, integrated into the reformer or water–gas shift reactor forming so-called membrane reactors, which were presented in Sections 7.1.4, 7.2.3 and 7.3.4. Some examples of fabrication technologies for membranes will be discussed in Section 10.2.5, and examples of membrane separation devices without the presence of chemical reactions will be presented below.

The permeation of a commercial palladium membrane containing 23 wt.% silver with 100-μm thickness was investigated by Emonts et al. [18]. Reformate from a methanol reformer (see also Section 9.1) was applied as feed for the separation device. A minimum operating temperature of 350 °C was determined, which was required to avoid poisoning of the membrane by the 2.6 vol.% carbon monoxide present in the reformate. The permeability of the membrane was determined to be 6 m^3 (m^2 h bar$^{0.5}$)$^{-1}$. For a power equivalent of 25 kW$_{el}$, the required palladium mass was calculated to be 1.74 kg for this membrane type [18]. A palladium membrane with much lower thickness of 4 μm, which was supported by ceramics, was then tested. Its permeability was determined to be 150 m^3 (m^2 h bar$^{0.5}$)$^{-1}$. The hydrogen to nitrogen selectivity of this membrane was 128 [553]. This reduction in membrane thickness reduced the required amount of palladium to only 10.5 g [553]. The supported membrane showed certain defects, which resulted in leakages. Thus, carbon monoxide passed through the membrane separation device. It could be removed by a methanation reactor (see Section 3.10.3). The permeability of the membrane was impaired by carbon monoxide [553]. The supported palladium membrane showed good adhesion for 1000 h of time on stream [553].

Kusakabe et al. prepared palladium membranes in micro-slits etched into a copper foil [554]. The foil had the dimensions 1 cm × 1 cm and 50-μm thickness. The palladium membrane was formed by electrodeposition in a bath containing 20 g L^{-1} palladium chloride, citric acid and ammonium sulfate, at pH 7. The electrolysis temperature affected the permeability of the membrane considerably and optimum separation factors were achieved at a temperature of between 35 and 45 °C. The thickness of the membrane was determined to be 4 μm. The micro-slits were introduced from the reverse side of the copper foil by photolithographic etching. They were about 100-μm wide. The total area of free standing membrane over the slits amounted to 5.4 mm^2. Permeation experiments were performed at ambient pressure using a mixture of nitrogen and hydrogen at the retenate side and argon as the sweep gas on the permeate side of the membrane. Hydrogen permeance increased with increasing temperature, whereas nitrogen permeance was not affected over the whole temperature range under investigation (200–400 °C). A separation factor of 2000 was determined at 300 °C. The hydrogen flux amounted to 1.3 L (m^2 s)$^{-1}$ under these conditions.

Han et al. developed a membrane separation module for a power equivalent of 10 kW$_{el}$ [402], shown in Figure 7.42. A palladium membrane containing 40 wt.% copper and of 25-μm thickness was diffusion bonded onto a metal frame. The separation module for a capacity of 10 Nm^3 h^{-1} hydrogen had a diameter of 10.8 cm and a length of 56 cm. The reformate fed to the modules contained 65 vol.% hydrogen

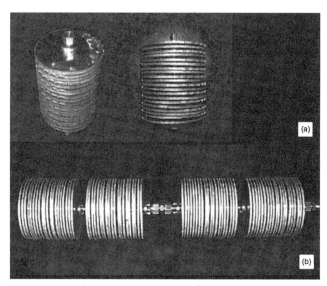

Figure 7.42 Membrane separation modules for a 10-kW$_{el}$ methanol fuel processor as developed by Han et al. [402].

and the hydrogen recovery through the membrane was in the range of 75%. The hydrogen flux through the membrane was determined to be about 22 cm^3 (min cm^2)$^{-1}$ for a hydrogen partial pressure difference of 6.5 bar and 300 °C temperature. Stable operation of the membrane separation was achieved for 750 pressure swing tests at 350 °C. The membrane separation devices were integrated into a methanol fuel processor, which is described in Section 9.1.

Gepert *et al.* prepared ceramic membrane supports [400] as inserts for an alcohol steam reformer with integrated membrane separation, the concept of which was discussed in Section 5.2.4. α-Alumina was chosen as the support. Capillaries of 20-cm length were produced, which had a wall thickness of from 100 to 500 μm and diameter from 500 μm to 3 mm. The porosity of the capillaries was in the range between 20 and 70%. The α-alumina was coated with γ-alumina to smooth the surface, resulting in a pore diameter of 4 nm. The palladium membrane was deposited onto the ceramics by electroless plating and had a thickness between 4 and 6 μm. The membranes were tested between 370 and 450 °C and up to an 8-bar pressure difference. The feed was a reformate surrogate composed of 75 vol.% hydrogen, 23.5 vol.% carbon dioxide and 1.5 vol.% carbon monoxide. The membranes showed a 60 m^3 (m^2 h)$^{-1}$ hydrogen flux at 6-bar pressure difference in pure hydrogen, while the nitrogen flux was negligible at 0.0032 m^3 (m^2 h)$^{-1}$. Material problems appeared through temperature cycling between the membrane and the glaze material, which the membranes had been coated with at the non-permeable zone. This was required for sealing to the environment. Thermal stress was a general problem, which originated from the different thermal expansion coefficients of palladium (11×10^{-6} K^{-1}) and α-alumina (6.5×10^{-6} K^{-1}). The permeance was calculated to be 20 m^3 (m^2 h bar$^{0.62}$)$^{-1}$, while the activation energy was at 19 kJ mol^{-1} in the usual range of palladium membranes. However, when the reformate surrogate was used as feed, the permeation rate decreased. At 4-bar hydrogen partial pressure difference, a 16% lower permeation rate was determined at 450 °C and 30% lower permeation rate at 370 °C. The reduced permeance was attributed to coke formation. Methane formation was observed in the presence of the reformate surrogate over the palladium by carbon oxide methanation (see Section 3.10.3). This was favoured by the absence of steam.

7.5
Catalytic Burners

Emonts reported the operation of catalytic burners for hydrocarbons and methanol/anode off-gas dedicated to fuel processing operations [555]. The burners were designed for thermal loads of up to 11.5 kW. Palladium catalyst deposited onto porous ceramic fibre was used for natural gas combustion. Only 0.5 g palladium was required for the burner, which was water-cooled by a concentric tube-bundle cooler mostly through radiation losses. At 11.5-kW power output, carbon monoxide emissions were below 10 mg kWh^{-1}, while NO$_x$ emissions were around 2 mg kWh^{-1}. The methanol burner worked with a platinum catalyst and showed lower NO$_x$ emissions of 0.4 mg kWh^{-1}.

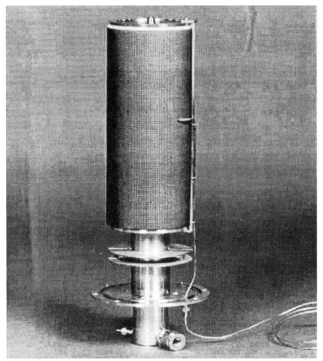

Figure 7.43 Permeate burner of the methanol fuel processor developed by Hansen et al. within the scope of a European Joule III project [553].

Within the scope of development work leading to a methanol fuel processor, a catalytic burner for anode off-gas combustion was built by Höhlein et al. [556]. This burner was designed as a separate device and not integrated into the reformer. The platinum/alumina catalyst [557] for the combustion reaction was deposited on wire mesh. The coating thickness on the wire mesh was calculated to be 260 µm [18]. The burner, which is shown in Figure 7.43, was cooled by water.

Arana et al. designed a small-scale catalytic combustor focussing on minimised heat losses, which could also be used for partial oxidation reactions [558]. For very small scale systems, heat conduction in heat-exchangers is dominated by heat transfer along the walls. This effect needs to be minimised to maintain efficiency. According to Arana et al., heat losses are even enhanced by insulation material as soon as the size of the insulated component is smaller than a few millimetres. Their so-called suspended tube reactor consisted of four rectangular ducts (called tubes) made from silicon nitride with a very low wall thickness of 2 µm in order to minimise axial heat transfer. The ducts had a width of 200 µm and a height of 480 µm. At each end, these ducts were embedded into a silicon substrate, which carried the connection to the external tubing at the front and formed a reaction zone at the end, as shown in Figure 7.44. The first and second couples of the tubes were

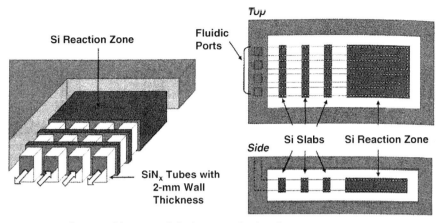

Figure 7.44 Schematic of the suspended tube reactor developed by Arana *et al.* [558]; the silicon slabs served for lateral heat transfer; homogeneous combustion took place in the silicon reaction zone.

connected to U-shaped reactors. The silicon reaction zone was positioned at the bend. In the middle of the tube length, silicon slabs served as heat transfer material and by these means the feed for the reaction zone was pre-heated with hot combustion off-gas. The slabs generated isothermal conditions between the inlet and outlet tubes, according to CFD simulations. The highest heat transfer resistance was calculated for the fluid itself. Thus, temperature profiles were only calculated across the width of the ducts, but not within the wall material, as shown in Figure 7.45. Posts made of silicon coated with silicon nitride were introduced into

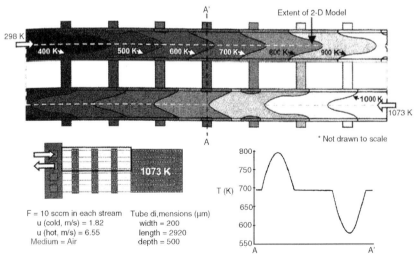

Figure 7.45 Temperature profile from 2-D CFD simulation for the recuperation ducts of the suspended tube reactor [558].

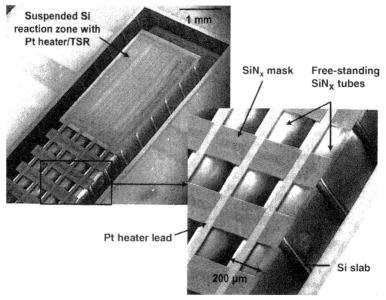

Figure 7.46 Suspended tube reactor developed by Arana et al. [558].

the silicon reaction zone. They improved the heat conduction and served as a support for the catalyst. By vacuum packaging of the device, steady state heat losses were reduced by 60%, the remaining losses were mainly attributed to radiation losses at temperatures above 500 °C. Meandering thin film platinum lines were coated onto the outer surface of the silicon reaction zone and on the supporting chip. They served both for resistance heating and temperature measurements. The reactor could be heated to 900 °C by these means. Platinum/alumina catalyst was introduced into the device for butane combustion experiments. After brief electrical heating for 10 s, the reaction ignited and proceeded without external heating. Full conversion of a stoichiometric butane/air mixture was achieved for 0.8 mL min^{-1} butane feed flow rate, which corresponded to an energy generation of 0.8 W. Figure 7.46 shows the reactor.

Microstructured reactors have high potential for the controlled combustion of all types of fuels, even for hydrogen at high temperatures due to the flame arresting features of small channel systems (see also Section 5.3).

Numerous examples of controlled hydrogen oxidation in microchannel systems do exist [424, 559–561] but will not be discussed in detail here.

8
Balance-of-Plant Components

Some examples of balance-of-plant components will be presented below, which does not claim to be a complete discussion by any means.

8.1
Heat-Exchangers

Castaldi *et al.* emphasised the need for compact-sized heat-exchangers in fuel cell systems [325]. However, the devices that are specifically dedicated to water condensation require excessive amounts of space in many instances. Nevertheless they are inevitable to close the water balance of the system.

For fuel processing applications, compact and high temperature resistant heat-exchangers do exist. However, these devices are still in the prototype and small series stage.

An important issue is regaining the low temperature heat in a fuel processor/fuel cell system, because the low temperature heat losses determine the system efficiency. Polymer heat-exchangers are possibly an alternative to conventional technology, which has the potential for weight and cost savings. Zaheed and Jachuk presented the development of a corrugated polymer cross-flow heat-exchanger made from poly (etheretherketone) (PEEK) films of 70-μm thickness [562]. The corrugations were of 2-mm width and 1-mm height and were alternated with uncorrugated films of 100-μm thickness for fluid flow separation. The maximum operating temperature of the heat-exchanger was 220 °C and maximum pressure difference 10 bar. Several fuel cell heat-exchanger applications in the kW range were discussed and weight savings of up to 96% compared with stainless steel heat-exchangers and up to 88% compared with aluminium heat-exchangers were calculated. Pressure drop savings were in the range of 75% for stainless steel and 40% for aluminium.

However, a counter-current design will increase the efficiency of the polymer heat-exchanger and allow for low temperature differences between the fluids, which would further increase their efficiency.

Fuel Processing for Fuel Cells. Gunther Kolb
Copyright © 2008 WILEY-VCH Verlag GmbH & Co. KGaA, Weinheim
ISBN: 978-3-527-31581-9

Figure 8.1 Micro-membrane liquid pump NF-5M from KNF Neuberger (photograph: courtesy of KNF Neuberger).

8.2
Liquid Pumps

The field of liquid pumps for process engineering applications is huge and its discussion is certainly not within the scope of this book. However, small pumps in the range of a few watts of power consumption are an issue in many portable applications. Therefore, an example of a pump of the smallest scale is shown below. It is a micro-membrane pump from the KNF Neuberger Company with a flow capacity of $50\,\text{cm}^3\,\text{min}^{-1}$ at atmospheric pressure. The pump is able to provide a maximum pressure of 2 bar for a power consumption of 0.66 W (Figure 8.1).

8.3
Blowers and Compressors

The air supply is provided either by blowers or compressors. The difference between these components is the pressure level provided. While compressors can achieve elevated pressures up to 10 bar or higher when designed with multiple stages, low power blowers are usually limited to fractions of a mbar. However, high pressure blowers do exist, which are able to achieve a supply pressure of 20 mbar or even up to several hundred mbar. This higher pressure is achieved by higher rotation speed and is accompanied by an increased power demand.

Some simplified equations are provided below for the estimation of the power consumption of compressors and blowers.

The power demand of a blower may be calculated by a simplified equation, according to Liu et al. [445]:

$$P_{\text{blower}} = \frac{2.67 \times 10^{-5}\, p_{\text{blower}}\, \dot{V}_{\text{air}}}{9.8\eta_{\text{blower}}} \tag{8.1}$$

where P_{blower} is the blower power demand in watts, p_{blower} is the pressure at the blower outlet in pascal, $\dot{V}_{\text{air}}$ the volume flow of air in $\text{L}\,\text{s}^{-1}$ and η_{blower} the blower efficiency as a fraction of unity. Efficiencies in the range of 50% at full load and 25% at partial load serve as a rough estimation for larger blowers [563].

Blowers on a larger scale can also provide much higher pressures. Clark and Arner reported on the development of a radial blower designed for a pressure build-up of 800 mbar at flow rate of $2\,m^3\,min^{-1}$, which worked with an extremely high speed of 150 000 rpm [563].

The power demand of an isothermal compressor (P_{comp}) can be calculated according to the following formula:

$$P_{comp} = \frac{\dot{m}_{gas} \dfrac{R}{M_{gas}} T_{in} \ln \dfrac{p_{out}}{p_{in}}}{\eta_{comp}} \qquad (8.2)$$

where $\dot{m}_{gas}$ is the mass flow of the compressed gas, R the universal gas constant, M_{gas} the molar mass of the compressed gas, T_{in} and p_{in} the inlet temperature and pressure, respectively, p_{out} the outlet pressure and η_{comp} the overall efficiency of the compressor.

Efficiency values of around 80% can be assumed for compressors [407], however, this value might well be lower at partial load of the device.

The development of a hybrid compressor/expander module was reported by Selecman and McTaggart [564]. It consisted of a scroll compressor that was fed by a separate compressor/expander module. The latter utilised the remaining compression energy of the fuel cell off-gas. However, the application of expander turbines, though attractive to reduce parasitic power losses of the compressor, is certainly limited to a minimum system power equivalent of 10 kW or higher due to weight, complexity and price. Gee reported on a turbo compressor/expander module developed by Honeywell for a 50-kW fuel cell [565]. The compressor, which worked with up to 110 000 rpm, had 8.0-kW input power with an expander and 14.3 kW

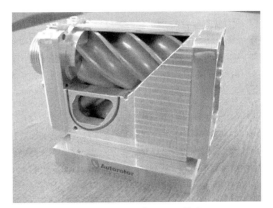

Figure 8.2 View into an Opcon Autorotor Twin-screw compressor providing oil-free air for use in pressurised fuel cell systems (photograph: courtesy of Opcon).

Figure 8.3 Micro-vane compressor G01–4EB from Rietschle Thomas (photograph: courtesy of Rietschle Thomas).

without [566]. The weight of the compressor and motor was 11 kg, the controller added another 6.5 kg, and the total volume was 15 L.

Other companies developing compressors for fuel cells are Vairex, who offer vane compressors, and Opcon, who have twin-screw compressors in their portfolio (see Figure 8.2).

For small scale applications, either miniaturised vane compressors, piston compressors or membrane compressors may be the preferred solution, depending on the application.

An example of a compressor on the smallest scale is shown in Figure 8.3. It is a micro-vane compressor from the Rietschle Thomas Company with a delivery rate of 0.4 L min^{-1} at a supply pressure of 200 mbar and power consumption of 2.2 W.

8.4
Feed Injection System

The addition point of gaseous fuels requires careful consideration to avoid homogeneous reactions upstream of the reformer with autothermal reforming and partial oxidation. Commercial flame arresters are normally not capable of operating under the elevated temperatures of the fuel processor. Microchannels are known to act as flame arresters (see Section 6.3.2) and may be inserted into the tubing system to avoid uncontrolled reaction of the fuel/air mixture. For liquid fuels, which are usually injected into the pre-heated steam feed or even into the air/steam feed mixture, either cooled injection nozzles [567] or the application of steam jackets may be used to ensure stable operation of the nozzle.

8.5
Insulation Materials

The application of appropriately chosen and efficient insulation materials is crucial for the efficient operation of the fuel processor. For very small scale fuel processors in the power range of a few watts, a vacuum packaging may be the only choice to avoid extensive heat losses. Another issue seems to be radiation losses for systems on the smallest scale even at temperatures well below 600 °C. Therefore, radiation shields [568] frequently surround the fuel processors in the low watt power range. Thin layers of gold deposited by vapour deposition are an efficient radiation shield.

Highly efficient insulation materials, based on silica, are provided by commercial suppliers, which have low heat conductivities in the range of from 0.025 to 0.05 W (m K)$^{-1}$ depending on temperature and the material. Another advantage of this type of material is its radiation reflection property, which becomes crucial at temperatures greater than at least 650 °C (see Section 5.4.1). These materials may be used as solid plate material or in the form of granules of different particle sizes. The advantage of the solid plates is that they may be milled to the appropriate shape, as shown in Figure 8.4, and quickly attached to the fuel processor. Alternatively, the fuel processor could be put into a housing that is then filled with granules of the insulation material.

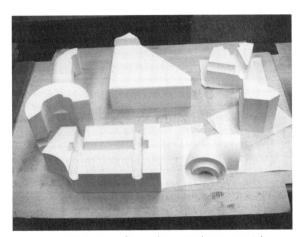

Figure 8.4 Milled plates of Microtherm insulation material (photograph: courtesy of Microtherm).

9
Complete Fuel Processor Systems

The requirements concerning volume, weight, efficiency and cost are most stringent with fuel processing for an automotive drive train. Below is an example of the specifications for a complete fuel cell/fuel processor system, which dates back to 2000 and had been specified by Docter *et al.*, from Daimler-Chrysler [489]:

- cost < 40 €/kW
- weight <4 kg kW^{-1} and volume < 3 L kW^{-1}
- system efficiency > 35%
- dynamics (idling to 90% of full power) < 1 s
- cold start < 1 min
- durability > 5000 h
- turn-down ratio 1:50.

The US Department of Energy has set key performance targets for drive train fuel processors, which are provided in Table 9.1.

9.1
Methanol Fuel Processors

Figure 9.1 shows a methanol steam reformer developed by the Kellogg Cy Company for the US Army. The system was designed for a 200-kW submarine fuel cell. The system worked with liquid oxygen as feed for the combustion chamber, in which off-gas from the membrane separation unit (L-1) was combusted during normal operation. During start-up, methanol was combusted to pre-heat the system. The oxygen was vaporised in evaporator C-3. The energy for evaporation came from the purified hydrogen, which had to be cooled before it entered the fuel cell at a temperature of 26 °C. The oxygen was then super-heated in the heat-exchanger C-4 by hot combustion off-gas. Methanol and water were fed by pumps J-1 and J-2 to the pre-heater C-1, which also utilised the heat of the purified hydrogen downstream of the separation unit. The methanol/water feed was further pre-heated by a split stream of the combustion off-gas in heat-exchanger C-2. The remaining combustion

Fuel Processing for Fuel Cells. Gunther Kolb
Copyright © 2008 WILEY-VCH Verlag GmbH & Co. KGaA, Weinheim
ISBN: 978-3-527-31581-9

Table 9.1 US Departments of Energy technical targets for an on-board fuel processor [448].

Characteristics	Calendar year		
	2003 Status	2005	2010
Energy efficiency[a] (%)	78	78	80
Power density (W/L)	700	700	800
Specific power (W/kg)	600	700	800
Cost[b] ($ per kW$_{el}$)	65	25	10
Cold start-up time to maximum power			
@ $-20\,°C$ ambient temperature (min)	TBD	2.0	1.0
@ $+20\,°C$ ambient temperature (min)	<10	<1	<0.5
Transient response (time for 10–90% power) (s)	15	5	1
Durability[c] (h)	2000[e]	4000[f]	5000[g]
Survivability[d] (°C)	TBD	-30	-40
CO content in product stream			
Steady state (ppm)	10	10	10
Transient (ppm)	100	100	100
H_2S content in product stream (ppb)	<200	<50	<10
NH_3 content in product stream (ppm)	<10	<0.5	<0.1

[a] Energy efficiency = total fuel cell system efficiency/fuel cell stack efficiency; or lower heating value of hydrogen × 0.95/lower heating value of fuel.
[b] 500 000 units/year.
[c] Time between catalyst and major component replacement; performance targets must be reached at the end of the durability period.
[d] Performance targets must be achieved at the end of an 8-h cold soak at a specified temperature.
[e] Continuous operation.
[f] Includes thermal cycling.
[g] Includes thermal cycling and realistic driving cycles.

off-gas was fed back to the combustor by the recycle compressor J-4. The feed entered the tubular reformer at a temperature of 178 °C, while the combustion gases were fed into the reformer vessel at a high temperature of 538 °C in co-current mode. The reformate left the reformer at a temperature of 273 °C and was purified in the membrane separator L-1 and then cooled in the heat-exchanger C-1 and in the oxygen evaporator C-3, as described above. The whole system was thus composed of two reactors, five heat-exchangers and four feed pumps, which does not seem to be too complex considering the size of the system.

In 1966, the Institut Francaise de Petrole (IFP) designed and built a methanol fuel processor of much smaller size for an alkaline fuel cell [12], which is shown in Figure 9.2. The methanol steam reformer was operated at 250 °C and contained a fixed bed of copper/chromium catalyst. The alkaline fuel cell required the removal of carbon dioxide from the reformate and thus a small diethanol-amine gas scrubber was built. Residual carbon oxides were removed by a methanation reactor, which contained a nickel/chromium fixed catalyst bed.

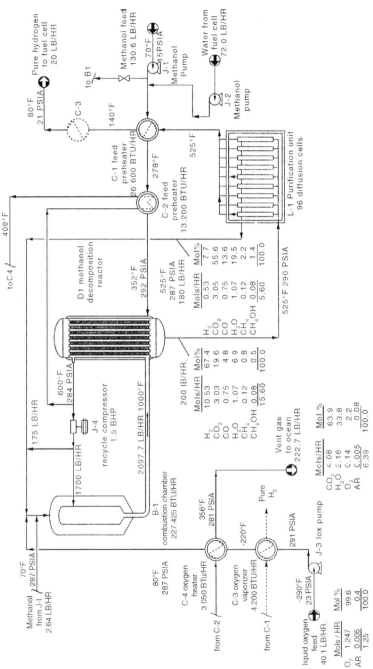

Figure 9.1 Process flow-scheme of a methanol fuel processor established for a 200-kW submarine fuel cell by Kellogg Cy [12].

Figure 9.2 Methanol fuel processor developed by the Institute Francaise de Petrole (IFP), www.ifp.fr for, a 2-kW alkaline fuel cell [12].

The development of a methanol fuel processor prototype was described by Höhlein et al. [556]. The methanol burner dedicated to this system has been described in Section 7.5. Later, a complete methanol reformer was developed by Wiese et al. [154]. It was operated at a S/C ratio of 1.5 and a pressure of 3.8 bar. The feed was evaporated and superheated to 280 °C. The reformer itself consisted of four pairs of concentric stainless steel tubes. In the annular gap between the tubes, steam was condensed at 65 bar and 280 °C for the heat supply, while the inner tube carried the copper/zinc oxide catalyst for steam reforming. The reformer response time to a load change from 40 to 100% was about 25 s, which was mainly attributed to the slow dynamics of the dosing pump. Because the dynamic behaviour of the reformer was too slow for an automotive drive system, which had been the target application of the work, an additional gas storage system was considered. To improve the system dynamics, Peters et al. considered the application of microreactor technology for a subsequent improved fuel processor [569].

Emonts et al. described the combination of the resulting 50-kW reformer with a hydrogen separation membrane system (see also Section 7.4) and a 1-kW Siemens PEM fuel cell to give a complete fuel processor/fuel cell system [50]. The breadboard system still required a footprint of 3 m^2. A flow scheme of the system along with a photograph is provided in Figure 9.3; 3 kg of a copper/zinc oxide catalyst were

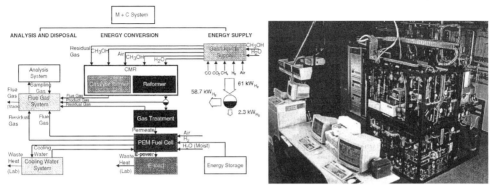

Figure 9.3 Flow scheme (left) and photograph (right) of the 50-kW methanol steam reformer/PEM fuel cell system developed by Forschungszentrum Jülich [50].

required for the steam reformer and the total weight of the fuel processor amounted to 130 kg. Therefore, an enlarged version of the burner described in Section 7.5 with a power of 12 kW was built for the system. It heated the fuel processor to the desired operating temperature of 260 °C within 36 min through methanol combustion [50]. The steam circuit alone, which was expected to be the heat supply of the reformer as described above, required 12 min to reach operating temperature and stable conditions. Problems with unstable operation of the reformer feed evaporation system made a hydrogen buffer tank mandatory. Specific hydrogen production of 6.7 m^3 H$_2$ (kg$_{cat}$ h)$^{-1}$ was achieved at 280 °C reformer temperature and 95% methanol conversion. More than 80% efficiency was calculated for the fuel processor at a 280 °C reforming temperature. At a 260 °C reformer temperature, efficiency was somewhat lower, between 77 and 83% in the load range between 25 and 100%, respectively. At 6% load, the efficiency decreased to 60%. An evaluation of the reformer efficiency over the New European drive cycle revealed a value of 77% [570]. A hydrogen storage vessel was proposed by Emonts *et al.* for their system as an intermediate energy buffer to make fast load changes possible [50].

Dams *et al.* described the overall results of the MERCATOX project funded by the European Community [571]. The target of the project was to set up a 20-kW methanol fuel processor for automotive applications based on plate heat-exchanger technology. The fuel processor consisted of different plate–fin heat-exchangers, namely of an evaporator with 5-L volume, and a heat-exchanger with 0.5-L volume for methanol combustion and superheating of the reformer feed. Both devices were fed with energy by catalytic combustion. A 0.5-L heat-exchanger served for pre-heating the reformer feed with its hot product gases. The reformer reactor itself had a volume of 5 L. The carbon monoxide clean-up was performed by preferential oxidation in two oil cooled plate–fin heat-exchangers, each of 2-L volume, which have been described in detail in Section 7.3.3. The single components could be operated at about 20–25% of their design capacity when tested in the complete system. Unfortunately little information on the overall performance of the system was provided by Dams *et al.*

Johnson Matthey developed the HotSpot™ fuel processor, a modular autothermal reformer. The basic idea of the HotSpot™ was that hydrogen back-diffusion to the reaction front, where it would be consumed by the oxygen feed, was prevented by the spot wise feed injection into the centre of the reactor. The heat of the exothermic reaction was also distributed from the centre of the reactor to its periphery, where it was required to supply the endothermic reactions with energy [572]. The original reactor had been invented in the 1980s [573]. In this early version of the reactor, the hot spot was a small spherical reaction space, in which most of the fuel was converted over a base metal catalyst such as 3 wt.% copper on silica [572], which had been designed for partial oxidation. The feed was injected into the centre of this space. The remaining reactor was filled with the same copper/silica catalyst, to which some 10 wt.% palladium catalyst was added. The latter contained 5 wt.% palladium supported by silica and served as a total oxidation catalyst. During cold start, total oxidation of the fuel took place in the outer reactor volume and the heat was transferred to the centre, where then partial oxidation started. The operating temperature at the hot spot was 600 °C and stable operation was achieved within 25–30 min [573]. The hydrogen output of this early design was 12 L $(L_{cat})^{-1}$ [574].

The unit was then improved for onboard methanol processing dedicated for the automotive drive train [573] but also for residential combined heat and power systems [575]. The catalyst bed was uniform and filled with a combined precious metal/base metal multi-component catalyst. The reaction produced 2.4 mol hydrogen per mole methanol [573], which corresponds to a S/C ratio of about 0.4 and an O/C ratio of about 0.6 in the feed. These conditions resulted in 58 vol.% dry hydrogen content of the reformate. The maximum temperature in the fixed bed was reduced to 400 °C [573]. The system start-up was performed under conditions of partial oxidation. The power density of the reactors was 3 kW L^{-1}. When the volume of the manifolds was taken into consideration this value was reduced to 1 kW L^{-1}, and to 0.75 kW L^{-1} when the carbon monoxide clean-up system was included. Owing to the high reaction temperature, the carbon monoxide content of the reformate supplied by the reformer ranged between 2 and 3 vol.%.

An eight unit HotSpot™ processor (see Figure 9.4) had a 6-L volume, 8.8-kg weight and produced 6 $m^3 h^{-1}$ hydrogen, [573]. The processor was started with partial oxidation of methanol. After 25 s, the reactors had been pre-heated and pre-heating of the manifolds could begin. After 100 s, when the manifolds had reached their operating temperature of 130–150 °C, the operating parameters could be changed to autothermal reforming [573], which finally reached steady-state conditions after 170 s [574]. The reformer was then operated at an O/C ratio of around 0.45 [574]. The start-up time demand could be reduced to 50 s by increasing the air feed during start-up by 20% [574], which corresponded to an O/C ratio of 0.55. Alternatively, hot off-gas from a separate anode off-gas afterburner was considered as the heat source for the manifold heating. It was emphasised that the S/C ratio of the fuel processor could be increased and the O/C ratio decreased by utilising the heat from such an afterburner [573]. This in turn increased system efficiency. At low system load some of the modules could be switched off, which increased the system efficiency under

Figure 9.4 8 unit HotSpot™ fuel processor producing 6 m³ h⁻¹ hydrogen [577].

these conditions [574]. The overall efficiency of the HotSpot reformer with anode off-gas combustion was determined to be 95.4% [574]. Reinkingh *et al.* reported on a methanol HotSpot processor which had been designed for a 20-kW$_{el}$ PEM fuel cell system [576].

The amount of carbon monoxide in the reformate was very low during start-up. However, about 2 vol.% carbon dioxide were still present in the reformate during normal operation, which made gas purification mandatory. The reformate left the manifold of the fuel processor at a temperature of 180 °C [574]. This temperature suggested the preferential oxidation of carbon monoxide as a clean-up process [574], because it is operated in the same temperature range (see Sections 3.10.2 and 4.5.2). To address this issue, a multiple stage catalytic carbon monoxide clean-up system (Demonox™) was developed, which worked, preferably, with preferential oxidation [575] but also alternatively with selective methanation [573]. The overall hydrogen loss was reported to be 6 vol.% for the removal of about 2 vol.% carbon monoxide [573]. Therefore, the equivalent overall O/C ratio for preferential oxidation was about 4. This clean-up system was up-scaled to an 18-kW$_{el}$ power equivalent [573]. Figure 9.5 shows a P2 prototype of the HotSpot™ fuel processor linked to a Demonox™ unit. The efficiency of this fuel processor system was calculated to be 89%. When the anode off-gas recycling to the fuel processor was taken into consideration for the calculations at 85% hydrogen utilization in the fuel cell [see Eq. (2.4) in Section 2.2], the fuel

Figure 9.5 P2 HotSpot™ fuel processor (top) linked to a Demonox™ carbon monoxide clean-up unit (bottom); the manifolds were placed in the centre [575].

processor efficiency was 76% [576]. For the entire fuel processor/fuel cell system, an overall efficiency of 40% was calculated [574].

Schuessler *et al.* [398] from XCellsis, subsequently BALLARD, presented an integrated methanol fuel processor system based on autothermal reforming, which is shown in Figure 9.6. The device had a size of 0.5 L and a weight of 1.8 kg at 1.2 m^3 h^{-1} hydrogen output [398]. Compared with most other approaches, the reactor technology was based on a sintering technique. This technology was chosen to achieve high heat conduction through the wall material

Figure 9.6 Single plate and complete integrated fuel processor for 20 L min^{-1} hydrogen output as developed by Schuessler from Ballard Power systems [398].

of the device. The fuel processor was built from copper powder, which could be sintered at temperatures between 500 and 700 °C. These temperatures were low enough to avoid damage of the catalyst during sintering, because the catalyst was incorporated into the reactor in the same fabrication stage. The catalyst was of the copper/zinc oxide type with a particle size of between 300 and 500 µm. Copper and aluminium powder were used as inert material for parts such as channels or diffusion layers of the fuel processor. The copper particles clung together after compression at only 1000 bar, which made the assembly of the fuel processor stack possible before sintering. The methanol and water evaporation zone of the fuel processor was closely coupled with the zone where the preferential oxidation of carbon monoxide took place at the reverse side of the same plate (see Figure 9.7). The evaporation was achieved by porous cylinders 7 mm in diameter and 1 mm in height introduced into each individual plate prior to the final sintering step that sealed the reactor. An air flow took up the evaporated feed components and the entire feed mixture was then mixed and pre-heated to

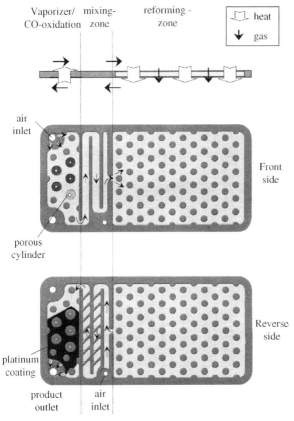

Figure 9.7 Front and reverse side of a single plate of the fuel processor developed by Schuessler et al. [398].

reaction temperature in a serpentine flow arrangement in the centre of the device, as shown in Figure 9.7.

Autothermal reforming took place, again unlike other approaches, not along the length of the plate, but rather perpendicularly through the porous catalyst mixed with sintered copper particles (pore diameter 2 µm, 60% porosity). As the whole reaction zone was only a few millimetres long, the hot spot formation was suppressed by thermal conduction through the wall material. The porous structure allowed a moderate gas hourly space velocity of $15\,000\,h^{-1}$ at a low flow rate of $0.1\,m\,s^{-1}$, which corresponded to a very low Re number of 0.01 [578]. Temperature gradients of less than 10 K were determined along the short reaction zone. Load changes did not change the temperature profiles in the reactor, only the residual methanol increased up to 1.5 vol.% [578]. The methane concentration was maintained below 100 ppm. Downstream the reformer outlet, air was added to the reformate in a mixing zone before the reformate entered the preferential oxidation zone (see Figure 9.7). The pressure drop over the whole coupled evaporator/reformer/preferential oxidation fuel processor was limited to 200 mbar at full load. At a 280 °C reaction temperature, $15\,mol\,h^{-1}$ methanol feed and a S/C ratio of 1, almost complete (>99%) conversion of the methanol could be achieved leaving just 0.9 vol.% methanol in the reformate. The carbon monoxide concentration was reduced from 1.3 vol.% at the reformer outlet to about 0.1 vol.% downstream of the preferential oxidation reactor. Around 50% selectivity was achieved in the preferential oxidation stage. The system was started by feeding with a surplus of air. Total combustion of methanol thus provided the energy for start-up. It took place specifically in the preferential oxidation stage. Owing to this indirect heating via the preferential oxidation stage, the system had a 10 min start-up time demand, which is still a fairly low value, especially when considering the power equivalent.

Certainly one of the most prominent fuel processor systems for automotive applications was implemented into the NECAR 3 system presented by Daimler-Benz at the Frankfurt automotive fair in August 1997 [579].

NECAR 3 (see Figure 9.8) was the first automobile that was driven by a methanol fuel processor/fuel cell system; 90% of the full system power was released by this system within 2 s, which is adequate performance compared with conventional

Figure 9.8 NECAR 3, the first vehicle powered by a fuel cell supplied by a methanol reformer [577].

9.1 Methanol Fuel Processors

Fuel cell system	Power (2Stacks)	50 kW
	Voltage range	185 - 280 V
Tank	Volume	38 l
Drive train	Electric drive cont.	33 kW
		max 45 kW
	Range	> 300 km
Gross vehicle weight		1750 kg

Figure 9.9 Cross-sectional view of NECAR 3 [70].

combustion engines. Carbon monoxide and nitrous oxide emissions of this vehicle were below the detection limits [580]. It had a range of 400 km [579]. However, there were only two seats in the vehicle, as shown in Figure 9.9. The electric engine had 33-kW power in permanent use and a 45-kW peak power [70]. The fuel processor of NECAR 3 was the ME 50-3 system from XCellsis [70]. It worked with methanol steam reforming supplied with energy by a thermal oil cycle from a catalytic afterburner, which also indirectly heated the evaporator [70]. The evaporator was a microstructured heat-exchanger [70]. The carbon monoxide removal was performed by a water-cooled preferential oxidation unit [70].

Cooper reported on development work by the former dbb Company aimed at developing a methanol-fuelled bus [581]. Two NECAR 3 fuel processors switched in parallel were coupled to six Ballard Mark 7 stacks resulting in 100 kW of net continuous power. A turn down ratio of 8:1 could be achieved. A two stage air compressor with heat and pressure energy recovery was used as the air supply.

Following on from NECAR 3, a second generation vehicle with a methanol fuel processor was built: NECAR 5 [70]. It was the first fuel cell vehicle with a methanol reformer to cross the USA. The ME 75-5 fuel processor of NECAR 5 is shown in Figure 9.10 [70]. Its evaporator was integrated into the catalytic afterburner, the reformer and preferential oxidation reactions were also combined. The start-up was performed in a novel reactor allowing exothermic, autothermal and endothermic operation according to zur Megede [70]. This means that the reformer could probably be operated in a combustion mode, autothermal mode and under the conditions of steam reforming.

Toyota developed a methanol-fuelled fuel cell vehicle that was presented shortly after the NECAR 3, in September 1997, which is shown in Figure 9.11. It had a range of 500 km and a maximum power of 50 kW. It worked with a methanol steam

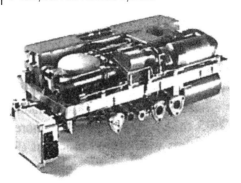

Figure 9.10 The ME 75-5 methanol fuel processor developed by XCellsis for NECAR 5 and other fuel cell vehicles [70].

reformer and catalytic carbon monoxide clean-up by water–gas shift and subsequent preferential oxidation [30].

An Opel Zanfira powered by a 30-kW fuel cell, which was supplied with hydrogen from a methanol fuel processor, was developed by General Motors [582]. The maximum efficiency of this system amounted to 30%, which considerably exceeded the efficiency of internal combustion engines. One of the development goals was to achieve a gasoline consumption equivalent of 3 L per 100 km [583].

The development of an even bigger methanol fuel processor/fuel cell system with a power output of 75 kW was reported by Yan *et al.* [584]. The autothermal fuel processor, which is shown in Figure 9.12, relied on catalytic carbon monoxide clean-up

Figure 9.11 Methanol-fuelled fuel cell electric vehicle (FCEV) developed by Toyota [30].

Figure 9.12 75 kW$_{el}$ methanol fuel processor developed by Yan et al. [584].

by a water–gas shift and preferential oxidation. It produced about 120 m^3 h^{-1} reformate containing 53 vol.% hydrogen and less than 30 ppm carbon monoxide during stationary operation. Load changes, however, led to carbon monoxide peak concentrations exceeding 100 ppm.

The development of a methanol fuel processor with a power equivalent of 240 kW$_{el}$ for naval applications was described by Sattler, from the HDW Company [585].

Membrane separation is a viable hydrogen purification process, especially for liquid fuels such as methanol.

A methanol fuel processor based on steam reforming in a fixed catalyst bed and membrane separation was described by Ledjeff-Hey et al. [401]. The system consisted of an evaporator, a steam reformer, which was supplied with heat by a catalytic burner, and a membrane separation module, which carried membranes of a very high thickness of 7.5 mm. At 5-bar system pressure and S/C ratio of 2.0, a hydrogen flow equivalent to 1.1-kW thermal power was generated by the system, which had an overall efficiency of 54%. Between 40 and 62% of the hydrogen produced by the reformer could be separated by the membrane module. Leakages in the sealing of the membrane module led to carbon monoxide spill-over to the permeate, but this was limited to carbon monoxide concentrations well below 100 ppm.

IdaTech LLC developed a 2-kW methanol steam reformer, which worked with a palladium membrane for reformate purification. The purified hydrogen then supplied a Ballard NEXA™ PEM fuel cell stack. Löffler et al. [105] and Edlund [419] described the reformer design of Idatech (see Sections 5.2.4 and 9.4). Later, IdaTech presented a 250-W methanol fuel processor/fuel cell system known as IGen®, which is shown in Figure 9.13. The system had a weight of 12 kg and a volume of about 30 L.

A 25-kW methanol fuel processor system based on membrane separation technology was described by Han et al. from the SK Corporation [587]. These workers claimed to produce high purity hydrogen (greater than 99.9995 vol.%) with their fuel processor. This had the consequence that the anode feed did not require an off-gas exit, but the

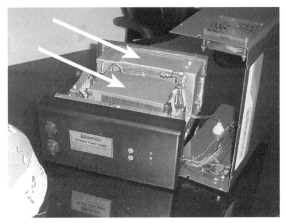

Figure 9.13 IGen 250-W methanol-fuelled portable fuel cell system with reformer and membrane separation developed by IdaTech: the upper arrow indicates the position of the fuel processor, the lower arrow the fuel cell [586].

anode could be designed with a dead-end, according to Han et al. This decreased the size of the anode gas distribution channels. The system worked with steam reforming and consequently only methanol and water were fed to the reformer [587]. Therefore, no compressor was required to generate the elevated system pressure that is mandatory for high purity membrane separation. Liquid pumps were sufficient.

This development work was performed in several stages; 2- and 10-kW systems [402] were built before the final size of 25 kW was achieved with the third generation prototype. The size of the 10-kW second generation methanol fuel processor was still fairly bulky at more than 860 L [402]. However, the efficiency of about 82% was relatively high already for membrane separation. The membrane separation modules have been described in Section 7.4. Figure 9.14 shows the gas flows and gas compositions for the 10-kW$_{el}$ system. About 95% methanol conversion was achieved and the reformate contained 3 vol.% carbon monoxide. The system had a high start-up time demand of between 30 and 60 min. The response to the load changes required between 2 and 3 min [402].

The 25-kW final methanol fuel processor, which was designed to meet the space limitations of a Hyundai "Santa Fe" passenger car is shown in Figure 9.15. Han et al. did not use a compressor/expander unit, but instead the air was fed to the system afterburner by a high performance blower. The blower had a supply pressure of 275 mbar, which decreased the parasitic power losses compared with a compressor. The steam reformer of Han et al. was operated with 7 kg of ICI 33-5 M commercial catalyst at a S/C ratio of 1.3 and 97% conversion was achieved [587]. The methanol and water feed was pre-heated in a heat-exchanger by the hot reformer product. The feed was then evaporated in a coupled evaporator/catalytic burner. According to Han et al., methane was formed in their steam reformer, in addition to the carbon oxides and hydrogen. The reformate was then purified by the membrane separation, which was operated at between 300 and 350 °C and 9-bar pressure. Between 70 and 75 vol.% of the hydrogen permeated through the membranes. Carbon monoxide, methane,

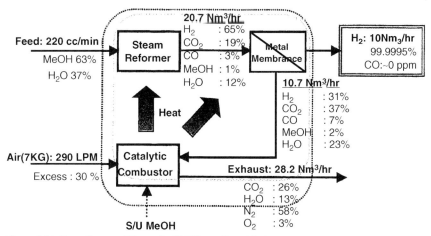

Figure 9.14 Operating conditions of a 10-kW$_{el}$ methanol fuel processor working with membrane separation as developed by Han et al. [402].

residual hydrogen and unconverted methanol contained in the retenate were combusted in the coupled evaporator/burner of the system. Methanol could be added to the afterburner feed for start-up purposes and temperature control of the evaporator. The combustion reaction also supplied the steam reforming and membrane separation with energy. Where a surplus of energy was generated by the system, the air feed of the catalytic burner was turned down. An additional catalytic afterburner then removed unconverted off-gases from the exhaust gas flow. Gas compositions and flow rates of the system are summarised in Table 9.2. The fuel processor had an efficiency of 75%. The purified hydrogen was then not fed to the fuel cell anode in a dead end arrangement, as had been proposed by Han et al. earlier. Only

Figure 9.15 25-kW methanol fuel processor developed by Han et al. [587].

Table 9.2 Gas composition and flow rates of the methanol fuel processor developed by Han et al. [587].

	Feed	Reformate	Permeate	Retenate
Flow rate	563 cm^3/min	53 Nm^3/h	25 Nm^3/h	28 Nm^3/h
Composition				
CH$_3$OH	63%	1%	—	2%
H$_2$O	37%	12%	—	23%
H$_2$	—	65%	99.9995%	31%
CO$_2$	—	19%	<5 ppm	37%
CO	—	3%	<1 ppm	7%

70% of the hydrogen was consumed by the fuel cell, which was connected to the fuel processor on a breadboard level. Because the fuel cell efficiency was rather low at 45%, a thermal power supply of 89 kW was required from the fuel processor. The start-up time demand of the system was estimated to be 15 min.

Methanol fuel processors are also under development for small scale applications in the power region below 100 W, and some of these systems will be described below. Since January 1st 2007 fuel cell cartridges containing flammable liquids or gases have appeared in international transportation regulations and their transportation is permitted in aircrafts when they comply with regulations [588]. The regulation does not apply to borohydride and metal hydride systems.

Target applications for micro fuel processors are cellular phones, laptops, camcorders and other types of portable electronic devices, but also military applications, because the power consumption of contemporary equipment for soldiers burdens them with considerable battery weight load – especially when the duration of the mission is long.

Detailed information about CASIO's reformer development was provided by Terazaki et al. [589]. The fuel processor was produced from 13 glass pieces connected by anodic bonding (see Figure 9.16). The fuel processor was composed of a gold thin film heater, a methanol steam reformer closely attached to a methanol burner, separate layers for the evaporation of the aqueous methanol solution that served as

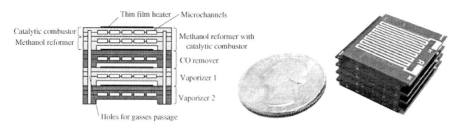

Figure 9.16 Small scale methanol fuel processor developed by CASIO: left, cross-sectional view; right, photograph of the fuel processor, which shows the gold thin film heater at the top of the fuel processor [589].

reformer feed and for the methanol fed to the burner, which supplied the steam reforming reaction with energy. Finally, one layer was dedicated to the preferential oxidation reaction to remove the carbon monoxide from the reformate. The thin film heater was sputtered onto the device, while the microchannels were introduced by sandblasting into glass layers. The anodic bonding between the Pyrex glass layers was performed with an intermediate tantalum film sputtered onto each glass substrate. The steam reformer was operated at 280 °C, while the preferential oxidation reaction was performed at 180 °C [589]. A vacuum insulation and high infrared reflective gold coating minimised heat losses from the system. The latter was required because the glass had high infrared emissivity. The fuel processor was 22-mm wide, 21-mm long and 10.7-mm thick. It was tested in connection with a miniaturised fuel cell [590]. It achieved 98% methanol conversion. About 2.5 W of electric energy were produced.

A similar methanol reformer with integrated heater was later presented by Kawamura *et al.*, also from CASIO [591]. It had been developed in cooperation with the University of Japan. Surprisingly, the reformer carried only one single meandering channel, which was 600-µm wide, 400-µm deep and 333-mm long but had only a 20-mbar pressure drop. The copper/zinc oxide catalyst was prepared from a boehmite alumina support by impregnation. It required reduction under a hydrogen flow and was operated at a S/C ratio of 2.0 and moderate weight hourly space velocity of 30 L $(h\, g_{cat})^{-1}$ as is usually required for copper/zinc oxide catalysts. Full methanol conversion could be achieved at 250 °C. The thermal power output of the hydrogen product was in the range of 3.3 W. Then a new generation fuel processor was designed [568]. It consisted of an anode off-gas burner, operated with a platinum catalyst at between 260 and 300 °C, which supplied a methanol reformer with energy. The reformer was operated in the same temperature range but now a palladium-based catalyst was used. The preferential oxidation reactor was operated between 110 and 130 °C. The fuel processor, which is shown in Figure 9.17, had 19-cm^3 volume and 30-g weight including the vacuum layer insulation and radiation shields. Through these means the heat losses of the system were reduced to 1.2 W. The

Figure 9.17 2.5-W methanol fuel processor developed by CASIO [568]: right, catalytic afterburner (top), methanol steam reformer (centre) and preferential oxidation reactor (bottom).

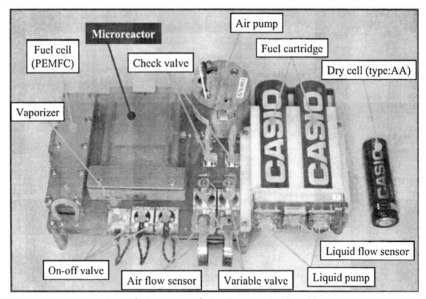

Figure 9.18 2.5-W methanol fuel processor/fuel cell system developed by CASIO [568].

electrical power consumption was around 70 mW. The fuel processor was combined with a fuel cell and balance-of-plant, which is shown in Figure 9.18. The system was operated at the Fuel Cell Seminar 2006, for demonstration purposes.

Motorola cooperated with Engelhard and the University of Michigan to develop a micro-structured steam reformer in a project funded by the US Commerce's Department Technology Administration [592]. The integrated fuel processor/fuel cell system consisted of an evaporator, a combustor, a reformer, heat-exchangers, insulation layers and the fuel cell. Ceramic technology was used. The maximum power output of this device, which was patented, amounted to 1 W [593].

Later, Hallmark et al. [594] presented a methanol fuel processor for the power range of from 10 to 100 W, which was under development at the Motorola laboratories, and was dedicated to operation in combination with a high temperature PEM fuel cell in order to decrease system complexity (see Section 2.3.1). The ceramic steam reformer worked with a copper/zinc oxide catalyst combined with a catalytic afterburner for anode off-gas combustion. A platinum/palladium catalyst was used in the afterburner. Very high catalyst loadings of up to 66 vol.% could be achieved by co-firing the catalyst onto the ceramic sheets, which was regarded as an advantage compared with slurry coating processes (see also Section 4.1.3).

Shah and Besser presented results from their development work that was aimed at a 20-W_{el} methanol fuel processor/fuel cell system [436]. The principle layout of the device consisted of a methanol steam reformer, preferential oxidation, a catalytic afterburner and an evaporator. The basic process and design parameters are summarised in Table 9.3. Nevertheless, it is obvious that the size of the steam reformer exceeded by far the size of all other components. The weight hourly space

Table 9.3 Overview process parameters and first layout results of the 20-W fuel processor under development by Shah and Besser [436].

Component	Process parameters and design results
SR	60% FC efficiency, 75% H_2 utilization CH_3OH conversion 100% $F_{MeOH} = 6.233\text{E-}5$ mol/s $F_{H_2O} = 8.10\text{E-}5$ mol/s $S/C = 1.3$ $T = 260\,°C$ $P = 1$ atm Microchannel: $50\,\mu m \times 400\,\mu m$ Parallel microchannels: 1000 Channel to channel spacing: $50\,\mu m$ Reaction zone width: 10 cm Catalyst film: $Cu/ZnO/Al_2O_3$, $5\,\mu m$ thickness (0.0478 g/cm) Reactor layers: 24 Total catalyst loading: 2.87 g
PrOx	$160\,°C$ Catalyst film: Pt/Al_2O_3, $5\,\mu m$ thickness (0.0478 g/cm) CO in reformate: 2.2% Reactor layers: 9 Total reactor loading: 1.1 g
Combustor	Off gas recycle to combustor: $2.5\text{E-}4$ mol/s Gas composition: H_2: 33.8 Vol.%, CO_2: 38.8 Vol.%, H_2O: 11.8 Vol.%, O_2: 0.6 Vol.% and N_2: 15.1 Vol.% Total catalyst loading: 1.2 mg

velocity that had been assumed by these workers was very low even for the copper/zinc oxide catalyst. Therefore, a packed catalyst bed was finally chosen by Shah and Besser to make the incorporation of sufficient catalyst possible.

The insulation strategy for the device was vacuum packaging, which is in line with other small scale system designs described above. A micro-fixed bed steam reformer coupled to a preferential oxidation reactor was then developed by Shah and Besser with a theoretical power output of 0.65 W. The heat supply of the combustion reaction was simulated by the platinum heater, as shown in Figure 9.19. The BASF commercial steam reforming catalyst was milled to 70 μm and introduced into the device by pneumatic transport, filter structures retained the particles within the device. Heating experiments revealed that the steam reformer could not be heated to more than 230 °C by the electric heater, which had reached a temperature of 550 °C even under these conditions. This was attributed to heat losses by various mechanisms. Insulation by fibreglass decreased the heat losses substantially. The conclusion of

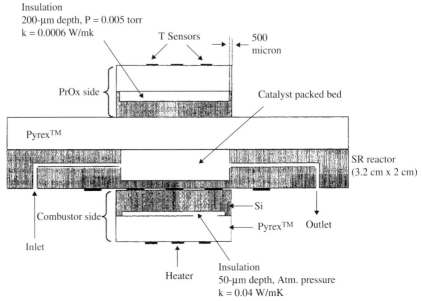

Figure 9.19 0.65-W methanol steam reformer and preferential oxidation reactor developed by Shah and Besser [436].

Shah and Besser was that it could be that it is impossible to build a self-sustaining fuel processor in the power region of a few watts, because minimisation of heat losses is limited. A minimum power of at least 10 W was required. However, the results from CASIO, described above, contradict these findings with practical examples.

Palo et al. [187] presented a concept of an integrated fuel processor for portable military applications. System specifications were less than 1 kg weight and less than 100 cm^3 volume, which corresponds to a power density of more than 0.15 kW L^{-1}. The concept of the system was to vaporise/pre-heat a methanol/air mixture, combust it in a separate combustor and feed the methanol steam reforming reaction with the energy of the hot combustion gases. Light-off of the combustion gases occurred at only 70 °C [595]. The combustion gases were further utilised to supply the fuel pre-heater/evaporator of the catalytic burner and finally the fuel pre-heater/evaporator of the reformer. Full conversion was achieved in the steam reformer, which was tested separately at a contact time of 300 ms and reaction temperature of 300 °C. At a very high reaction temperature of 375 °C, even 50 ms contact time was sufficient to achieve full conversion of the steam reforming reaction. Methane formation was below the detection limit of 100 ppm under these conditions [595]. Results generated with a partially integrated demonstrator, which was still heated by electricity during start-up, were presented. At 350 °C reaction temperature, 140-ms contact time and S/C ratio 1.8, full conversion of the methanol was achieved. The carbon monoxide concentration of 0.8% detected at this high reaction temperature was surprisingly low. The efficiency of the fuel processor was calculated to be 45% at an equivalent electrical power output of 14 W.

Kwon et al. presented a miniaturised methanol fuel processor/fuel cell system [596]. The reformer, a preferential oxidation reactor and the fuel cell were fabricated separately from silicon wafers by photolithographic methods. The wafers were sealed by anodic bonding of a Pyrex glass cover. Commercial copper/zinc oxide catalyst from Süd-Chemie was used for methanol steam reforming, while a platinum catalyst from Johnson Matthey served for the carbon monoxide removal. The fuel cell carried carbon monoxide tolerant platinum/ruthenium catalyst. The devices were heated by electrical thin film resistance heaters. The small reformer was $1\,cm^3$ in size and converted the methanol feed completely at a 280 °C reaction temperature and low S/C ratio of 1.0 and a weight hourly space velocity of 87.5 L $(h\,g_{cat})^{-1}$. The hydrogen output of the fuel processor corresponded to a thermal power of 3.2 W. The carbon monoxide concentration of the reformate was reduced from 3170 to almost 0 ppm at 220 °C reaction temperature by the preferential oxidation reactor, which had a size of only $0.57\,cm^3$. When the whole system was combined and tested, the fuel cell performance with reformate hydrogen was comparable to operation with pure hydrogen and no degradation of the fuel cell performance was observed for 20-hours duration.

Men et al. reported the operation of a small scale, breadboard fuel processor composed of electrically heated reactors [163]. A methanol steam reformer, two stage preferential oxidation reactors and a catalytic afterburner were switched in series. A small scale 20-W fuel cell equipped with a reformate tolerable membrane was connected to the fuel processor and operated for about 100 h.

Jones et al. [597] presented an integrated and miniaturised device for methanol steam reforming composed of two evaporators/pre-heaters, a reformer and a burner with a total volume of less than $0.2\,cm^3$ for a power range between 50 and 500 mW. This is probably the smallest fuel processor that has been presented in the open literature to-date. The energy for the steam reforming reaction was supplied by the burner, which had 3-W maximum power. It became obvious that the power generation of the burner had to exceed the heat consumption of the reformer by far, to compensate for the extensive heat losses of this smallest scale system. However, Holladay et al. [188] calculated that even a fuel processor efficiency as low as 5% would outperform an Li-ion battery. The burner was fed by a hydrogen/oxygen mixture for start-up and later on by up to $0.4\,mL\,h^{-1}$ methanol and up to $14\,mL\,min^{-1}$ air. The burner was claimed to have a volume of less than $1\,mm^3$, the reformer volume was higher ($5\,mm^3$). In this later paper, more detailed data were provided. The flow rate of methanol for the reforming reaction was in the range between 0.02 and $0.1\,mL\,h^{-1}$ and more than 99% conversion were achieved at a S/C ratio of 1.8 and 325 °C reaction temperature. A fixed bed containing 14 mg palladium/zinc oxide catalyst particles of 200 μm in size [598] was applied for steam reforming [185]. Upstream, methanol was burnt in a porous disk [598]. The burner temperature exceeded 400 °C. Downstream of the reformer catalyst bed a second bed followed, which was filled with 13 mg ruthenium catalyst for selective methanation (see Sections 3.10.3 and 4.5.3) of the carbon monoxide produced. The thermal power of the hydrogen product was 200 mW at 9% efficiency. Assuming 60% fuel cell efficiency and 80% hydrogen utilisation, the net efficiency of the fuel processor/fuel cell system was calculated to be 4.5% and the theoretical electrical power output to 100 mW. In a second prototype,

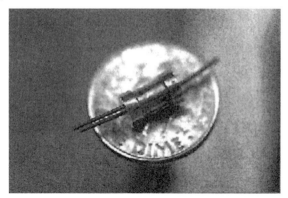

Figure 9.20 500-mW methanol fuel processor developed by Holladay et al. [601].

which is shown in Figure 9.20, the reaction temperature was reduced to 250 °C, which reduced the carbon monoxide concentration from 1.2 vol.% to less than 1 vol.% [599]. The efficiency of the fuel processor could be increased to 33%.

Cao *et al.* performed numerical calculations for the system described above [598]. Assuming adiabatic conditions and despite large temperature gradients in the region of 90 K, which were estimated for the steam reforming catalyst bed, 100% methanol conversion could be achieved, however, at low weight hourly space velocity of only 3 L $(h\,g_{cat})^{-1}$. Then heat losses were taken into consideration for the calculations, which revealed that 1.39 W of heat had to be produced by methanol combustion to compensate for the power demand of evaporation, steam reforming and 0.58 W heat losses to the environment through 12.7-mm thick insulation by glass wool [598]. The heat losses were, however, regarded as an optimistic value by these workers.

Subsequently, the first generation fuel processor prototype described above was linked to a meso-scale high temperature PEM fuel cell developed at Case Western University by Holladay *et al.* [600], which was astonishingly tolerant towards carbon monoxide up to 10 vol.%. Thus, no CO clean-up was necessary to run the fuel processor together with the fuel cell. A power output of 23 mW was demonstrated by Holladay *et al.* [599]. This value was lower than the calculated value of 100 mW, which was due to a lower hydrogen supply of the reformate (more than 99% conversion of 0.03 mL h^{-1} methanol could be achieved at a 420 °C reformer temperature), lower cell voltage of the fuel cell by the 2 vol.% carbon monoxide present in the reformate and a dilution effect by carbon dioxide at the gas diffusion layer material of the fuel cell.

9.2
Ethanol Fuel Processors

Very few fuel processors that utilise ethanol as a fuel have been described in the open literature. However, ethanol is of course very attractive from an environmental and sustainability point of view, because its production and consumption cycle may well be performed with reduced, or even zero, net carbon dioxide emissions.

Mitchell *et al.*, at the company Arthur D. Little, described early projections for an ethanol fuel processor based upon partial oxidation [475]. The efficiency of the fuel processor was calculated to be 80% and power densities of 1.44 L kW^{-1} and 1.74 kg kW^{-1} were achieved.

9.3
Natural Gas Fuel Processors

Natural gas reforming plays a dominant role especially in the domestic sector: 75% of the fuel cells existing today that are in the power region below 10 kW are fuelled by natural gas [16].

Reinkingh *et al.* from Johnson Matthey reported on a 5-kW$_{el}$ HotSpot natural gas fuel processor [576]. The multi-step unit produced reformate containing 43 vol.% hydrogen, 1 vol.% methane and less than 10 ppm carbon monoxide. The reformer itself produced between only 1 and 2 vol.% carbon monoxide. The natural gas conversion exceeded 90%.

A methane or natural gas fuel processor with 2.5-kW thermal energy output was described by Heinzel *et al.* [17]. It consisted of a pre-reformer, which made future multi-fuel operation possible, the reformer itself, which carried a nickel catalyst [433], it was operated between 750 and 800 °C, and had catalytic carbon monoxide clean-up. The preferential oxidation reactor was operated at an O/CO ratio of 3.5 [433]. A carbon monoxide content of between 20 and 50 ppm could be achieved during steady state operation. An external burner supplied the steam reforming reaction with energy. The natural gas was desulfurised by a fixed bed of impregnated charcoal. Figure 9.21

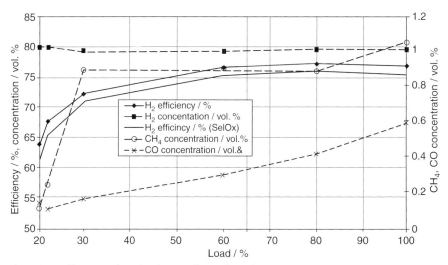

Figure 9.21 Efficiency with and without preferential or selective oxidation (SelOx) and reformate composition of the methane fuel processor as described by Heinzel *et al.* [17].

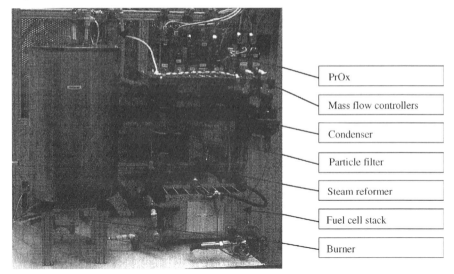

Figure 9.22 Coupled breadboard 1-kW$_{el}$ methane fuel processor/ fuel cell system as developed by Mathiak et al. [433].

shows the efficiency and reformate composition of the fuel processor. The efficiency was, of course, lowered slightly due to the hydrogen losses in the preferential oxidation reactor. Anode off-gas recirculation to the burner supplied some 500 W of energy to the fuel processor, which increased the reformer efficiency from 75 to 88.5% at full load. At 20% load the efficiency decreased to 64%. The fuel processor was then connected to a 1-kWel fuel cell, from the Proton Motor Company, Starnberg, Germany, which was tolerant to carbon monoxide up to 50 ppm (see Figure 9.22). An overall electrical efficiency of 33% was determined for the complete system. However, the parasitic losses to the balance-of-plant components were not taken into consideration for these calculations. Load changes were performed from 20% load to 100% load. The carbon monoxide content of the reformate fed to the fuel cell could be maintained below 100 ppm (see Figure 9.23). Turn down from 100% load to 66% load could be achieved in about 5 min.

Numerous natural gas fuel cell/fuel processor systems are under development for residential applications. Because off-heat could well be used for heating purposes and hot water generation in residential applications, the overall system efficiency is usually high for such systems.

Seris et al. presented a prototype methane fuel processor based upon the Heatric Printed Circuit Technology, which uses plates fabricated by wet chemical etching (see Sections 5.1.1 and 10.2.2) [602]. The system, which was designed to feed a 15-kW fuel cell with hydrogen, worked with steam reforming. The basic layout of the system is shown in Figure 9.24. It consisted of pre-reforming, steam reforming coupled to anode off-gas combustion, high and low temperature water–gas shift in multiple adiabatic fixed beds, because the fuel processor was sealed by diffusion bonding. This sealing technique is performed at high temperature, greater than 1000 °C to which

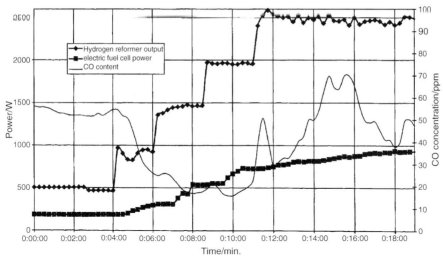

Figure 9.23 Fuel processor thermal power output, fuel cell electrical power output and CO content of the reformate as determined by Mathiak et al. for a 1-kW$_{el}$ fuel processor/fuel cell system [433].

the catalyst must not be exposed, because it would lead to irreversible damage. As an alternative, monoliths coated with catalyst could be introduced into the device to reduce the pressure drop. The steam reforming side of the fuel processor was operated at a pressure of 2 bar, while the catalytic burner was operated close to ambient pressure. The prototype fuel processor is shown in Figure 9.25. In the foreground the pre-reformer and the high temperature water–gas shift reactors are

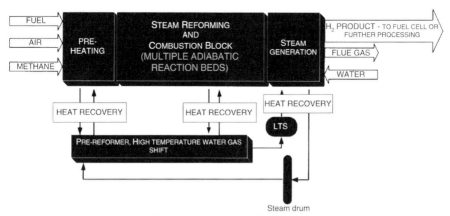

Figure 9.24 Schematic diagram of the 15-kW prototype methane fuel processor developed by Seris et al. [602].

Figure 9.25 15-kW prototype methane fuel processor developed by Seris et al. [602].

visible. The rectangular plates on top of the reformer are the covers of the adiabatic reactor beds, which were sealed by conventional welding.

Lee et al. presented a natural gas fuel processor for the thermal power range up to 16 kW, which was developed by the Argonne National Laboratory along with the H2fuel LLC Company [603]. Natural gas containing between 20 and 50 ppm sulfur was fed to the monolithic autothermal reformer containing commercial catalyst, which was operated between 450 and 750 °C. The feed composition was a S/C ratio of around 1.2 and high O/C ratio of about 1.2, which shifts the operating conditions into the field of partial oxidation. More than 98% methane conversion was achieved and the reformate contained about 40 vol.% hydrogen on a dry basis and 10 vol.% carbon monoxide on a dry basis, which agreed well with the thermodynamic equilibrium. The reformate was desulfurised at 350 °C by zinc oxide downstream of the reformer. Precious metal catalysts were used in annular fixed beds for high and low temperature water–gas shift, which surrounded the reformer. The reformer was started by electrical heating. The fuel processor was operated for more than 2000 h with stable performance at different load levels after an initial period of catalyst deactivation. The efficiency, which was defined as the ratio of the lower heating value of the hydrogen in the product of the water–gas shift reactor to the lower heating value of the natural gas feed, was in the range of between 80% at 20% load and 90% at full load.

The FLOX® steam reforming technology, developed by the German company WS Reformer, allows stable anode off-gas utilisation and a system efficiency of greater than 40%. The size range covered by the technology extends from 1 to 50 kW$_{el}$ [16]. The FLOX® heating concept is a flameless oxidation process, which facilitates ultra-low NO$_x$ emissions (see Figure 9.26). The water evaporation for the reformer feed is supplied with heat from the hot reformate. The FLOX® heating systems supplies heat to the reformer and to the steam super-heater. The air feed of the heating system is pre-heated by the heat of the exhaust gas of the heating system in a heat-exchanger. Natural gas is fed not only to the reformer, but also to the heating system as an

9.3 Natural Gas Fuel Processors

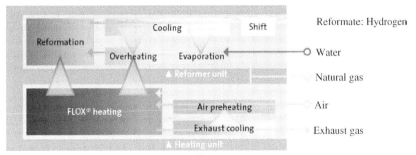

Figure 9.26 Schematic principle of the FLOX® reformer concept [16].

additional heat source [16]. The heating process works at a temperature of between 900 and 1100 °C. The internals of a FLOX® reformer and its outer appearance are shown in Figure 9.27. The gas distribution system is positioned at the top, followed by the evaporator and the reformer, which is surrounded by flow ducts for the off-gas of the heating system. The steam reformer contains conventional nickel catalyst

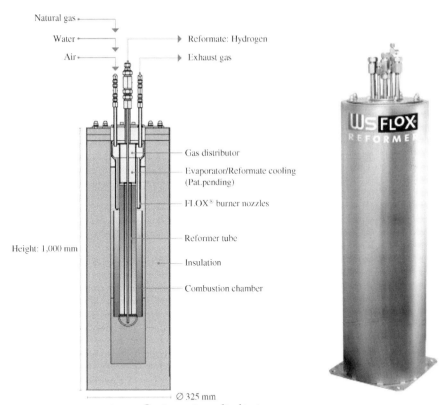

Figure 9.27 Internals of a FLOX® reformer for 1 m³ h⁻¹ hydrogen production (left) and the complete reformer/heater/evaporator system (right) [16].

operated at S/C ratios between 2.5 and 3.5 with sulfur-free natural gas. The system requires de-ionised water at 6-bar pressure and generates a steam jet, which drags the natural gas out of the supply grid without additional equipment. The steam jet generates an over-pressure, which is just sufficient to supply the reformate to the fuel cell downstream. The system has a size of approximately 19 L with insulation and water–gas shift but without the carbon monoxide fine clean-up. However, even more compact systems with a power equivalent of 20 kW$_{el}$ and more are also offered by the company. The system has a start-up time demand of less than 30 min, and can be modulated from 30 to 100%. Its efficiency was determined to be 83% at full load and 73% at 40% load. A 4-kW$_{el}$ version of the fuel processor was coupled to a selective methanation reactor for carbon monoxide fine clean-up and to a fuel cell, both developed by the German research centre ZSW. The prototype system had an electrical efficiency of greater than 30%. The weight amounted to 200 kg with a volume of 1 m^3, including the balance-of-plant components such as pumps, valves and electrical converters. The parasitic losses to the system periphery ranged between 300 and 500 W depending on the load.

Baxi Innotech develops a fuel cell heating system for natural gas with 1.5 electrical and 3.0 kW thermal power output [604]. Figure 9.28 shows the BETA 1.5 PLUS system. The design operating capacity of the auxiliary boiler is in the range between 3.5 and 15 kW. The overall system efficiency exceeds 80%. The system has a weight of 350 kg and a volume of 1.3 m^3. The company cooperates with the heating system manufacturer August Brötje.

Figure 9.28 BETA 1.5 PLUS fuel cell heating system of Baxi Innotech GmbH [604] (photograph: courtesy of Baxi Innotech).

Figure 9.29 4.5 kW$_{el}$ combined heat and power system developed by Vaillant and the Plug Power company [605.]

Vaillant developed a multi-family home fuel cell system, along with the US company Plug Power, with an electric power output of between 1.5 and 4.5 kW$_{el}$. The thermal power output is in the range from 1.5 to 7.0 kW$_{th}$. The system uses PEM fuel cell technology and a fuel processor and has an electrical efficiency of 35%, while the overall efficiency amounts to 85%. So far 60 fuel cell systems have already been installed for field trials. By the end of 2006 1 000 000 kWh of electricity had been produced by these systems [605] (Figure 9.29).

The Japanese combined heat and power systems are usually smaller. As one prominent example, from numerous systems under development in the Japanese and Asian markets, Figure 9.30 shows the combined heat and power system working with a natural gas fuel processor and a PEM fuel cell, which was developed by Ballard for Tokyo gas [606]. The system achieved an efficiency of 38% at an electrical power output of 800 W$_{el}$.

In the first half of 2005, 175 fuel cell systems were installed in Japan, 67 of them by Tokyo Gas.

Echigo *et al.* from Osaka gas reported on highly elaborate 500-W and 1-kW natural gas fuel processors [607]. The composition of their natural gas feed was 88 vol.% methane, 6 vol.% ethane, 3 vol.% propane, 3 vol.% butane and 5 ppm sulfur as odorant. Anode off-gas surrogate was fed to the fuel processor as an additional heat source, which was composed of 49 vol.% hydrogen, 42 vol.% carbon dioxide, 6 vol.% nitrogen and 3 vol.% methane, the last as a surrogate of unconverted reformer feed.

Figure 9.30 Combined heat and power system developed by Ballard for Tokyo gas [606].

The anode off-gas composition resulted from 75% hydrogen utilisation in the fuel cell. The fuel processor size was 280-mm width, 440-mm length and 395-mm height, which corresponded to 48.6 L at 1 kW$_{el}$ power output. The fuel processor was composed of a desulfuriser, an evaporator, clean-up reactors and a burner. The reformer was operated at a S/C ratio 2.5 and the preferential oxidation reactor at an O/CO ratio 3.0. Self-developed ruthenium catalyst was used for preferential oxidation, which reduced the carbon monoxide to below 0.5 ppm. An efficiency of 78% of the fuel processor could be achieved. This value decreased only slightly to 72% at 30% load. The fuel processor showed stable performance for more than 10 000 h. More than 25 000 load changes with a total test duration of 6800 h were performed with one fuel processor from 100% load instantaneously to 50% load and gradually back to 100% load, while the carbon monoxide content in the reformate was always maintained below 10 ppm. Up to 1000 start/stop cycles were performed without degradation of the fuel processor. A total operation time for 94 fuel processors had already exceeded 94 000 h in 2004, which is an impressive durability and reliability demonstration.

A system of a larger size is the Wärtsila WFC20 solid oxide fuel cell system, which has an electrical net power output of 18.6 kW. The targeted application are auxiliary power systems for marine applications. Figure 9.31 shows a flow scheme of the system, which consists of a zinc oxide sulfur trap, a pre-reformer, which is fed partially by recycled anode off-gas, and a catalytic afterburner, which delivers the heat for steam generation and natural gas pre-heating. The afterburner is fed with oxygen contained in the cathode off-gas. Figure 9.32 shows the system housing and its internal workings. The company provided fairly detailed information on the energy balancing of their system. About 40 kW of thermal energy are contained in the natural gas feed; 3 kW of heat are lost to the environment, while the power conversion, automation, control and other balance-of-plant consumes sum up to 5.4 kW power; 18.6 kW is the net electrical power output of the system, which corresponds to a high efficiency of 46%. The overall efficiency target of this system is 80% when off-heat is utilised. The system has a volume of 5.7 m^3 and a weight of 3 tons.

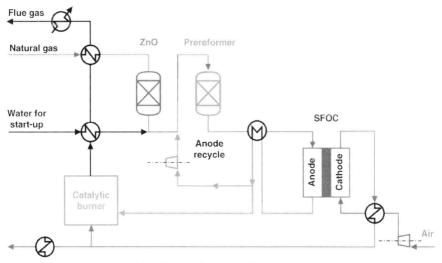

Figure 9.31 Wärtsila WFC20 solid oxide fuel cell system with pre-reformer and catalytic afterburner [608].

Seo *et al.* reported on the development and operation of a 100-kW natural gas fuel processor, which was developed for a molten carbonate fuel cell [609]. The molten carbonate fuel cell does not require any carbon monoxide clean-up (see Section 2.3.2), and thus the system consisted merely of a burner to supply the steam reformer, a compressor, heat-exchangers, the desulfurisation stage and the reformer itself. The reformer was built by relying on conventional technology with tubular reactors top-fired externally from the natural gas burner. The 16 steam reformer tubes shown in Figure 9.33 were operated at a S/C ratio of 2.6 and 3-bar pressure, while the design operating temperature was 700 °C. Seo *et al.* reported that the efficiency of their system was still too low. Therefore, an improved version of the fuel processor is under development.

Figure 9.32 Wärtsila WFC20 solid oxide fuel cell system: left, housing; right, internal arrangement of the components [608].

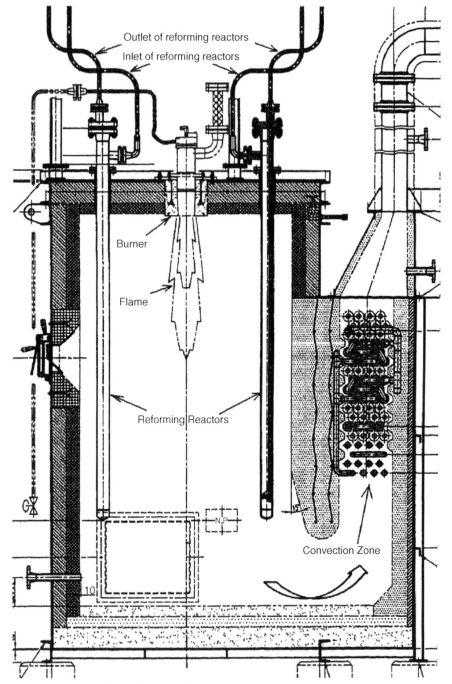

Figure 9.33 100-kW$_{el}$ tubular natural gas steam reformer dedicated for a molten carbonate fuel cell [609].

9.4
Fuel Processors for LPG

Because of the existing distribution grid for Liquefied Petroleum Gas (LPG) and its widespread application in caravans and trailers, it is an attractive fuel for the electrical power supply of such vehicles. The German company TRUMA, Europe's largest manufacturer of heating systems for caravans and trailers, has developed a fuel processor/fuel cell system together with IMM (Institut für Mikrotechnik Mainz GmbH). The utilisation of microstructured plate heat-exchanger technology made it possible to produce a compact design of the fuel processor. The proprietary fully integrated system is shown in Figure 9.34.

The concept of a 300-W_{el} LPG based auxiliary power unit was presented by Dokupil et al. and Beckhaus et al. [542,610]. The target applications were caravans and sailing yachts. A fuel supply of 5 kg of LPG was calculated for 2-weeks operation of the proposed system. The system concept was composed of a desulfurisation unit, a steam reformer, two water–gas shift stages, a preferential oxidation reactor operated at an O/CO ratio of 2 and a conventional air cooled PEM fuel cell stack operated with 70% hydrogen utilisation. All reactors were of a monolithic design. The reformer off-gas was cooled by a heat-exchanger, which in turn pre-heated the reformer feed. The reformer was operated at 630 °C, a S/C ratio around 3 and 15 000 h^{-1} gas hourly space velocity. It was supplied with energy from an external catalytic burner, which combusted both LPG and anode off-gas at 800 °C and 60 000 h^{-1} gas hourly space velocity. The reformate was cooled with air in extra heat-exchangers downstream of each of the carbon monoxide purification reactors, which generated about 100 W off-heat. A heat-exchanger was integrated into the preferential oxidation reactor. Calculations revealed a maximum efficiency of 75% for this concept. The overall system efficiency was calculated to be 25%.

A breadboard laboratory fuel processor was then set up, which had an efficiency of between 56 and 60%. These values were lower than the calculations despite the

Figure 9.34 The 250-W_{el} fuel cell/fuel processor system VEGA developed by a cooperation of TRUMA and IMM (photograph: courtesy of TRUMA).

utilisation of anode off-gas, which was attributed to heat losses to the environment. The system had a weight of 3.6 kg (without balance-of-plant) and a volume of 17 L.

The system was started by hydrogen combustion from ambient temperature, however, the source of this hydrogen was unclear for a practical system. Therefore, these workers planned to combust LPG in a future modification of the fuel processor. The start-up was performed in four steps. Firstly, the burner was started at 50% load and switched to 100% load after 2 min. At the same time, an air flow pre-heated the reactors of the fuel processor downstream of the reformer. After 16 min water was added to the reformer feed and after 21 min LPG was fed to the reformer at full load. At the same time, air was fed to the low temperature water–gas shift reactor to combust some of the hydrogen and carbon monoxide, similar to the procedure described by Springmann et al. [389] (see Section 5.4.5). The system then stabilised after 4 min, and therefore the total start-up time demand of the device amounted to 25 min.

This system set-up consisting of a total of 4 reactors and 3 heat-exchangers on top of an evaporator seems to be rather complex for such a low power device (Figure 9.35).

Then Spitta et al. presented a modified system with 300-W_{el} power output, which was connected to a low temperature PEM fuel cell [611]. The monolithic steam reformer of the system was again supplied with energy from an external burner, which also served as a start-up burner (started by a spark plug) and anode off-gas burner. Further components of the fuel processor were an evaporator, a superheater and a heat-exchanger. The reformer was operated at a S/C ratio of 3.3 and between 640 and 650 °C. The peak burner temperature was maintained below 1000 °C. The carbon monoxide clean-up was, similar to the former system, composed of two stage adiabatic water–gas shift reactors and a cooled preferential oxidation reactor operated at an O/CO of 1.8. A fuel processor efficiency of 61.8% and 25-min start-up time demand were reported for the system. A volume of less than 8 L and a weight of less than 3.6 kg were reported, including insulation. The fuel cell stack was air cooled and had a volume of 5 L and weight of 5 kg. Condensation problems occurred in the fuel

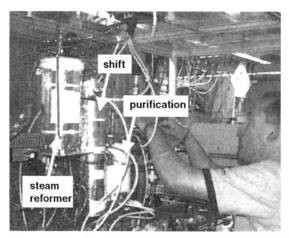

Figure 9.35 Breadboard 1-kW$_{th}$ LPG fuel processor as developed by Beckhaus et al. [610].

cell stack due to the high water content of the reformate. This originated from the relatively high S/C ratio of 3.3. Therefore the fuel cell operating temperature was increased to 85 °C to widen the operating window without condensation. Alternatively, water was removed from the reformate by condensation. The overall efficiency of the system was calculated to be 28.5% by assuming 50-W parasitic power losses to periphery components such as pumps and blowers.

Recupero *et al.* described development work for a fuel processor for liquefied petroleum gas [612]. Autothermal reforming was chosen and a platinum/ceria catalyst was used, but the reformer operating conditions were set to S/C 3.6 and O/C 1.3. A medium temperature water–gas shift reactor was put downstream of the reformer. A ruthenium/alumina catalyst was chosen for the preferential oxidation stage and the catalyst was operated at relatively high O/CO ratio of 4.0. Despite this high value, the carbon monoxide content in the test reactor effluent showed peaks at 250 ppm during dynamic operation, even though the feed concentration of the preferential oxidation stage was only 5000 ppm. Then Cipiti *et al.* presented an integrated 2-kW$_{el}$ fuel processor/fuel cell system [613]. The system, which is shown in Figure 9.36, was of 870-mm width, 880-mm length and 970-mm height. The reformer was operated at a S/C ratio of 3.0, O/C ratio of 1.0 and at 600 °C. Commercial catalysts were used for medium temperature water–gas shift and preferential oxidation. Surprisingly the water–gas shift reactor had only 8% of the size of the preferential oxidation reactor. The reformer was pre-heated to 300 °C with nitrogen within 50 min, the nitrogen in turn was heated with electricity. Owing to the low reforming temperature, about 6–8 vol.% of methane were present in the reformate. The preferential oxidation reactor failed to convert carbon monoxide to a level of less than 2000 ppm and thus connection of the system to the fuel cell was postponed.

A completely different system, which worked with cracking of liquefied petroleum gas was presented by Ledjeff-Hey *et al.* [79]. The system had the charm of being, at first glance, much simpler compared with conventional reforming coupled with catalytic carbon monoxide clean-up. It was very similar to the concept presented

Figure 9.36 2-kWel fuel processor/fuel cell system for liquefied petroleum gas developed by CNR-ITAE [613].

earlier by Poirier *et al.* [80] (see also Section 5.4.4). It required only the LPG fuel as feed, no water or air. The low content of carbon monoxide formed by the cracking process could be removed by a methanation stage, which also required no addition of air. The methanation was less complicated in the present case compared with the methanation of reformate originating from reforming reactions because much less carbon dioxide was present. Therefore, the selectivity of the reaction was not an issue (see Section 3.10.3). However, the carbon formed during the cracking process (see Section 3.4) required removal by combustion, and this complicated the process considerably. A portion, 55%, of the lower heating value of the propane was stored in the carbon [401], thus creating a heat surplus of the overall system. Two cracking reactors had to be operated in parallel, one reactor worked in the cracking mode and the other in the carbon combustion mode in an alternating manner. During carbon combustion the catalyst was exposed to high temperatures of around 1200 °C. This led to stability issues of the catalyst.

However, a 120-W_{el} prototype cracker was built by Ledjeff-Hey *et al.* [79]. A homogeneous propane burner with 1.5-kW thermal power served for rapid system start-up. A high initial carbon monoxide concentration was observed during the cracking step after the start-up. Therefore, the cracking product could not be supplied to the fuel cell but was rather combusted in the homogeneous burner. Owing to the high operating temperature, a high concentration of carbon monoxide was also formed during the combustion phase. Depending on the catalyst technology applied, up to 35 vol.% of carbon monoxide were detected in the combustion gases by Ledjeff-Hey *et al.* [79]. The theoretical efficiency of this cracker fuel processor was as low as 47%, which is only about half the value that could be achieved by reforming (see Section 3). The theoretical overall efficiency of this concept in combination with a fuel cell was calculated to be 9% [79], which is fairly poor. According to Ledjeff-Hey *et al.*, this value could have been improved to 23% by reducing the parasitic losses to electrical consumers and by utilisation of the combustion heat of the process.

A much more elaborated system developed by the US company IdaTech was presented by Löffler *et al.*, which was suitable for liquefied petroleum gas, and also for methane fuel [105]. Because the system worked with membrane purification, steam reforming was chosen (see Section 5.2.4). The desulfurisation was performed upstream of the reformer, steam was then added downstream the desulfurisation. The reformer was composed of a pre-reformer and the main steam reformer. Packed bed reactor technology was applied, more specifically tubular reactors placed as a bundle around a homogeneous burner, as shown in Figure 9.37. Alternatively, a concentric annular design of the reformer was presented, which is also shown in Figure 9.37. Rhodium catalysts were used in the reformer and the operating temperature of the pre-reformer was around 500 °C, while the reformer itself had a wide temperature gradient of between 600 and 800 °C [105]. The S/C ratio amounted to 2. The fuel conversion was low in this reformer arrangement, specific values for various system sizes are provided in Table 9.4. Downstream of the reformer, the membrane separation followed, which consisted of planar palladium membranes. The thickness of the membranes had been minimised to reduce cost. Thus leakages and imperfections of the membrane occurred and therefore some carbon oxides were

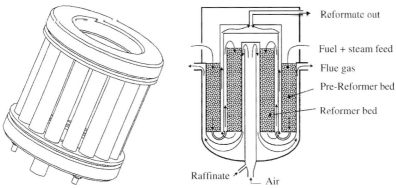

Figure 9.37 Design concepts of IdaTech steam reformers: left, tubular fixed bed steam reformer reactors are placed around a central burner; right, heat-exchange reformer; the pre-reformer is placed in the outer area of the device while the reformer is more in the centre; the combustion gases of the homogeneous burner pass through several annular gaps between the annular catalyst beds for heating [105].

present in the reformate, which made a methanation unit mandatory [105]. Both carbon monoxide and carbon dioxide were converted by methanation into methane. Owing to the low concentration of carbon oxides the hydrogen losses remained low. A certain residual hydrogen partial pressure was required as the driving force for the membrane separation. Hydrogen was also required in the retenate of the membrane separation to supply the steam reformer with energy. Therefore, the efficiency of this

Table 9.4 Performance data of the IdaTech natural gas/liquefied petroleum gas fuel processor [105].

	Fuel	
	Natural Gas	LPG
Hydrogen production (L/min)	25	25
	80	85
Thermal efficiency (%)	55	54
	69	67
Outlet temperature (°C)	715	750
	800	785
Fuel conversion (%)	65	73
	83	80
Approach to equilibrium (°C)	10	10
	15	15

fuel processor was limited. Depending on the fuel this varied between 54 and 69%, as shown in Table 9.4 [105]. A turn-down ratio of 3 could be achieved. The differences between the natural gas and liquefied petroleum gas fuel processors were only determined by the choice of balance-of-plant components.

9.5
Gasoline Fuel Processors

The reasons for worldwide development efforts in the field of gasoline fuel processors for automotive applications are two-fold. One area of application are auxiliary power units, which are envisaged as providing electrical energy to a car during normal operation and especially when the engine is turned off. The energy demand of passenger vehicles has grown exponentially over the last 80 years as shown in Figure 9.38 and today power supply by conventional batteries is now becoming insufficient. Another possible application of auxiliary power units is an electric vehicle supplied by a lithium ion battery with a capacity of 10 kWh [614]. Such a battery has a weight of about 100 kg at present [614]. A 10 kW$_{el}$ auxiliary power unit could charge the battery and extend the range of the vehicle considerably.

The other, more ambitious, goal is to supplement the power supply of the drive train of the vehicle completely. In both instances fuel processing technology allows the introduction of fuel cell technology into the market place in the short term, because the hydrogen infrastructure required to run fuel cells without fuel processor is becoming obsolete. The cost of such a hydrogen filling station was estimated to be ten times higher compared with a conventional gasoline station [70].

An early status report on development work for a gasoline fuel processor for automotive applications was provided by Flynn et al. from McDermott Technology

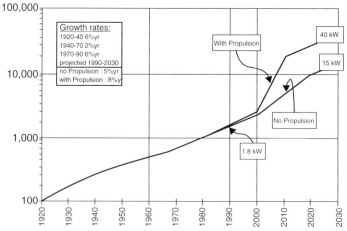

Figure 9.38 Increasing on-board power demand of passenger cars in watts versus time [442].

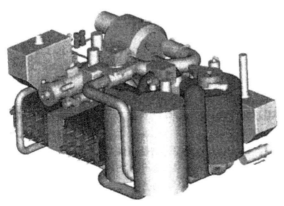

Figure 9.39 Design concepts of a 50-kW gasoline fuel processor/fuel cell system as developed by International Fuel Cells [616].

Inc. and Catalytica Advanced Technologies [615]. The concept was fairly conventional, consisting of desulfurisation, autothermal reforming and catalytic carbon monoxide clean-up. Desulfurisation by adsorption in the liquid phase was combined with a zinc oxide bed for residual hydrogen sulfide removal. The zinc oxide was positioned downstream of the high temperature water–gas shift reactor. Only 225 g of zinc oxide were required for a residual sulfur content of 5 ppm during a 24 000-km annual driving cycle according to the calculations.

A design study for a gasoline fuel processor/fuel cell system was presented by King and O'day [616]. It consisted of an autothermal reformer, sulfur removal, water–gas shift and two stage preferential oxidation. The system pressure was close to ambient to reduce parasitic power losses of the compressor (Figure 9.39).

A breadboard gasoline fuel processor was assembled by Moon *et al.* [67]. Fixed bed reactors served for reforming by steam supported partial oxidation (see Section 7.1.1), followed by high and low temperature water–gas shift. Commercial iron oxide/chromium oxide catalyst was applied for high temperature shift at a 4200 h^{-1} gas hourly space velocity and 450 °C reaction temperature, while the copper/zinc oxide low temperature water–gas shift catalyst was operated at 250 °C and 5600 h^{-1} gas hourly space velocity.

A breadboard gasoline fuel processor relying on a "shift-less" concept, in other words omitting the water–gas shift purification steps, was developed by Bosco *et al.* [617]. The basic idea behind the development was to reduce the reformer temperature to a range between 550 and 650 °C, thus reducing the carbon monoxide content of the reformate. Partial oxidation supported by significant steam addition was performed in a fixed catalyst bed. The gasoline feed was sulfur-free. The reforming catalyst was 1 wt.% rhodium supported by ceria/zirconia. An annular reactor concept was applied for the preferential oxidation reaction, because heat removal problems had occurred previously in a two stage fixed bed catalyst system. This originated from the high carbon monoxide concentration in the reformate feed. The catalyst chosen for preferential oxidation was 5 wt.% ruthenium/5 wt.% cerium supported by alumina. The catalysts beds for reforming and clean-up were diluted with quartz sand, which would be unsuitable for a practical application, which has to be optimised for size and weight. However, the reformate was fed to a single PEM fuel cell equipped with reformate tolerant platinum/ruthenium anode catalyst and the fuel cell performance was also validated. The reformer was operated at a S/C ratio of between 2.6 and 3.6 and O/C ratio of between 0.5 and 0.7. The outlet temperature of the reformate downstream of the reformer ranged between 545 and 610 °C, while the maximum temperature in the catalyst bed was between 625 and 650 °C. Owing to this low reformer temperature, the carbon monoxide concentration of the reformate downstream of the reformer was low, between 4 and 5 vol.%. The dry carbon monoxide content of the reformate after the preferential oxidation stage could be reduced to values below 50 ppm, while 10–20 vol.% of methane were present, which is a consequence of the thermodynamics at the low reformer temperature. About 27% of the hydrogen was lost in the preferential oxidation reactor due to the oxidation by excess air, which was required to remove the unusually high content of carbon monoxide. Along with the high methane content this decreased the efficiency of the approach to very low values.

Qi et al. presented a 1-kW breadboard gasoline fuel processor [451]. The device consisted of a concentric reactor arrangement, similar to the design developed by Ahmed et al. [448], see Section 5.4.5. The overall dimensions were very low, a diameter of 150 mm and length 150 mm were reported by these workers [451]. The preferential oxidation reactor was a separate device, but the autothermal fixed bed reformer was positioned in the centre of the fuel processor and surrounded by annular high and low temperature water–gas shift fixed bed reactors, as shown in Figure 9.40. The feed

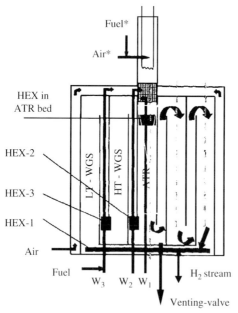

Figure 9.40 Integrated gasoline fuel processor as developed by Qi et al. [451] HEX stands for heat-exchangers, W_1 to W_3 for water injection systems.

temperatures of the water–gas shift reactors were controlled by heat exchangers, which were cooled by water evaporation. Self-developed ruthenium/ceria/alumina catalyst was used for reforming, commercial iron oxide/chromium oxide catalyst and copper/zinc oxide catalyst served for the high and low temperature water–gas shifts, respectively. Platinum/alumina catalyst was used for preferential oxidation.

The system was started by a homogeneous burner in which fuel and air were combusted over a platinum filament. The burner served as the feed injection system during normal operation. The combustion gases were fed through the autothermal reformer and left the system through a by-pass valve. The high operating temperature of this valve would create an issue in a practical system. It took 50 min to reach the operating conditions of the reformer and shift reactors, but about 4 h to reach stable performance and 1000 ppm carbon monoxide downstream of the preferential oxidation, which is a particularly high time demand. The autothermal reactor suffered from a low outlet temperature of between 450 and 550 °C, which is likely to lead to high methane selectivity. Unfortunately this was not discussed by Qi *et al.* However, the fuel processor was operated for 85 h and the hydrogen content in the reformate was between 38 vol.% for n-octane and 33 vol.% for gasoline.

Kolb *et al.* presented a breadboard isooctane fuel processor designed for an electrical power equivalent of 5 kW$_{el}$ [543]. It was, to their knowledge, the first complete hydrocarbon fuel processing system in the kilowatt range that was based on microchannel plate technology. The fuel processor was constructed of an autothermal reformer, a 1-kW heat-exchanger for cooling the reformate downstream of the reformer, a first stage water–gas shift reactor, which had cross-flow cooling capabilities, a second stage water–gas shift reactor and a preferential oxidation reactor, also with three stage cross-flow cooling. At a power equivalent of 3.65 kW, the isooctane was converted completely at a S/C ratio 3.3, O/C ratio 0.68 and pressure of 3 bar. An 88% conversion of the carbon monoxide was achieved in the first stage water–gas shift, with 40% in the second stage. The reformate still contained about 1 vol.% carbon monoxide on a dry basis (water-free) when it entered the preferential oxidation reactor. The latter was operated at an O/CO ratio of 2.1 and converted the carbon monoxide to about 200 ppm. The heat removal by cooling air from the first stage preferential oxidation was around 170 W under these conditions. The fuel processor efficiency was calculated to be 74%.

A 1–3 kW$_{el}$ autothermal gasoline fuel processor was presented by Severin *et al.* [618]. The flow scheme of the breadboard system is shown in Figure 9.41, the system itself is shown in Figure 9.42. The fuel was injected into the mixing chamber upstream of the autothermal reformer (ATR). The reformate leaving the ATR was cooled in a heat-exchanger by the steam and water feed. Then two water–gas shift reactors (HTS and LTS) followed downstream. A heat-exchanger was placed between them to cool the reformate with part of the reformer air feed. Downstream of the water–gas shift there followed a preferential oxidation reactor (SelOx). Finally, a heat-exchanger cooled the purified reformate and water was removed by condensation before it entered the fuel cell. The remaining hydrogen in the fuel cell off-gas was combusted in an afterburner. The afterburner off-gases pre-heated the air feed of the reformer in a heat-exchanger downstream of the afterburner. Obviously the feed temperature of the reformer air

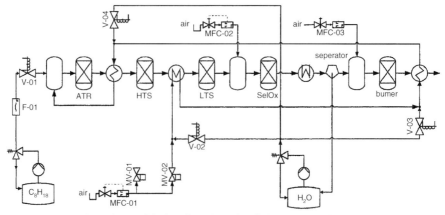

Figure 9.41 Flow scheme of the breadboard gasoline fuel processor as built by Severin et al. [618] MFC stands for mass flow controllers.

was high enough to evaporate the water feed completely through simple injection. The reactor temperatures were controlled by water injection into the cooling gas flows of the heat-exchangers. All reactors were of monolithic design.

The autothermal reformer was operated at temperatures between 600 and 800 °C [618]. According to the stoichiometry and thermodynamics, the highest efficiency of the reactor was achieved at the lowest O/C ratio of 0.75 (here expressed as an air to fuel ratio of 0.25) as shown in Figure 9.43. This was almost independent of the S/C ratio, which was varied from 1.0 to 2.5, and was also independent of the inlet temperature of the feed, which had been changed in parallel.

However, a higher steam to carbon ratio inevitably impairs the overall system efficiency of a practical system (see Section 5.4.4).

The high and low temperature water–gas shift reactors were operated at around 450 and 300 °C, respectively. At partial load of the system, the operating temperature of the shift reactors could be decreased. This increased the conversion, according to the thermodynamic equilibrium. The preferential oxidation reactor reduced the

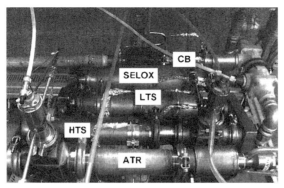

Figure 9.42 Breadboard gasoline fuel processor as built by Severin et al. [618].

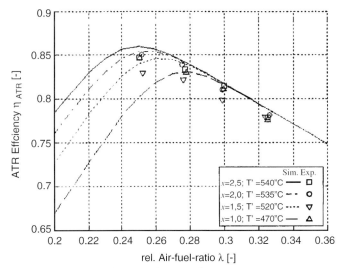

Figure 9.43 Efficiency of the autothermal reformer versus air to fuel ratio as calculated and determined experimentally by Severin et al. The parameters that are varied are: feed inlet temperature T' and S/C ratio (expressed as χ) [618].

carbon monoxide concentration to approximately 160 ppm. The start-up procedure was not totally suitable for a future practical system and the start-up time demand was approximately 30 min. The main origin of the start-up time demand was again the subsequent heating of a consecutive chain of reactors as described previously for other systems (see Sections 5.4.5 and 9.4). Electrical pre-heating of the inlet section of the reformer monolith as described by Springmann et al. [389] actually reduced the time demand for pre-heating. The experimental prototype had a weight of 18 kg and a volume of 40 L (without insulation).

A design concept was then developed for a future fuel processor, which is shown in Figure 9.44. To allow for more compact packaging, a rectangular rather than cylindrical reactor and heat-exchanger geometry was chosen.

General Motors developed a rapid start-up gasoline fuel processor, which was presented by Goebel et al. [619]. The fuel processor consisted of an autothermal reformer operated at a S/C ratio of 1.8 and an O/C ratio of 0.83, two water–gas shift stages, two reactors for preferential oxidation of carbon monoxide, each operated at an O/CO ratio of 3. The total amount of precious metals loaded onto the reactors was 44.5 g, more than 75% of which were used in the water–gas shift reactors. Heat-exchangers were positioned after each reactor. The reactors were of monolithic design and 144 mm diameter apart from the reformer, which had only an 80-mm diameter. In the high-temperature part, the heat-exchanger between the reformer and the first water–gas shift reactor was used for pre-heating air and steam, whereas the remaining heat-exchangers downstream were mainly used as evaporators as shown in Figure 9.45. The fuel processor contained two burners with direct steam generation by injection. They had dimensions of 235-mm length and 92-mm

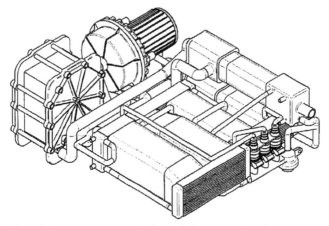

Figure 9.44 Design concept of a future fuel processor based on rectangular components, according to Severin et al. [618].

diameter and each had a heating power of 18.6 kW. The weight of the whole fuel processor was only 34 kg without insulation [619]. Under normal operating conditions the system had a power output of 68 kW of the lower heating value of the hydrogen produced. This corresponded to an electrical power output of 27 kW for the corresponding fuel cell. The efficiency of the fuel processor was calculated to be 78%, however, significant amounts of methane and higher hydrocarbons, mostly unconverted toluene, were contained in the fuel processor product. This decreased the efficiency to below 70%, regardless of the system load chosen.

The main problem during normal operation of the fuel processor was a temperature mal-distribution in the autothermal reformer, which could not be fixed by using different injection nozzles [619].

The start-up procedure relied on steam generation by water injection into the start-up burners. The steam generators of the fuel processor were positioned in the area of the carbon monoxide clean-up equipment and could not provide steam during start-up. However, steam addition was regarded as mandatory even during start-up, because the adiabatic temperature rise of the reformer under steam-free conditions of partial

Figure 9.45 Schematic (left) and photograph (right) of the gasoline fuel processor as developed by Goebel et al. [619]; burn 1 and burn 2 stands for the two start-up burners of the system.

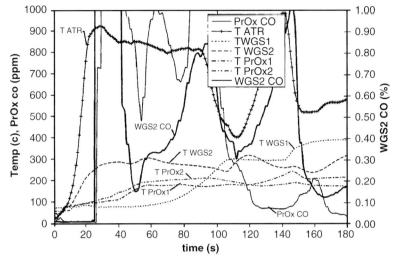

Figure 9.46 Fuel processor reactor temperatures and carbon monoxide concentration after the second stage water–gas shift (WGS2 CO) and after the second stage preferential oxidation reactor (PrOx CO) as determined during start-up of a gasoline fuel processor to full power [619].

oxidation was too high, causing catalyst stability issues. Hydrogen combustion was used to light-off the catalysts of the reformer and preferential oxidation reactors. Electrical heating of the inlet section of the reactors was regarded as an alternative to hydrogen combustion. The two burners heated the reformer and first water-gas shift stage (burner 1) and the remaining reactors (burner 2) separately [619].

The start-up procedure was initiated by air addition to the burners and the preferential oxidation reactors. Next, the burners were started and steam generation initiated. Then light-off hydrogen was added to speed up the pre-heating of the autothermal reformer and of the preferential oxidation reactors. All these measures were complete after about 20 s and the reforming process was begun at an increased O/C ratio. Burner 2 could be shut off after 40 s and burner 1 after 100 s. The reactor temperatures and the carbon monoxide content determined downstream of the second water–gas shift reactor and downstream of the second preferential oxidation reactor are shown in Figure 9.46. The start-up power demand determined during the different phases is shown in Figure 9.47. After only 140 s full hydrogen output could be achieved with this fuel processor.

When the system was started at half power, the carbon monoxide content after the preferential oxidation reactor could be kept to below 100 ppm and the time demand was also only 180 s [619].

The total start-up energy demand of the fuel processor was calculated to be 4.2 MJ. It was calculated that this start-up energy demand increased the fuel consumption of a 24-km driving cycle by 14%. The start-up hydrogen demand was very low, a tank volume of only 12 cm^3, pressurised to 340 bar, was required to power one start-up.

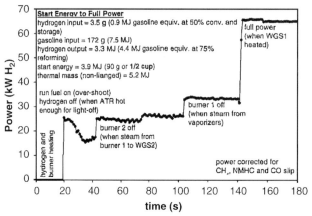

Figure 9.47 Start-up power and transition test as determined during start-up of a gasoline fuel processor to full power [619].

Thus, a replaceable tank was taken into consideration for a future system. Attempts to run the burners under conditions of partial oxidation resulted in soot formation in the reactors.

In August 2001, General Motors presented a Chevrolet S-10 pick-up with the world's first fuel cell vehicle supplied by a gasoline fuel processor. The electrical power of the system amounted to 25 kW and the start-up time demand was in the order of 3 min. The peak efficiency of the reformer was 80%.

Toyota presented a fuel cell vehicle with on-board reforming of clean hydrocarbon fuel at the Tokyo Motor show in 2001 (see Figure 9.48).

The most promising advanced fuel processing system at least for gasoline fuel, which was dedicated to the automotive drive train application, was developed by Nuvera Fuel Cells, Inc., for the car manufacturer Renault. The target carbon dioxide emissions of the system (tank to wheel) were set to $80 \, g \, km^{-1}$, which corresponded to a gasoline fuel consumption of $3.2 \, L \, 100 \, km^{-1}$. The fuel processor was designed for a fuel cell system with a 70-kW power output and an efficiency target of 40%, which transferred to a thermal power of $200 \, kW_{th}$ for the fuel processor. Table 9.5 summarises the objectives for the power plant and Table 9.6 the design targets for the third generation of the fuel processor, as set by Renault [620]. The most important issues were the utilisation of sulfur-free gasoline (below 2 ppm sulfur), the small volume target (below 80 L) and the relative high pressure level of the system, which amounted to 3 bar at the fuel processor outlet. Owing to the low size, a relatively high pressure drop of 0.7 bar was assumed for the fuel processor. The start-up time demand was set to 4 min in order to reduce the size of the battery, which was anticipated as being used to move the vehicle until the fuel cell power plant was in operation. The immediate start of the system was expected to compete with conventional internal combustion engines. However, future end-users should change their expectations from a novel technology for the sake of environmental benefits, according to the opinion of the author of this book.

9.5 Gasoline Fuel Processors | 341

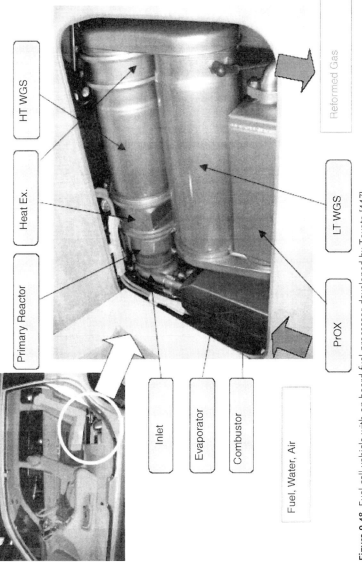

Figure 9.48 Fuel cell vehicle with on-board fuel processor developed by Toyota [447].

Table 9.5 Targets for the fuel processor/fuel cell power plant as set by Renault for their gasoline system, developed together with Nuvera [620].

Characteristic	Objective
Fuel Type	Sulfur Free Gasoline, Ethanol
Electric Power	70 kW$_{el}$
Efficiency	30% @ mid load
	35% @ full load
Volume	280 L
Mass	235 kg
Start-up time	2 min
Water requirements	No water addition to the vehicle
Transient (idle to 90%)	7 s

The fuel processor worked with autothermal reforming [620] and with conventional carbon monoxide clean-up, namely high and low temperature water–gas shift and preferential oxidation. Substrates coated with catalyst that were not specified further were used to build the reactors. A catalytic afterburner was used to utilise

Table 9.6 Target for the third (A) generation fuel processor as set by Renault for their gasoline system, developed together with Nuvera [620].

Characteristic	Design Goal	Comments
Start-up time	4 min	<100 ppm CO; <1 min needed in future
Maximum hydrogen in reformate	1.4 g/sec	~168 kWth based on LHV
Durability	3000 hours steady	Long term goal of 5000 hours of cycling
	500 hours cycle	
Fuel Processor Volume* *without balance of plant components or plumbing	≤80 L	Includes everything between cold feed streams and 100 °C fuel-cell-quality reformate outlet stream
Height	<23 cm	"Flat" aspect ratio for vehicle installation
Catalysts	Non-pellet-catalyst	Minimize volume
Fuel Type	Sulfur Free Gasoline	<2 ppm-weight sulfur
Full power hydrogen Efficiency	≥77%	LHV H$_2$/LHV ATR fuel
Residual fuel (as CH$_4$)	<1% (dry)	At PrOx exit
Assumed FC H$_2$ utilization	80%	At peak power
CO – steady state	≤100 ppm (dry)	At PrOx exit
CO – transient	<1000 ppm (dry)	
Reformate pressure	3 bar	At PrOx exit
Pressure loss	≤0.7 bar	ATR air inlet – PrOx exit

Figure 9.49 Second generation fuel processor developed by Nuvera [620].

the energy remaining in the anode off-gas. Figure 9.49 shows the second generation of the fuel processor, which had been equipped with automotive style control components [620]. The third generation fuel processor system (Figure 9.50) had a size of 150 L, including balance-of-plant.

The fuel processor could be operated in a turn-down ratio of 6.5:1, which corresponded to a thermal output of between 215 and 33 kW$_{th}$. The key performance data of the 3A generation fuel processor as provided by Bowers *et al.* are summarised in Table 9.7. The pressure drop of the fuel processor increased linearly with the system load and did not exceed the target value of 0.7 bar [621].

The start-up time demand was demonstrated to be kept below 4 min, which was impressive, for the second generation fuel processor, as shown in Figure 9.51 [620]. The start-up procedure was described. Gasoline was fed to the autothermal reformer at 7 kW$_{th}$ for 1.5 min. Then the gasoline flow was increased to 30 kW$_{th}$. The carbon

Figure 9.50 3A generation fuel processor developed by Nuvera [620].

Table 9.7 Key performance data of the third (A) generation fuel processor as set by Renault for their gasoline system, developed together with Nuvera [620].

Thermal Power		33	100	150	215
	[kWth]				
Hydrogen efficiency	[%]	81	80	78.5	80–82
Hydrogen content at PrOx outlet (dry base)	[%]	41.5	40	40	39
CO content at PrOx outlet	[ppm]	<50	60–77	70–104	100

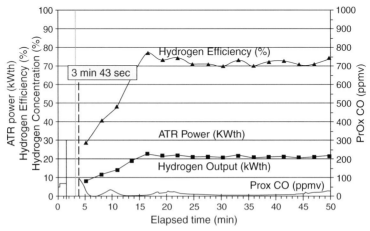

Figure 9.51 Start-up time demand from ambient (25 °C) as determined for the second generation fuel processor developed by Nuvera [620].

monoxide concentration at the outlet of the preferential oxidation stage dropped below 100 ppm after 3 min 43 s [620]. However, the hydrogen efficiency was below 30% at that time and it took the system 20 min to reach efficiency values greater than 70%. Full power was also not achieved after 4 min, and it actually took 15 min to reach this value. The start-up time demand could be further reduced below 90 s for the third generation prototype when started from a temperature slightly above ambient (40 °C) [620].

A cost analysis breakdown was performed for the fuel processor, which will be discussed in Section 10.1.

9.6
Diesel and Kerosine Fuel Processors

There are various applications for diesel fuel processors. Apart from the energy supply for diesel fuelled passenger cars and military applications, the most prominent is certainly the auxiliary power supply to trucks. In the United States in particular, idling of trucks is a common practice for generating power for the driver cabin during breaks. This is a pollution problem, because it is very inefficient and generates about twice as much NO_x as compared with driving at 90 km h^{-1} [622]. The idling time amounts to more than 1800 hours per truck a year and 3.8 billion L of diesel are consumed in the United States this way every year [622]. Another target market is recreational vehicles.

Piewetz *et al.* [623] described a 32-kW$_{el}$ diesel fuel processor prototype designed for a molten carbonate fuel cell. It consisted of a hydrotreater unit operated at 45 bar and 380 °C for conversion of the 0.2 wt.% sulfur components present in the feed. A zinc oxide bed was positioned downstream, which adsorbed the hydrogen sulfide generated in the hydrotreater. The sulfur free feed then entered the reformer, which was operated at 25 bar and 480 °C.

Delphi Corp. US, develops solid oxide fuel cell based stationary and mobile auxiliary power units. Delphi is developing reforming technology for natural gas, gasoline and diesel/JP-8 fuel for their applications. Catalytic partial oxidation is under development as well as autothermal reforming. While partial oxidation is addressed owing to its simplicity, oxidative steam reforming is regarded as the more efficient solution. Part of the anode tail-gas is recycled to the reformer to supply steam (see Section 2.3.2). Cathode and anode off-gas are mixed and the combustion supplies energy to the reformer. Figure 9.52 shows the Delphi steam reformer and the 5-kW SPU 2 stationary power unit.

Since 2004 Volkswagen has been developing a diesel fuel processor/fuel cell system with IdaTech LLC, US. The system works with a palladium membrane for reformate purification.

The Webasto Thermosystems GmbH company is developing a 5-kW solid oxide auxiliary power unit. A partial oxidation reactor supplies the fuel cell with reformate.

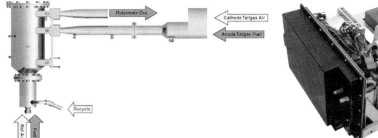

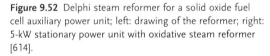

Figure 9.52 Delphi steam reformer for a solid oxide fuel cell auxiliary power unit; left: drawing of the reformer; right: 5-kW stationary power unit with oxidative steam reformer [614].

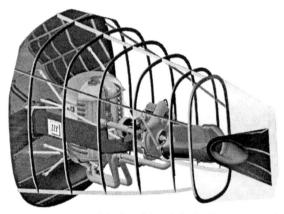

Figure 9.53 CAD model of a solid oxide fuel cell APU mounted in the aircraft APU cavity [625].

Airbus and Boeing have initiated the development of auxiliary power units for aircrafts. One advantage of fuel cell systems is that they produce water. This could be used as service ("grey") water, reducing the weight demand of the airplane water storage system. Up to 2000 L of water are required in an airplane during a long-term flight [624].

Daggett et al., from Boeing, presented their concept of a fuel cell APU for a commercial aircraft [625]. The main driver for this development was the reduction of NO_x emissions, noise and the low thermal efficiency of current APU systems, which is around 15%. The system design presented consisted of a solid oxide fuel cell along with a pre-reformer working either with partial oxidation or autothermal reforming. An afterburner was shown to combust the fuel cell anode off-gas and the combustion gases drove an expander, which in turn propelled the compressor of the system. A CAD (computer-aided design) model is shown in Figure 9.53. Heat-exchangers and a water separator to produce grey water (e.g., toilet water) from the combustion gases completed the design concept. The market for commercial aircraft APU's was estimated to about 440 devices per year, however, a modular design of the fuel cell APU might even triple this number.

Rosa et al. [251] set up a complete 5-kW diesel fuel processor based on autothermal reforming and catalytic carbon monoxide clean-up, which was dedicated to a low temperature PEM fuel cell. The breadboard system was composed of the autothermal reformer operated between 800 and 850 °C with a ruthenium/perovskite catalyst (see Section 4.2.8), a single water–gas shift reactor containing platinum/titania/ceria catalyst operated between 270 and 300 °C (see Section 4.5.1), and a preferential oxidation reactor containing platinum/alumina catalyst operated between 165 and 180 °C. Figure 9.54 shows the gas composition and reactor temperatures achieved. The hydrogen content of the reformate was in the range from 40 to 44 vol.% on a dry basis. The carbon monoxide content of the reformate was 7.4 vol.% and could be reduced to values of between 0.3 and 1 vol.% after the water–gas shift reactor and to below 100 ppm after the preferential oxidation reactor.

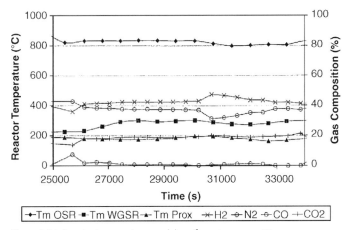

Figure 9.54 Reactor temperatures and dry reformate composition of the diesel reformer developed by Rosa et al. The gas analysis shows the gas composition of the fuel processor product [251].

A fuel processor for diesel, JP-8 and Jet-A fuels dedicated to feed a solid oxide fuel cell was developed by Roycoudhury et al. [150]. Because no carbon monoxide clean-up was required in this instance, the system was composed only of the reformer, a heat-exchanger to evaporate and pre-heat the steam with the hot reformate gases and finally of a sulfur trap (zinc oxide operated between 300 and 350 °C) downstream of the reformer. Similar to the propane ATR reformer presented by Rampe et al. [151] (see Section 7.1.2), the reformer was operated at an O/C ratio of between 1.1 and 1.2 but at a higher S/C ratio of 2.2. Therefore, partial oxidation rather than oxidative steam reforming took place in the reactor. The efficiency of the reformer was consequently in the same region, around 75%. Pre-heating of the catalyst, which was coated onto a metal-foam like structure (Microlith, see Section 6.2) was performed with a glow plug. The reactor is shown in Figure 9.55. The conversion of the fuel, however, was incomplete in the reactor and never exceeded 95%. It increased with increasing O/C ratio due to increasing total oxidation of the feed carbon but was

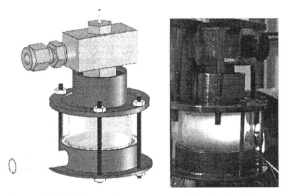

Figure 9.55 ATR diesel reformer developed by Roychoudhury et al. [150].

independent of the S/C ratio, because the steam only contributed to the water–gas shift but not to steam reforming in the reactor. The power density of the device amounted 1 kg kW^{-1}.

Another application for diesel fuel processors is the propulsion of naval systems. Krummrich et al. [626] reported from a conceptual study of a 2.5-MW fuel processor/fuel cell system, which was dedicated to submarine applications for the German ship manufacturer HDW. The system consisted of a desulfurisation step, an adiabatic pre-reformer operated between 400 and 550 °C, steam reforming at 800 °C and catalytic carbon monoxide clean-up. The critical step turned out to be the desulfurisation of F76 diesel fuel, which in Europe contains as much as 0.2 wt.% sulfur, world-wide as much as 1 wt.%. These workers then set up and operated a 25-kW demonstration model of the fuel processor, which achieved an efficiency of 82%.

9.7
Multi-Fuel Processors

Some workers and companies are also developing fuel processors suitable for a variety of fuels. If these systems were as reliable and efficient as those fuel processors that are tailor made for a specific fuel, multi-fuel processors would of course be a superior solution.

The fuel processor that was developed by Nuvera and described in detail in Section 9.5, was capable of working not only with gasoline, but also with alcohol fuels and natural gas [620], diesel and jet fuel [621] at comparable efficiency and carbon monoxide content after the preferential oxidation stage, as shown in Figure 9.56.

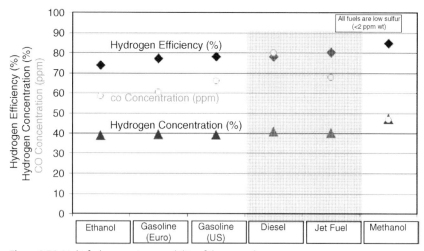

Figure 9.56 Multi-fuel conversion capability of the second generation fuel processor developed by Nuvera [621].

Real-time switching between gasoline and diesel fuel could actually be demonstrated at different load levels without any apparent change to the performance of the system [621]. However, low sulfur fuels were utilised exclusively. The fuel processor was tested in combination with an Andromeda fuel cell stack developed by Nuvera. No apparent difference could be observed when operating the combined system with ethanol, gasoline or diesel fuel [621]. However, recent activities of these workers have been directed towards a 10 kW$_{el}$ on-board power plant or auxiliary power unit [621]. This fuel processor had a volume of 35 L [621].

Meyer et al. described the development of a multi-fuel processor by International Fuel Cells, LLC [627]. Methanol and gasoline (quality: California reformulated gasoline grade II) were the major fuel alternatives. The technology chosen consisted of feed desulfurisation, autothermal reforming and catalytic carbon monoxide removal by two water–gas shift stages and two preferential oxidation reactors. The system had a power equivalent of 50 kW. However, performance data were only provided with respect to the autothermal reformer. Desulfurisation proved to increase the reformer conversion up to 98%. No residual heavy hydrocarbons then remained in the product. The hot spot of the autothermal reformer approached 1000 °C.

Meyer et al. also described the full range of hydrocarbon fuelled fuel cell systems that International Fuel Cells, LLC had developed by the year 2000 [627]. The fuels ranged from methane to heavy hydrocarbons and the system size from 500 W to 11 MW. The fuel cell technology, which was supplied with hydrogen from the fuel processors, were reported to cover alkaline, proton exchange, molten carbonate and phosphoric acid fuel cells. However, no details were provided concerning the specific applications or performance of these systems.

Irving et al. [628] presented a microreactor with fixed catalyst beds that was capable of reforming gasoline, diesel, methanol and natural gas by steam reforming at temperatures up to 800 °C. Not just reforming, but also mixing of fuel and steam, heat exchange, evaporation and feed pre-heating were covered by the integrated device, which was made from both stainless steel and ceramics. Gasoline steam reforming at flow rates of 0.1 g min^{-1} gasoline feed was performed at high S/C ratios of between 5 and 8. The dry reformate contained 70 vol.% hydrogen. Methane formation and catalyst deactivation were negligible due to the high S/C ratio applied. Sulfur contained in the gasoline led to hydrogen sulfide formation. Lowering the S/C ratio with isooctane steam reforming increased the methane concentration in the reformate to 5 vol.%. At a feed rate of 0.3 g min^{-1} and S/C ratio of 4, 100% conversion and a methane content well below 5 vol.% was achieved for a mixture of 60% isooctane, 20% toluene and 20% dodecane, which also contained 476 ppm sulfur. However, the weight hourly space velocity was rather low at about 30 L (h g$_{cat}$)$^{-1}$. A multi-fuel processor developed by Innovatec was later presented by Irving et al. [629], which is shown in Figure 9.57. It was sized for a 1-kW$_{el}$ PEM fuel cell and again microstructured components were used to build the system. The fuel processor was claimed to operate with a wide variety of fuels, from methane to bio-fuels and JP-8 fuel. However, no further specifications were provided for the device.

Figure 9.57 Multi-fuel processor developed by Irving et al. for a 1-kW PEM fuel cell [629]; the device contains microstructured components.

9.8
Fuel Processors Based on Alternative Fuels

Metkemeijer and Archard have set up a combined ammonia cracker/alkaline fuel cell system [630]. The cracking reactor was a tubular fixed catalyst bed of 4-cm diameter and 75-cm length filled with commercial magnetite ammonia synthesis catalyst from Atochem. The particle size of the catalyst was 600 µm and the reactor was heated electrically to supply the endothermic cracking reaction with energy (see Section 3.8). 99.9% conversion could be achieved at a reaction temperature of 627 °C and a high ammonia feed rate of 2.4 m^3 h^{-1}. The pressure drop was below 300 mbar under these conditions, and a hydrogen output of 3.6 m^3 h^{-1} was also achieved, which would be sufficient to generate approximately 3.5 kW of electrical energy. However, the alkaline fuel cell stack (see Section 2.3.2) connected to the cracking reactor was composed of 24 cells and had a nominal power of only 432 W [630]. Compared with the operation with pure hydrogen, performance losses of no more than 3% were observed for the fuel cell when operated with the reformate, which contained only 40% hydrogen [630].

A 1-kW$_{el}$ sodium borohydride hydrogen generation system was developed by Zhang et al. [101], from Millennium Cell Inc. Figure 9.58 shows the setup, which consisted of the storage tank, the fixed bed catalytic reactor, a gas–liquid separator to separate the liquid product (sodium borate solution) from the hydrogen, a plate heat-exchanger to remove the product heat and condense the steam still present in the gas phase, a second gas–liquid separator for water removal and a dryer for the hydrogen, which then entered the fuel cell. The sodium borohydride was stored in a tank and tested at various concentrations as discussed below, the solution was stored at pH 14 through the addition of 3% sodium hydroxide. The reactor had a volume of 78 cm^3 and was filled with 56 g of 3 wt.% ruthenium catalyst supported by carbon, which had a particle size of 2 mm (Johnson Matthey). The liquid feed flow rate was varied from

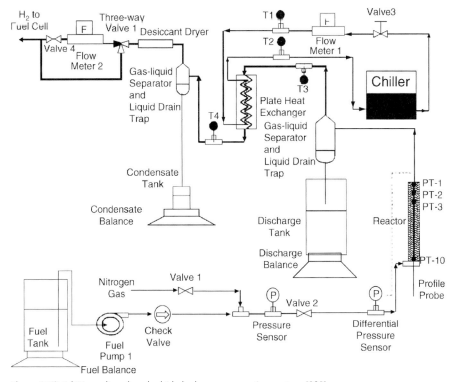

Figure 9.58 1-kW$_{el}$ sodium borohydride hydrogen generation system [101].

10 to 60 mL min^{-1} and full conversion of the fuel could be achieved up to a feed rate of 30 mL min^{-1}. Figure 9.59 shows the temperature distribution along the small reactor as determined for various flow rates. The rising temperature gradient moved further into the catalytic bed with increasing feed rate and some of the heat of reaction was removed by water evaporation. The small reactor produced 8.8 L min^{-1} hydrogen at full fuel conversion, which corresponded to a power equivalent of 500 W$_{el}$ [101]. A higher feed concentration of 15 wt.% and higher feed flow rate of 80 mL min^{-1} increased the hydrogen production rate up to 22.9 L min^{-1}, however, fuel conversion remained incomplete under these conditions. The higher feed concentration had a positive effect on the activity of the catalyst. The temperature plateau, which corresponded to full conversion, was thus achieved earlier and therefore a smaller reactor was required [101]. Feed pre-heating increased the conversion, while a higher system pressure had a negative effect probably due to the thermodynamic equilibrium [101]. The catalyst degraded significantly within 4 h. The systems under development had fast dynamic responses within seconds and generated up to 120 L min^{-1} hydrogen [631]. The issue of catalyst deactivation was addressed by including the reactor in the fuel cartridge unit [632]. Thus the reactor and catalyst would be exchanged together with the cartridge.

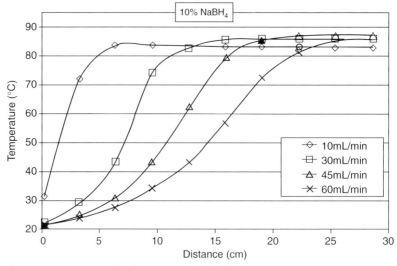

Figure 9.59 Temperature distribution versus length of reactor axis for various feed flow rates at 10 wt.% sodium borohydride concentration in the feed [101].

Ammonia hydride generators are relatively simple systems [104]. Ammonia was fed into the lithium aluminium hydride reactor and hydrogen was generated until a pressure equal to the vapour pressure of ammonia had built up. The fuel cell connected to the system (not shown here) consumed the hydrogen, which decreased the pressure. This caused the ammonia to flow into the reactor again. An ammonia getter had to be placed downstream of the reactor to remove traces of unconverted ammonia. While the ammonia reservoirs could be exchanged, the reactor was of course exhausted when all of the lithium aluminium hydride was consumed.

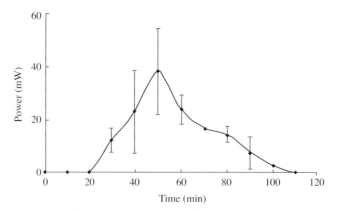

Figure 9.60 Electricity generation by a bio-fuel processor/fuel cell system [95].

Ammonia, 145 g, was stored in the reservoir, while the fresh reactor contained 324 g lithium aluminium hydride. The total weight of the system amounted to 830 g and produced 400 Wh energy.

Finally, a portable bio-fuel processor/fuel cell system will be mentioned, which was presented by Ishikawa et al. [95]. Micro-organisms were immobilised in gel beads and packed into a 20 cm^3 glass bottle along with a glucose solution and incubated at 37 °C. Up to 9 cm^3 h^{-1} hydrogen could be produced by these means, which was used to generate electricity with a small fuel cell. Up to 40 mW of power were generated for 2 h duration, as shown in Figure 9.60.

10
Introduction of Fuel Processors Into the Market Place – Cost and Production Issues

The technical performance issues of fuel processors such as efficiency, start-up time demand, size, weight, responsiveness to load changes, lifetime and reliability have been discussed in the sections above. Unlike chemical reactors for chemical industry, which are frequently fabricated with a few or even as single devices, energy generation reactors and fuel processors for future distributed power generation are heading for mass production. Consequently, fabrication costs become the most crucial issue as soon as the technology has proven its technical feasibility.

10.1
Factors Affecting the Cost of Fuel Processors

According to studies of Arthur D. Little Inc. [633] and targets set by the US Partnership of New Generation Vehicles (PNGV), the cost for a drive train fuel processor had been limited to 30 US$ kW^{-1} for the year 2000. Another study even limited the total cost of the whole drive train including the fuel cell to below the current value of a diesel turbo engine (50 US$ kW^{-1}) [634]. However, these ambitious goals are still far from being attainable. They do not apply to residential applications [633] and auxiliary power units (APU). For these applications values higher by one to three orders of magnitude or more could be competitive depending on the application. The target costs for residential fuel processor/fuel cell systems were around 1500 US$, but today's actual investment costs are about 15 000–20 000 US$ [635]. Fuel processor costs depend on the main factors summarised below:

- fabrication technique
- type of fuel
- catalyst cost
- production quantity.

Cheap fabrication techniques that are suitable for mass production need to be applied for fuel processors. The various techniques available for fuel processor production will be discussed in Section 10.2.

Fuel Processing for Fuel Cells. Gunther Kolb
Copyright © 2008 WILEY-VCH Verlag GmbH & Co. KGaA, Weinheim
ISBN: 978-3-527-31581-9

The type of fuel employed has a significant impact on the overall fuel processor cost. Fuels such as methanol, for which no distribution grid exists, would lead to considerable extra costs. As an example, to replace 10% of the existing gasoline demand of the USA with methanol would add 65 billion US$ in costs (status: 1998) [633], these charges would be passed on to the corresponding 25 million vehicles, resulting in additional costs of approximately 65 US$ kW^{-1}. This is a strong argument for the utilisation of fuels with existing infrastructure that has already been amortised. In particular, for first niche applications the infrastructure will play a dominant role. On the other hand, sustainability of power systems is a strong political argument, which will of course drive the development of energy supply systems in the direction of renewable sources more and more in the future, especially for mobile systems, where carbon dioxide sequestration seems to be a less viable option.

Catalyst cost may play a significant role in the overall fuel processor cost and could reach values as high as 38% [633]. In this situation, tailor-made catalyst formulations of enhanced activity are required along with measures to increase the utilisation of the catalyst. This may be achieved by coating the catalyst into small channel systems of ceramic or metallic monoliths or into microstructured plate heat-exchangers, which improves the mass transfer, as described in Chapter 6.

For the gasoline fuel processor developed for the automotive drive train by Nuvera, for Renault, (see Section 9.5), a cost breakdown was performed both for the overall system and for the individual reactors, which is shown in Figure 10.1 [636]. It is obvious that (precious metal) catalyst costs dominated at 61% the system cost and thus required optimization for this specific system. The application of a fuel cell tolerant to elevated amounts of carbon monoxide would further reduce the system cost, because the catalytic carbon monoxide clean-up reactors contributed 59% to the overall cost for the reactors, as shown in Figure 10.2. Similar results were found much earlier by Teagan *et al.* [633] for a 50-kW_{el} gasoline fuel processor as shown in Figure 10.3. Again, the catalyst costs dominated the overall costs of the system for high volume production (10 000 units).

Finally, the quantity of fuel processors produced has a strong impact on costs. Fuel processor fabrication for residential, mobile and portable applications is heading

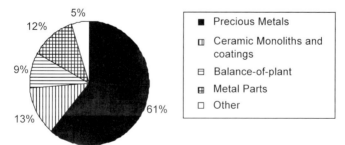

Figure 10.1 Cost breakdown for the major components of the third generation gasoline fuel processor developed by Nuvera, for Renault [636].

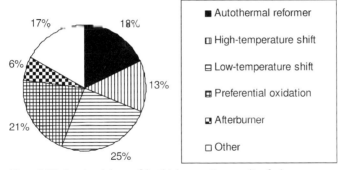

Figure 10.2 Cost breakdown of the third generation gasoline fuel processor developed by Nuvera, for Renault, by reactors [636].

towards high production quantities. Starting with some hundreds to a thousand devices per year for first and niche applications, hundreds of thousand of fuel processors are the final goal.

As a rule of thumb, increasing the production quantity from 100 to 10 000 devices reduces the fabrication cost by one order of magnitude [633]. Another rule of thumb for decreasing the fabrication costs of high-tech products is that doubling the cumulated production costs brings about a reduction in the fabrication costs per unit by a factor of 0.8.

Below some rules for the cost estimation of up- or downscaling of fuel processors and balance-of-plant equipment are provided.

The unknown fuel processor cost C_x of a system of size E_x may be calculated from the known cost C_k of a system of size E_k according to [622]:

$$C_x = C_k \left(\frac{E_x}{E_k}\right)^y \tag{10.1}$$

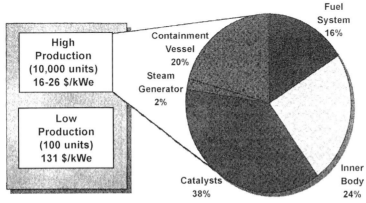

Figure 10.3 Cost breakdown of a 50-kW$_{el}$ gasoline fuel processor [633].

Barrato et al. determined a value of 0.94 for the cost-capacity factor y, however, its value is usually in the region of 0.6 [622].

The cost of a compressor C_{comp} increases with increasing power P demand according to [622]:

$$C_{comp} = AP^{0.82} \tag{10.2}$$

where A is a multiplicative factor, which needs to be determined previously for a specific compressor size and technology.

Changing the size of a heat-exchanger with known heat-exchange area A_{HX} results in a cost change according to the following rule [622]:

$$C_{HX} = BA_{HX}^{0.65} \tag{10.3}$$

B is again a pre-determined factor.

For a combined heat and power (CHP) system for a one-family house, an optimum configuration with respect to cost effectiveness was determined by Schmid and Wünning [16]. The fuel cell system was shown to provide about 70% of the average annual heat demand of the house, while peaks in demand were buffered by a conventional burner with an efficiency of 95%. The relative energy savings P of the fuel processor/fuel cell system over a centralised power supply with conventional steam gas turbines (SGT) was calculated as follows:

$$P = \left(1 - \frac{GU_{CHP}}{GU_{SGT+CBT}}\right) 100\% \tag{10.4}$$

where GU_{CHP} is the gas consumption of the combined heat and power fuel cell system and $GU_{SGT+CBT}$ is the gas consumption of the steam gas turbine and of the conventional burner technology (CBT),

$$GU_{CHP} = \frac{1 - f_{CHP}}{\eta_{th,PB}} + \frac{f_{CHP}}{\eta_{th,CHP}} \tag{10.5}$$

f_{CHP} is the ratio of heat generation of the CHP system to the overall heat demand of the house and is equal to 0.7, while $\eta_{th,PB}$ and $\eta_{th,CHP}$ correspond to the thermal efficiency of the CHP system and of the peak burner (PB), respectively.

$$GU_{SGT+CBT} = \frac{1}{\eta_{th,CBT}} + \frac{f_{CHP} \, SK_{CHP}}{\eta_{el,SGT}} \tag{10.6}$$

where $\eta_{th,CBT}$ and $\eta_{el,SGT}$ correspond to the thermal efficiency of the CBT system and the electrical efficiency of the steam gas turbine [16]. SK_{CHP} is the CHP coefficient:

$$SK_{CHP} = \frac{\eta_{el,CHP}}{\eta_{th,CHP}} \tag{10.7}$$

Figure 10.4 shows the primary energy savings of the CHP system versus its electrical efficiency for two different values of the total efficiency of the CHP system, namely 80 and 90%. Because current PEM fuel cell systems were assumed to have an electrical efficiency of 30%, only an energy saving of about 15% relative to conventional

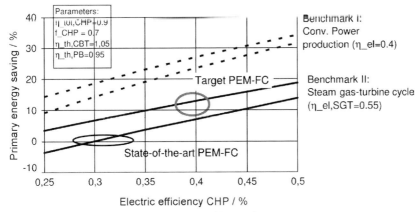

Figure 10.4 Primary energy savings of a combined heat and power (CHP) system for a one-family house compared with conventional power production ($\eta_{el} = 40\%$) and the steam gas turbine process ($\eta_{el} = 55\%$); upper lines correspond to an overall efficiency of the CHP system of 90%, lower lines to 80% [16].

power generation was to be expected. No advantage was apparent if compared with steam gas turbines [16]. Schmid and Wünning concluded from these results that an electrical efficiency of 35% and an overall efficiency of 90% is required to introduce fuel cell CHP systems into the market place.

10.2
Production Techniques for Fuel Processors

This section discusses various fabrication techniques that are suitable for different fuel processor technologies taking the cost issues into consideration.

10.2.1
Fabrication of Ceramic and Metallic Monoliths

The most widely used construction material for ceramic monolith carriers is cordierite [130], alumina being an alternative. The ceramic monolith is produced from the cordierite by mixing with water and binder material, such as methylcellulose, and extruding the resulting paste. Heat resistant inorganic fibres may be added to the paste [136]. The resulting green monolith body is then dried and calcined at 1300–1400 °C for 3–4 h [130]. Alternatively, because cordierite is not usually readily available, different source materials may be mixed, in order to achieve the composition of cordierite ($5SiO_2:2Al_2O_3:2MgO$). However, numerous other formulations, additives and production procedures can be applied, which will not be discussed further as they are not within the scope of this book.

Once the ceramic monolith is fabricated and coated with catalyst, it needs to be canned in a metallic reactor body. A mat material is used to bridge the gap between

Figure 10.5 View into metallic monoliths produced by EMITEC (photograph: courtesy of Emitec).

the monolith and the reactor shell. It holds the monolith and prevents by-pass of gases through the gap [57]. A popular ceramic mat used in automotive exhaust systems is Interam™ produced by 3M™ [57]. It degrades at temperatures above 800 °C, and therefore a high temperature ceramic fibre material such as CC-Max® from Unifrax must be used for monolithic reformer reactors, which are operated at higher temperature [57].

The most prominent production technique for metallic monoliths is the rolling up of a flat and of a corrugated metal foil. In this way, an alternating arrangement of both foils is achieved, as shown in Figure 10.5.

The construction material of metallic monoliths are alloys of iron, about 15–20 wt. % chromium and 5 wt.% aluminium (Fecralloy). The unique feature of these alloys is the formation of a thin alumina layer (0.5 μm) on their outer surface when treated at temperatures above 850 °C [130]. Long alumina whiskers are formed at temperatures of around 900 °C and an oxidation time of greater than 12 h (see Figure 10.6). This alumina layer then acts as an adhesion layer for the catalyst coating. On top of this, the alumina layer protects the alloy from corrosion, which allows the operation of ultra-thin Fecralloy foils at high temperatures.

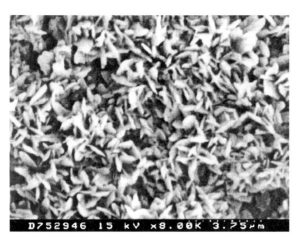

Figure 10.6 Alumina whiskers produced on the surface of Fecralloy by treatment in air at a temperature of 900 °C for 22 h [130].

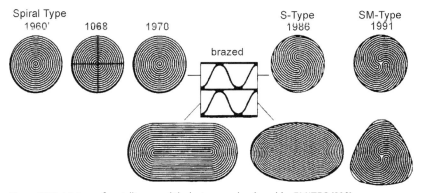

Figure 10.7 History of metallic monolith designs as developed by EMITEC [130].

Foils with a thickness as low as 20 µm are used in metallic monoliths. Increasing the aluminium content of the Fecralloy above 6 wt.% increases the resistance of the material towards corrosion, but also its brittleness. Different methods are applied to obtain the axial mechanical stability of the metallic monolith. Amongst these are forcing bins, welding and brazing techniques [130]. The company EMITEC applies sandwich foils of Fecralloy and aluminium. The aluminium then melts during operation and produces a type of welding of the monolith structure [130]. Figure 10.7 shows various metallic monolith structures as developed by EMITEC. Stacks of corrugated and flat foils are wound alternatively around two ("S"-type in Figure 10.7) and three ("SM"-type in Figure 10.7) mandrels. Torsional deformation then stabilises the monolith structure.

An alternative to Fecralloy is aluminium when the operating temperature of the monolith is limited to below about 450 °C. An alumina layer may be formed on the aluminium, for example, by anodic oxidation.

Ceramic foams are produced from organic precursor foams such as polyurethane or polyolefins. Their pores are then filled with an aqueous slurry of the ceramic typically containing 20 wt.% of ceramic particles in the size range of from 0.1 to 10 µm [461]. Wetting agents, dispersion stabilisers and viscosity modifiers are added to the slurry. Suitable ceramics are alumina, alumina silicates, zirconia, stabilised zirconia and titania, amongst others. The pores of the precursor foam may be filled completely or only coated on their surface by the ceramic particles. The foam is then dried and calcined at 1000 °C, which removes the polymer and sinters the ceramic. Metallic foams have similar properties compared with ceramic foams, but superior mechanical stability and improved heat conductivity.

10.2.2
Fabrication of Plate Heat-Exchangers/Reactors

Plate heat-exchangers are usually made from stainless steel, however, to deal with the demands of high temperature applications, such as hydrocarbon steam reforming and partial oxidation reactions, nickel-based alloys may be considered, despite their higher price.

Another option are Fecralloys, which are also used for fabrication of metallic monoliths and are mechanically stable up to very high temperatures exceeding 1200 °C. However, the material is brittle, which generates practical problems with respect to mechanical stability.

Ceramics offer advantages for the high temperature regime, as proven by the widespread use of ceramic monoliths. Interesting work has been performed by Wang et al. [637] on a ceramic microreactor made from ceramic tapes, which might be considered in the future for certain applications.

Copper, and particularly aluminium, are alternative metals for low temperature processes such as methanol reforming and carbon monoxide purification. The higher heat conductivity of these metals, 401 and 236 W $(m\,K)^{-1}$, respectively, compared with stainless steel [around 15 W $(m\,K)^{-1}$] makes them attractive when isothermal conditions are required, which may well be the situation for evaporators or reactors operating within narrow temperature windows.

On the other hand, the efficiency of small-scale counter-flow heat-exchangers suffers from heat transfer through the wall material. Corrosion issues need to be addressed, especially with aluminium. Comparing the energy demand for fuel processor start-up, which results from the product of specific heat capacity and density, aluminium is the most favourable material as opposed to copper and stainless steel (as long as the geometry and consequently the volume of the device remains the same).

Polymers are an interesting construction material for the fabrication of highly efficient low-temperature heat-exchangers, which withdraw the last portion of energy from the fuel processor off-gases before releasing them to the environment. This arises because of the low heat conductivity of the polymers.

Various production technologies for heat-exchangers and heat-exchanger/reactors were compared by Dudfield et al. [550]. Plate heat-exchangers sealed by gaskets are established products of process engineering. This technique makes the coating of the catalyst feasible, because the devices can be disassembled. Their surface to volume ratio is typically in the region of 200 $m^2\,m^{-3}$ according to Dudfield et al. [550]. However, the gasket materials available, namely nitrile, neoprene and viton, limit the operating temperature of the devices to 200 °C [550]. For temperatures greater than 200 °C, metallic gaskets and sealing techniques should be applied. Heat-exchangers sealed by gaskets require screws and bulky housing, which becomes an issue the smaller the devices are, because the thermal mass is increased considerably. This increases start-up time demand.

Alternative sealing techniques are performed at elevated temperatures and are more or less irreversible. For process engineering devices such as heat-exchangers and evaporators, the processing temperature of the sealing technique is not critical because the construction material, such as stainless steel, can withstand the temperature. The same applies for chemical reactors, provided the catalyst or the catalyst coating is introduced after the sealing procedure. However, where the catalyst coating needs to be introduced before sealing the reactor, the temperature of the sealing process has to be well below the maximum operating temperature of the catalyst, which may range from about 300 to more than 800 °C depending on the catalyst formulation. This limits considerably the applicability of many of the sealing techniques that are described below.

Shell and tube heat-exchangers are all exclusively metallic constructions, which means that a much wider operating temperature range of up to 650 °C and more is feasible. To incorporate catalyst coatings into such devices, the tubes can either be coated with catalyst or catalyst supports can be incorporated into the tubes [550]. The surface to volume ratio of the shell and tube design is in the range of 740 m^2 m^{-3}. The devices are usually sealed by conventional welding techniques.

Laser welding is a viable option for the fabrication of plate heat-exchanger/reactors. For devices on a larger scale the length of the weld needs to be minimised to achieve competitive prizing. An advantage of laser welding is the limited local energy input, which protects incorporated catalyst from damage. Figure 10.8 shows a 10-kW counter-flow heat-exchanger, which was sealed by laser welding. Electron beam welding, already established in the automotive industry, is an alternative to laser welding with similar limited local energy input.

Diffusion bonding is an alternative sealing technique, which requires a high temperature and high vacuum. The material is compressed and heated to temperatures approaching its melting point, which generates a quasi boundary-free single workpiece. Takeda *et al.* described the procedure for diffusion bonding of nickel-based alloys (Hastelloy) [638]. A plate and fin heat-exchanger (40 × 40 × 3 mm^3 dimensions, channel dimensions 1000 × 1000 µm^2) was cleaned with a 1% solution of nitric acid and hydrogen fluoride to remove the oxidation layer. The operating pressure during bonding amounted to 6 mPa. For NiCroFer and other materials, some workers have claimed that even lower pressures of 1 µPa are required. A 1150 °C bonding temperature and 440-bar contact pressure for 30-min duration were reported by Takeda *et al.* as the optimised conditions [638]. They claimed that 3% deformation was sufficient for bonding. Leak tightness was verified at ambient temperature up to 630 bar. Yu *et al.* reported

Figure 10.8 10 kW counter-flow heat-exchanger sealed by laser welding (source: IMM).

successful diffusion bonding of Fecralloys at 900 °C and a mechanical pressure of 200 bar [502].

Heatric, a member of the Meggit Group, is one of the companies that has developed of a proprietary diffusion bonding process. The company applies diffusion bonding for fabrication of large meso-scaled heat-exchangers with exchange capabilities in the MW range, mainly for off-shore applications. The Karlsruhe Research Center frequently applies diffusion bonding in their cross-flow microstructured heat-exchangers (see Figure 10.9) [639,640].

Compact plate heat-exchangers are frequently sealed by brazing techniques. However, the brazing compounds often contain heavy metals such as cadmium and tin, which are known catalyst poisons [57]. Therefore, it is advisable to finish the brazing procedure prior to a thorough cleaning step and the catalyst coating procedure [57]. Migration of metal or metal alloy compounds into the catalyst is less of an issue with stainless steels than for the braze compounds. In addition, the melting temperature of brazing compounds needs to be compatible with the catalyst coatings.

Sintering was applied by Schuessler et al. [398] to seal their compact methanol fuel processor (see Section 9.1).

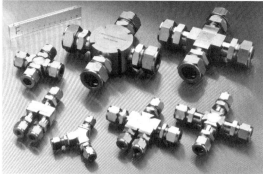

Figure 10.9 Top, diffusion bonded cross-flow heat-exchanger bodies developed by the Karlsruhe Research Center; bottom, ready made heat-exchangers (source: FZK).

10.2.3
Fabrication of Microchannels

Techniques for the fabrication of microstructured channel systems will be discussed below, where the focus is on their applicability for future mass production rather than providing detailed information about the techniques.

Micro-milling was frequently applied in many early applications and it is definitely a useful tool for experimental work and rapid prototyping, but is certainly not a choice when considering future mass production.

There are two types of Electro Discharge Machining (EDM), that is, controlled spark micromachining under a dielectric fluid. They are namely EDM die-sinking, were an electrode is moved into the work-piece creating a reverse shaped ablation, and EDM wire-cutting, where the discharge process is achieved by a wire (minimum diameter about 20 µm) moving through the work-piece. Both techniques are, similar to milling, excellent tools for rapid prototyping and fabrication of mold inserts. However, they are not suitable for mass production of microstructured devices.

Wet chemical etching, which applies a photo-resist for masking and usually iron chloride solution for etching, is an established, although not widespread, technique that is industrially available for many applications. It is competitive for mass production and covers a wide range of channel depths from about 100 up to 600 µm, which is the channel size usually applied in microstructured reactors for fuel processing applications. However, for applications in the kW range, large microstructured foils are required and costs become a critical issue when etching is applied. Thus, there is a need for cheaper alternatives in some cases (Figure 10.10).

Punching is a cheap technique suitable for mass production. To achieve a sealed microchannel system, the punched plates need to be separated from each other by unstructured plates. To the knowledge of the author of this book only cross-flow heat-exchangers can be fabricated out of punched plates, because the channel systems

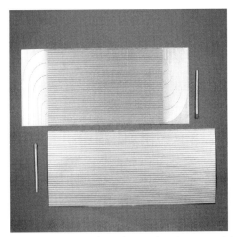

Figure 10.10 Microchannels introduced into a stainless steel plate by wet chemical etching (source: IMM).

Figure 10.11 Left, microstructures introduced into a stainless steel plate by embossing; right, microstructured and embossed heat-exchanger (source: IMM).

need to be designed without any bends. Coating of punched heat-exchangers with catalyst remains a challenging task.

Embossing is another cheap technique for introducing microstructures into metal foils and is certainly particularly suitable for mass production (see Figure 10.11).

At present, corrugated metal foils fabricated by rolling are already being applied in the field of automotive exhaust gas systems. The metallic monoliths fabricated this way achieve channel dimensions in the sub-millimetre range and are actually "microtechnology".

Laser ablation is frequently applied in many industrial applications. However, for the fabrication of microchannels of several hundred micrometre depth the method appears not to be cost competitive. For smaller channel dimensions in a region well below 100 µm, laser ablation might well be a cost competitive option, especially for small scale applications.

Schuessler et al. [398] used sintering of copper and aluminium powder to form microstructured plates for their integrated autothermal methanol reformer (see Section 9.1). The powders were compressed before sintering at a pressure of 1000 bar, which provided plate-type elements for assembly. The sintering of the copper was then performed at temperatures between 500 and 700 °C, which bonded the plates to a stack-like reactor.

Wang et al. reported the construction of a microreactor from Low Temperature Co-fired Ceramic tapes (LTCC DuPont 951 AX), which are inexpensive and also allow the integration of resistors and conductors [637]. The minimum thickness of the tapes available started at 111 µm. They consisted of oxide particles, glass frit and organic binder and could be machined by laser, milling, chemical treatment or even by photolithography. On firing, the organic binder combusted and the oxide particles were joined by sintering. The tapes could be stacked and co-fired to form monoliths. Shin et al. also reported the application of low temperature co-fired ceramics (LTCC) for the fabrication of a methanol fuel processor [641].

Finally, polymer heat-exchangers may be fabricated by injection molding, and microchannels have been produced by micro-injection molding [642] for numerous applications.

10.2.4
Fabrication of Chip-Like Microreactors

Techniques for the production of chip-like systems will be discussed below. These techniques should allow the cost-efficient mass production of the smallest scale (sub-Watt) fuel processors in the future.

Figure 10.12 shows the fabrication steps of a chip-like microreactor made from a silicon wafer as described by Pattekar and Kothare [532]. Firstly, a 10-µm thick photoresist layer was coated onto the silicon chip, which served as the etch mask for Deep Reactive Ion Etching (DRIE). Then the photoresist was removed and Pyrex glass was connected to the chip by anodic bonding to seal the channels. Teflon capillaries were then bonded as inlet/outlet connectors to the chip. On the reverse side of the chip, the photoresist was patterned, which served as an etch mask for a

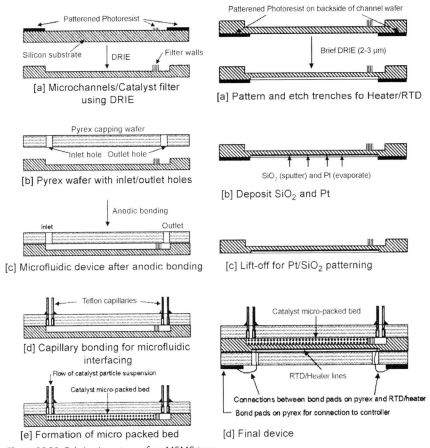

Figure 10.12 Fabrication steps of an MEMS-type (microelectromechanical system) methanol steam reformer by deep reactive ion etching [532]. Left: treatment of the top side of the substrate; right: bottom side.

brief DRIE step. Silica was sputtered onto the etched surface, for electrical insulation of the platinum layer, which served as a temperature measurement device and electrical resistance heater. The platinum was coated onto the surface by physical vapour deposition. The photoresist was then removed and Pyrex glass also bonded to the reverse side of the device.

Another fabrication procedure was described by Kim and Kwon [143]. Photosensitive glass was covered with a 1-nm chromium layer by sputtering and with 1-μm photoresist by spin coating (see Figure 10.13). The photoresist was treated by a typical lithographic process and the glass was subsequently treated with UV light, which crystallised the part of the glass not protected by the chromium layer about 1-mm deep. A bottom layer was attached to the glass by fusion bonding at 550 °C and 1000 N m^{-2}. The crystallised glass was then removed by etching with hydrofluoric acid. The catalyst coating technique shown in Figure 10.13 will be discussed briefly in Section 10.2.6.

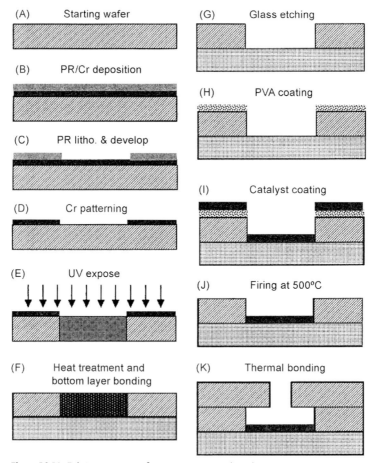

Figure 10.13 Fabrication steps of an MEMS-type methanol steam reformer by hydrofluoric acid etching [144].

10.2.5
Fabrication of Membranes for Hydrogen Separation

Thin metallic membranes can be produced by techniques such as rolling. Tosti *et al.* described the fabrication of palladium/silver membranes by cold rolling [643]. Several steps of rolling and annealing were required and eventually resulted in membranes of 50-μm thickness. The Vickers hardness of the membranes was increased by the rolling procedure. Cylindrical membrane separators were fabricated from the membranes by welding.

However, the generation of thin palladium membranes onto ceramic surfaces is more complicated. Methods such as spray pyrolysis (see also Section 4.1.3), chemical vapour deposition and sputtering are used. Another method commonly applied [400] is electroless plating [408]. Palladium particles are produced from palladium solution containing amine complexes of palladium in the presence of reducing agents. Palladium nuclei need to be seeded onto the surface prior to the coating procedure [408]. Ceramic surfaces such as α-alumina are first sensitised in acidic tin chloride and then palladium is seeded from acidic palladium ammonia chloride [408].

Combination with osmosis can improve the plating procedure and repair defects as described by Li *et al.* [408]. It was performed by feeding sodium chloride through the tube side of the membrane, which allowed the water of the electroless plating solution to diffuse through the membrane defects and increase the palladium concentration locally, and healed the defects.

Franz *et al.* developed a palladium membrane microreactor for hydrogen separation based on MEMS technology, which carried integrated devices for heating and temperature measurement [644]. The reactor was composed of two channels separated by the membrane, which was composed of three layers. Two of these layers, which were made of silicon nitride introduced by low pressure chemical vapour deposition (0.3-μm thick) and silicon oxide by temperature treatment (0.2-μm thick), served as a perforated support for the palladium membrane. Both layers were deposited on a silicon wafer (see Figure 10.14). They were removed from one side completely and from the other side partially to achieve the support function. Then one channel was etched out of the wafer with potassium hydroxide. The platinum/titanium films were deposited by an electron beam onto the device, which served as the heaters and temperature sensors. Then a "blanket" of palladium was deposited onto one side of the support using a titanium film as the adhesion layer. In this way, a 700-μm wide and 17-mm long membrane of 200-nm thickness was realised. An aluminium plate sealed the permeate channel underneath the membrane. The second (retenate) channel was fabricated from poly(dimethylsiloxane) applying a moulding process. The membrane showed a high mechanical stability. When the device was pressurised below the support structure on the permeate side, rupture occurred at a pressure of more than 5 bar. Surprisingly, the pressure tolerance was much lower (1.4 bar) when the membrane was pressurised from the opposite direction, on the retenate side. A pressure drop of 1 bar was even tolerated at a temperature of 500 °C, when the higher pressure was placed below the support structure. The performance of the membrane was tested with a mixture of 90%

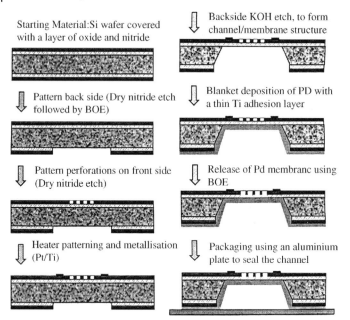

Figure 10.14 Microfabrication sequence for the silicon component of the palladium membrane reactor developed by Franz et al. [644.] BOE stands for Buffered Oxide Etching.

nitrogen and 10% hydrogen at atmospheric pressure and 500 °C. The permeate side of the membrane was evacuated. A hydrogen flux of 100 L $(m^2 s)^{-1}$ was determined at a pressure drop of 1 bar.

Gielens et al. described the fabrication of a palladium/silver membrane which contained 77 wt.% palladium and was 750-nm thick. The membrane was sputtered onto a silicon chip [645]. At 450 °C a hydrogen flux of 22.4 L $(m^2 s)^{-1}$ was determined for the membrane. Karnik et al. reported the fabrication of a palladium membrane of 50-nm thickness over a support structure which had holes that were 5.5-μm thick. This membrane could withstand a pressure difference of 690 mbar without rupture. Wilhite et al. [526] reported the preparation of a palladium/silver membrane only 20-nm thick. A 50 L $(m^2 s)^{-1}$ hydrogen flux was determined for the membrane for a hydrogen partial pressure of only 35 mbar.

10.2.6
Automated Catalyst Coating

Dip-coating is the automated procedure wide-spread established for ceramic and metallic monoliths.

Wash-coating techniques for metal foils exist in automotive exhaust treatment technology [646]. Adomaitis et al. has presented the Metreon process, which was developed for coating of unstructured Fecralloy metal foils [646]. The foils wrapped up as coils were processed through rolls, where they received a corrugated structure.

Figure 10.15 Semi-automated continuous coating machine for catalyst coating. The procedure is demonstrated here using a microstructured metal foil unwound from a roll. The wash-coat slurry is fed through a slot die onto the metal foil. After coating, the foil is sent through a drying compartment, an infrared calciner (not shown here) and then wound up again (source: IMM).

A heat treatment followed to generate the alumina layer (see Sections 4.1.3 and 10.2.1). Then the foils were wash-coated with the catalyst up to four times. Each coating step included drying and calcination. Finally, the metal foils were recoiled. The coatings prepared by this method were highly resistant to thermal shocks.

Continuous wash-coating of microstructured metal foils is possible using coating machines, as shown in Figure 10.15. In a continuous process, a microstructured steel coil structured by wet-chemical etching is unwound from a roll and continuously coated with the catalyst slurry. If required, the metal foil could be singularised later, to any plate size by laser-cutting at defined positions.

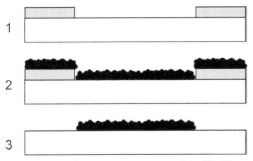

Figure 10.16 Schematic representation of the lift-off process for catalyst coatings developed by Thybo et al. [139]: (1) deposition of photoresist; (2) deposition of catalyst; (3) lift-off of the photoresist by dissolving.

Thybo *et al.* proposed a method to create well-defined patterns of catalyst deposits by application of a photoresist [139]. As shown in Figure 10.16, the photoresist was deposited on the surfaces not to be coated by spin-coating, subsequent UV-photo-lithographic treatment and removal of the undesired resist with sodium hydroxide solution. Then the catalyst was deposited on the entire surface and finally the photoresist was removed by dissolving in acetone. Kim and Kwon [144] used a similar method for catalyst coating but applied poly(vinyl alcohol) binder and a mask rather than a photoresist, as shown in Figure 10.13 (H–J) Section 10.2.4, and also Section 4.1.3. The binder along with catalyst, which were at undesired positions, were then removed by temperature treatment.

References

1 Song, C. (2002) Fuel processing for low-temperature and high-temperature fuel cells. Challenges and opportunities for sustainable development in the 21st century. *Catal. Today*, **77**, 17–49.
2 www.fuelcelltoday.com, Fuel cell market survey: portable applications, (01 November 2003).
3 Heurtaux, F., Meille, V., Pitault, I., Santa-Cruz, G., Swesi, Y., Rouveyre, L., Soares, E. and Forissier, M. (2004) Storage in cycloalkanes and *in situ* production of hydrogen in vehicles, in Proceedings of the 15th World Hydrogen Energy Conference, Yokohama, Japan.
4 Delsman, E.R., Uju, C.U., de Croon, M.H.J.M., Schouten, J.C. and Ptasinski, K.J. (2005) Exergy analysis of an integrated fuel processor and fuel cell system. *Energy*, **31**, 3300–3309.
5 Dicks, A.L. (1996) Hydrogen generation from natural gas for the fuel cell systems of tomorrow. *J. Power Sources*, **61**, 113–124.
6 Muradov, N. (2003) Emission-free fuel reformers for mobile and portable fuel cell applications. *J. Power Sources*, **118**, 320–324.
7 Docter, A. and Lamm, A. (1999) Gasoline fuel cell systems. *J. Power Sources*, **84**, 194–200.
8 Cheekatamarla, P.K. and Lane, A.M. (2005) Catalytic autothermal reforming of diesel fuel for hydrogen generation in fuel cells I. Activity tests and sulfur poisoning. *J. Power Sources*, **152**, 256–263.
9 Campbell, T.J., Shaaban, A.H., Holcomb, F.H., Salavani, R. and Binder, M.J. (2004) JP-8 catalytic cracking for compact fuel processors. *J. Power Sources*, **129**, 81–89.
10 Lenz, B. and Aicher, T. (2005) Catalytic autothermal reforming of jet fuel. *J. Power Sources*, **149**, 44–52.
11 Mengel, C., Konrad, M., Hartmann, L., Lucka, K. and Köhne, H. Liquid logistic fuels for fuel cells - requirements and prospects, in Proceedings of the International Symposium and workshop on fuel cells and hydrogen for aerospace and maritime applications, pp. 95–109; (September 16–17, 2004) Shaker, Hamburg, Germany.
12 Prigent, M. (1997) On board hydrogen generation for fuel cell powered electric cars. *Revue de l'institut francais du prétrole*, **52** (3), 349–359.
13 Hagh, B.F. (2004) Stoichiometric analysis of autothermal fuel processing. *J. Power Sources*, **130**, 85–94.
14 Lutz, A.E., Bradshaw, R.W., Keller, J.O. and Witmer, D.E. (2003) Thermodynamic analysis of hydrogen production by steam reforming. *Int. J. Hydrogen Energ*, **28**, 159–167.
15 Feitelberg, A.S., Rohr Jr., D.F. (2005) Operating line analysis of fuel processors for PEM fuel cell systems. *Int. J. Hydrogen Energy*, **30**, 1251–1257.
16 Schmid, H.P. and Wünning, J.A. (2007) Flox steam reforming for PEM fuel cell systems. *Fuel cells*, **4**, 4–10.

Fuel Processing for Fuel Cells. Gunther Kolb
Copyright © 2008 WILEY-VCH Verlag GmbH & Co. KGaA, Weinheim
ISBN: 978-3-527-31581-9

17 Heinzel, A., Roes, J. and Brandt, H. (2005) Increasing the electric efficiency of a fuel cell system by recirculating the anodic off-gas. *J. Power Sources*, **145**, 312–318.

18 Emonts, B., Hansen, J.B., Jörgensen, S.L., Höhlein, B. and Peters, R. (1998) Compact methanol reformer test for fuel cell powered light -duty vehicles. *J. Power Sources*, **71**, 288–293.

19 Grove, W.R. (1842) *Phil. Mag. (III)*, **21**, p. 417.

20 Schoenbein, C.F. (1838) *Schweiz. Ges.*, **82**.

21 Jones, M. (1995) Hybrid vehicles - the best of both worlds. *Chem. Ind. (London)*, **15**, 589–592.

22 Springer, T.E., Rochward, T., Zawodzinski, T.A. and Gottesfeld, S. (2001) Model for polymer electrolyte fuel cell operation on reformate feed. Effects of CO, H_2 dilution and high fuel utilization. *J. Electrochem. Soc.*, **148**, A11–A23.

23 Narusawa, K., Hayashida, M., Kamiya, Y., Roppongi, H., Kurashima, D. and Wakabayashi, K. (2003) Deterioration in fuel cell performance resulting from hydrogen fuel containing impurities: poisoning effects by CO, CH_4, HCHO, and HCOOH. *J. Soc. Automotive Eng. Jpn.*, **24**, 41–46.

24 Cheng, X., Shi, Z., Glass, N., Zhang, L., Zhang, J., Song, D., Liu, Z.-S., Wang, H. and Shen, J. (2007) A review of PEM hydrogen fuel cell contamination: impacts, mechanisms and migitation. *J. Power Sources*, **165**, 739–756.

25 Schmidt, V.M., Bröckerhoff, P., Höhlein, B., Menzer, R. and Stimming, U. (1994) Utilization of methanol for polymer electrolyte fuel cells in mobile systems. *J. Power Sources*, **49**, 299–312.

26 Hoogers, G. and Thompsett, D. (1999) Releasing the potential of clean power. *Chem. Ind. (London)*, 18 October, 796–799.

27 de Brujn, F.A., Papageorgopoulos, D.C., Sitters, E.F. and Janssen, G.J.M. (2002) The influence of carbon dioxide on PEM fuel cell anodes. *J. Power Sources*, **110**, 117–124.

28 Amphlett, J.C., Mann, R.F. and Peppley, B.A. (1996) On board hydrogen purification for steam reformation/PEM fuel cell vehicle power plants. *Int. J. Hydrogen Energ.*, **21** (8), 673–678.

29 Amphlett, J.C., Mann, R.F. and Peppley, B.A. (1996) Performance and operating characteristics of methanol steam reforming catalyst for on-board fuel cell hydrogen production, in Proceedings of the 11th World Hydrogen Energy Conference, pp. 1737–1743; Stuttgart, Germany.

30 Kawatsu, S. (1998) Advanced PEFC development for fuel cell powered vehicles. *J. Power Sources*, **71**, 150–155.

31 Shi, Y., Khan, R. and Cross III, J.C. (2005) Steam reforming of liquid hydrocarbon fuels for fuel cell applications, Fuel Cell Seminar November 15–17, Palm Springs, California.

32 Wang, J.T., Savinell, R.F., Wainright, J., Litt, M. and Yu, H. (1996) A H_2/O_2 fuel cell using acid doped polybenzimidazole as polymer electrolyte. *Electrochim. Acta*, **41** (2), 193–197.

33 Carrette, L., Friedrich, K.A. and Stimming, U. (2001) Fuel cells - Fundamentals and applications. *Fuel Cells*, **1**, 5–39.

34 Metkemeijer, R. and Achard, P. (1994) Comparison of ammonia and methanol applied indirectly in a hydrogen fuel cell. *Int. J. Hydrogen Energ.*, **19** (6), 535–542.

35 Park, S., Gorte, R.J. and Vohs, J.M. (2000) Application of heterogeneous catalysis in the direct oxidation of hydrocarbons in a solid-oxide fuel cell. *Appl. Catal. A*, **200**, 55–61.

36 Singhal, S.C. (2002) Solid oxide fuel cells for stationary, mobile and military applications. *Solid State Ionics*, **152–153**, 405–410.

37 Belyaev, V.D., Politova, T.I., Mar'ina, O.A. and Sobyanin, V.A. (1995) Internal steam reforming of methane over Ni-based electrode in solid oxide fuel cells. *Appl. Catal. A*, **133**, 47–57.

38 Peters, R., Dahl, R., Klüttgen, U., Palm, C. and Stolten, D. (2002) Internal reforming

of methane in solid oxide fuel cell systems. *J. Power Sources*, **106**, 238–244.

39 Meusinger, J., Riensche, E. and Stimming, U. (1998) Reforming of natural gas in solid oxide fuel cell systems. *J. Power Sources*, **71**, 315–320.

40 Chan, S.H. and Wang, H.M. (2000) Effect of natural gas composition on autothermal fuel reforming products. *Fuel Process. Technol.*, **64**, 221–239.

41 Costamagna, P., Cerutti, F., Di Felice, R., Agnew, G., Bozzolo, M., Collins, R., Cunningham, R. and Tarnowski, O. (2004) Production and utilization of H_2 in a reforming reactor coupled to an integrated planar solid oxide fuel cell (IP-SOFC): Simulation analysis. *Chem. Eng. Trans.*, **4**, 81–86.

42 Borup, R.L., Inbody, M.A., Tafoya, J.I., Guidry, D.R. and Jerry, W. (2004) SOFC anode recycle effect on diesel reforming, in Proceedings of the AIChE Spring Meeting, New Orleans.

43 Cavallaro, S. and Freni, S. (1998) Syngas and electricity production by an integrated autothermal reforming/molten carbonate fuel cell system. *J. Power Sources*, **76**, 190–196.

44 Freni, S., Maggio, G. and Cavallaro, S. (1996) Ethanol steam reforming in a molten carbonate fuel cell: a thermodynamic approach. *J. Power Sources*, **62**, 67–73.

45 Maggio, G., Freni, S. and Cavallaro, S. (1998) Light alcohols/methane fuelled molten carbonate fuel cells: a comparative study. *J. Power Sources*, **74**, 17–23.

46 Joensen, F. and Rostrup-Nielsen, J.R. (2002) Conversion of hydrocarbons and alcohols for fuel cells. *J. Power Sources*, **105**, 195–201.

47 Avci, A.K., Omsan, Z.I. and Trimm, D.L. (2001) On-board fuel conversion for hydrogen fuel cells: comparison of different fuels by computer simulations. *Appl. Catal. A*, **216**, 243–256.

48 Cavallaro, S. and Freni, S. (1996) Ethanol steam reforming in a molten carbonate fuel cell. A preliminary kinetic investigation. *Int. J. Hydrogen Energ.*, **21** (6), 465–469.

49 Amphlett, J.C., Creber, K.A.M., Davis, J.M., Mann, K.F., Peppley, B.A. and Stokes, D.M. (1994) Hydrogen production by steam reforming of methanol for polymer electrolyte fule cells. *Int. J. Hydrogen Energ.*, **19**, 131–137.

50 Emonts, B., Hansen, J.B., Schmidt, H., Grube, T., Höhlein, B., Peters, R. and Tschauder, A. (2000) Fuel cell drive system with hydrogen generation in test. *J. Power Sources*, **86**, 228–236.

51 Takahashi, K., Kobayashi, H. and Takezawa, N. (1985) *Chem. Lett.*, 759.

52 Jiang, C.J., Trimm, D.L. and Wainwright, M.S. (1993) Kinetic mechanism for the reaction between methanol and water over a Cu-ZnO-Al_2O_3 catalyst. *Appl. Catal. A*, **97**, 145–158.

53 Lwin, Y., Daud, W.R.W., Mohamad, A.B. and Yaakob, Z. (2000) Hydrogen production from methanol steam reforming: thermodynamic analysis. *Int. J. Hydrogen Energ.*, **25**, 47–53.

54 Fishtik, I., Alexander, A., Datta, R. and Geana, D. (2000) A thermodynamic analysis of hydrogen production by steam reforming of ethanol via response reactions. *Int. J. Hydrogen Energ.*, **25**, 31–45.

55 Rostrup-Nielsen, J.R., Catalytic Steam Reforming, in: Catalysis, Science and Technology (eds. Anderson, J.R., Boudert, M.), Elsevier, Amsterdam, Vol. 5, 3–115.

56 Springmann, S., Friedrich, G., Himmen, M., Sommer, M. and Eigenberger, G. (2002) Isothermal kinetic measurements for hydrogen production from hydrocarbon fuels using a novel kinetic reactor concept. *Appl. Catal. A*, **235**, 101–111.

57 Giroux, T., Hwang, S., Liu, Y., Ruettinger, W. and Shore, L. (2005) Monolithic structures as alternatives to particulate catalysts for the reforming of hydrocarbons for hydrogen generation. *Appl. Catal. B*, **56**, 95–110.

58 Williams, K.A. and Schmidt, L.D. (2006) Catalytic autoignition of higher alkane

partial oxidation on rhodium coated foams. *Appl. Catal. A*, **299**, 30–45.
59 Pennemann, H., Hessel, V., Kolb, G., Löwe, H. and Zapf, R. (2008) Washcoat-based catalysts for the partial oxidation of propane using a micro structured reactor. *Chem. Eng. J.* **135**, 66–73.
60 Tornianien, P.M., Chu, X. and Schmidt, L.D. (1994) Comparison of monolith-supported metals for the direct oxidation of methane to syngas. *J. Catal.*, **146**, 1–10.
61 Heitnes Hofstad, K., Sperle, T., Rokstad, O.A. and Holen, A. (1997) Partial oxidation of methane to syngas over a Pt/10% Rh gauze. *Catal. Lett.*, **45** (1–2), 97–105.
62 Ioannides, T. and Verykios, X.E. (1998) Development of a novel heat-integrated wall reactor for the partial oxidation of methane to synthesis gas. *Catal. Today*, **46**, 71–81.
63 Lyubovsky, M., Roychoudhury, S. and LaPierre, R. (2005) Catalytic partial oxidation of methane to syngas at elevated pressure. *Catal. Lett.*, **99** (3–4), 113–117.
64 Specchia, S., Negro, G., Saracco, G. and Specchia, V. (2007) Fuel processor based on syngas production via short contact time catalytic partial oxidation reactors. *Appl. Catal. B*, **70**, 523–531.
65 Panuccio, G.J., Williams, K.A. and Schmidt, L.D. (2006) Contributions of heterogeneous and homogeneous chemistry in the catalytic partial oxidation of octane isomers and mixtures on rhodium coated foams. *Chem. Eng. J.*, **61**, 4207–4219.
66 Seo, Y.S., Shirley, A. and Kolaczkowski, S.T. (2002) Evaluation of thermodynamically favourable operating conditions for production of hydrogen in three different reforming technologies. *J. Power Sources*, **108**, 213–225.
67 Moon, D.J., Sreekumar, K., Lee, S.D., Lee, B.G. and Kim, H.S. (2001) Studies on gasoline fuel processor system for fuel-cell powered vehicles application. *Appl. Catal. A*, **215**, 1–9.
68 Ahmed, S. and Krumpelt, M. (2001) Hydrogen from hydrocarbon fuels for fuel cells. *Int. J. Hydrogen Energ.*, **26** (4), 291–301.
69 Hartmann, L., Lucka, K. and Köhne, H. (2003) Mixture preparation by cool flames for diesel-reforming technologies. *J. Power Sources*, **118**, 286–297.
70 zur Megede, D. (2002) Fuel processors for fuel cell vehicles. *J. Power Sources*, **106**, 35–41.
71 Villegas, L., Guilhaume, N., Provendier, H., Daniel, C., Masset, F. and Mirodatos, C. (2005) A combined thermodynamic/experimental study for the optimisation of hydrogen production by catalytic reforming of isooctane. *Appl. Catal. A*, **281**, 75–83.
72 Aicher, T., Lenz, B., Gschnell, F., Groos, U., Federici, F., Caprile, L. and Parodi, L. (2006) Fuel processors for fuel cell APU applications. *J. Power Sources*, **154**, 503–508.
73 Kolb, G., Baier, T., Schürer, J., Tiemann, D., Ziogas, A., Ehwald, H. and Alphonse, P. (2007) A micro-structured 5 kW complete fuel processor for iso-octane as hydrogen supply system for mobile auxiliary power units. Part I - development of the autothermal catalyst and reactor. *Chem. Eng. J.*, in press.
74 Flytzani-Stephanopoulos, M. and Voecks, G.E. (1981) Catalytic autothermal reforming increases fuel cell flexibility. *Energy Progress*, **1** (1–4), 52–58.
75 Ellis, S.R., Golunski, S.E. and Petch, M.I. (2001) *HotSpot processor for reformulated gasoline* Report No. ETSU F/02/00143 REP DTI/PUB URN 01/958 Johnson Matthey Technology Centre, Sonning, UK.
76 Chan, S.H. and Wang, H.M. (2000) Thermodynamic analysis of natural-gas fuel processing for fuel cell applications. *Int. J. Hydrogen Energ.*, **25**, 441–449.
77 Pacheco, M., Sira, J. and Kopasz, J. (2003) Reaction kinetics. and reactor modeling for fuel processing of liquid hydrocarbons to produce hydrogen:

78 Hagh, B.F. (2003) Optimization of autothermal reactor for maximum hydrogen production. *Int. J. Hydrogen Energ.*, **28**, 1369–1377.

79 Ledjeff-Hey, K., Kalk, T., Mahlendorf, F., Niemzig, O., Trautmann, A. and Roes, J. (2000) Portable PEFC generator with propane as fuel. *J. Power Sources*, **86**, 166–172.

80 Poirier, M.G. and Sapundzhiev, C. (1997) Catalytic decomposition of natural gas to hydrogen for fuel cell applications. *Int. J. Hydrogen Energ.*, **22** (4), 429–433.

81 Christensen, T.S. (1996) Adiabatic prereforming of hydrocarbons - an important step in syngas production. *Appl. Catal. A*, **138**, 285–309.

82 Mator da Silva, J., Hermann, I., Mengel, C., Lucka, K. and Köhne, H. (2004) Autothermal reforming of gasoline using a cool flame vaporizer. *AIChE J.*, **50** (5), 1042–1050.

83 Naidja, A., Krishna, C.R., Butcher, T. and Mahajan, D. (2003) Cool flame partial oxidation and its role in combustion and reforming of fuels for fuel cell systems. *Prog. Energ. Combust. Sci.*, **29**, 155–191.

84 Borup, R., Inbody, M., Semelsberger, T., Perry, L. and Parkinson, J. (2002) Testing of fuels in fuel cell reformers. Progress report hydrogen, fuel cells and infrastructure, technologies, US Department of Energy.

85 Cohn, D.R., Rabinovich, A., Titus, C.H. and Bromberg, L. (1997) Near-term possibilities for extremely low emission vehicles using onboard plasmatron generation of hydrogen. *Int. J. Hydrogen Energ.*, **22** (7), 715–723.

86 Bromberg, L., Cohn, D.R. and Rabinovich (1997) Plasma reformer-fuel cell system for decentralised power applications. *Int. J. Hydrogen Energ.*, **22** (1), 83–94.

87 Bromberg, L., Cohn, D.R., Rabinovich, A., Alexeev, N., Samokhin, A., Ramprasad, R. and Tamhankar, S. (2000) System optimisation and cost analysis of plasma reforming of natural gas. *Int. J. Hydrogen Energ.*, **25**, 1157–1161.

88 Benilov, M.S. and Naidis, G.V. (2006) Modeling of hydrogen-rich gas production by plasma reforming of hydrocarbon fuels. *Int. J. Hydrogen Energ.*, **31**, 769–774.

89 Sekiguchi, H. and Mori, Y. (2003) Steam plasma reforming using microwave discharge. *Thin Solid films*, **435**, 44–48.

90 Jiang, T., Li, Y., Liu, C.-J., Xu, G.-H., Eliasson, B. and Xue, B. (2002) Plasma methane conversion using dielectric-barrier discharges with zeolite A. *Catal. Today*, **72**, 229–235.

91 Lesueur, H. and Czernichowski, A. (1994) Electrically assisted partial oxidation of methane. *Int. J. Hydrogen Energ.*, **19** (2), 139–144.

92 Paulmier, T. and Fulcheri, L. (2005) Use of non-thermal plasma for hydrocarbon reforming. *Chem. Eng. J.*, **106**, 59–71.

93 Biniwale, R.B., Mizuno, A. and Ichikawa, M. (2004) Hydrogen production by reforming of iso-octane using spray-pulsed injection and effect of non-thermal plasma. *Appl. Catal. A*, **276**, 169–177.

94 Cortright, R.D. (2006) Hydrogen generation via aqueous-phase reforming of glycerol, Small Fuel Cells Conference, April 2–4, Washington DC, US.

95 Ishikawa, M., Yamamura, S., Takamura, Y., Sode, K., Tamiya, E. and Tomiyama, M. (2006) Development of a compact high-density microbial hydrogen generator for portable bio-fuel cell system. *Int. J. Hydrogen Energ.*, **31**, 1484–1489.

96 Semelsberger, T.A., Borup, R.L. and Greene, H.L. (2006) Dimethyl ether (DME) as an alternative fuel. *J. Power Sources*, **156**, 497–511.

97 Taube, M. and Taube, P. (1981) A liquid organic carrier of hydrogen as a fuel for automobiles. *Adv. Hydrogen Energ.*, **2**, 1077–1085.

98 Wee, J.-H. (2006) A comparison of sodium borohydride as a fuel for proton

exchange membrane fuel cells and for direct borohydride fuel cells. *J. Power Sources*, **155**, 329–339.

99 Amendola, S.C., Sharp-Goldman, S.L., Saleem Janjua, M., Spencer, N.C., Kelly, M.T., Petillo, P.J. and Binder, M. (2000) A safe, portable hydrogen gas generator using aqueous borohydride solution and Ru catalyst. *Int. J. Hydrogen Energ.*, **25**, 969–975.

100 Kojima, Y., Suzuki, K., Fukumoto, K., Sasaki, M., Yamamoto, T., Kawai, Y. and Hayashi, H. (2002) Hydrogen generation using sodium borohydride solution and metal catalyst coated on metal oxide. *Int. J. Hydrogen Energ.*, **27**, 1029–1034.

101 Zhang, J., Zheng, Y., Gore, J.P. and Fisher, T.S. (2007) 1 kWe sodium borohydride hydrogen generation system. Part I: Experimental study. *J. Power Sources*, **165**, 844–853.

102 Choudhary, T.V. and Goodman, D.W. (2002) CO-free fuel processing for fuel cell applications. *Catal. Today*, **77**, 65–78.

103 Li, L. and Hurley, J.A. (2007) Ammonia-based hydrogen source for fuel cell applications. *Int. J. Hydrogen Energ.*, **32**, 6–10.

104 Sifer, N. and Gardner, K. (2004) An analysis of hydrogen production from ammonia hydride hydrogen generators for use in military fuel cell environments. *J. Power Sources*, **132**, 135–138.

105 Löffler, D.G., Taylor, K. and Mason, D. (2003) A light hydrocarbon fuel processor producing high-purity hydrogen. *J. Power Sources*, **117**, 84–91.

106 Wang, W., Turn, S.Q., Keffer, V. and Douette, A. (2007) Study of process data in autothermal reforming of LPG using multivariate data analysis. *Chem. Eng. J.*, **129**, 11–19.

107 Farrauto, R., Hwang, S., Shore, L., Ruettinger, W., Lampert, J., Giroux, T., Liu, Y. and Ilinich, O. (2003) New material needs for hydrocarbon fuel processing: Generating hydrogen for the PEM fuel cell. *Annu. Rev. Mater. Res.*, **33**, 1–27.

108 Hennings, U. and Reimert, R. (2007) Behaviour of sulphur-free odorants in natural gas fed PEM fuel cell systems. *Fuel Cells*, **7** (1), 63–69.

109 Song, C. (2003) An overview of new approaches to deep desulfurization for ultra-clean gasoline, diesel fuel and jet fuel. *Catal. Today*, **86**, 211–263.

110 Ahmed, S., Kumar, R. and Krumpelt, M. (1999) *Fuel Cells Bull.*, **2**, 4–7.

111 Hartvigsen, J.J., Elangovan, S. and Khandkar, A.C. (2003) System design, in *Handbook of fuel cells*, vol. 4 (eds W. Vielstich, A. Lamm and H.A. Gasteiger), Wiley-VCH, Weinheim, pp. 1072.

112 Wang, X., Krause, T. and Kumar, R. (2002) Sulphur removal from reformate. Progress report hydrogen, fuel cells and infrastructure, technologies, US Department of Energy.

113 Moe, J. (1962) *Chem. Eng. Process.*, **58**, 33.

114 Kahlich, M.J., Gasteiger, H.A. and Behm, R.J. (1998) Preferential oxidation of CO over Pt/γ-Al$_2$O$_3$ and Au/a-Fe$_2$O$_3$: Reactor design calculations and experimental results. *J. New Mat. Electrochem. Syst.*, **1**, 39–46.

115 Nibbelke, R.H., Campman, M.A., Hoebink, J.H.B.J. and Marin, G.B. (1997) Kinetic study of the CO oxidation over Pt/γ-Al$_2$O$_3$ and Pt/Rh/CeO$_2$/γ-Al$_2$O$_3$ in the presence of H$_2$O and CO$_2$. *J. Catal.*, **171**, 358–373.

116 Nikolaidis, G., Kolb, G., Baier, T., Maier, W.F., Zapf, R., Hessel, V. and Löwe (2008) H. Kinetic study of CO preferential oxidation over Pt/Rh/γ-Al$_2$O$_3$ catalyst in a microstructured recycle reactor, *Cat. Today*, submitted for publication.

117 Ouyang, X., Bednarova, L., Besser, R.S. and Ho, P. (2005) Preferential oxidation (PrOx) in a thin-film catalytic microreactor: advantages and limitations. *AIChE J.*, **51** (6), 1758–1771.

118 Redenius, J.M., Schmidt, L.D. and Deutschmann, O. (2001) Millisecond catalytic wall reactors: I Radiant burner. *AIChE J.*, **47** (5), 1177–1184.

119 Wauters, S. and Martin, G.B. (2002) *Ind. Eng. Chem. Res.*, **41**, 2379.

120 Sone, Y., Kichida, H., Kobayashi, M. and Watanabe, T. (2000) A study of carbon deposition on fuel cell power plants - morphology of deposited carbon and catalytic metal in carbon deposition reactions on stainless steel. *J. Power Sources*, **86**, 334–339.

121 Bartholomew, C.H. (2001) Mechanisms of catalyst deactivation. *Appl. Catal. A*, **212**, 17–60.

122 Meille, V. (2006) Review of methods to deposit catalysts on structured surfaces. *Appl. Catal. A*, **315**, 1–17.

123 Ganley, J.C., Riechmann, K.L., Seebauer, E.G. and Masel, R.I. (2004) Porous anodic alumina optimized as a catalyst support for microreactors. *J. Catal.*, **227**, 26–32.

124 Wunsch, R., Fichtner, M., Görke, O., Haas-Santo, K. and Schubert, K. (2002) Process of applying Al_2O_3 coatings in microchannels of completely manufactured microstructured reactors. *Chem. Eng. Technol.*, **25** (7), 700–703.

125 Gorges, R., Käßbohrer, J., Kreisel, G. and Meyer, S. (2002) Surface-functionalization of microstructures by anodic spark deposition, in Proceedings of the 6th International Conference on Microreaction Technology, IMRET 6, pp. 186–191; (11–14 March); *AIChE Pub. No. 164*, New Orleans, USA.

126 Presting, H., Konle, J., Strakov, V., Vyatkin, A. and König, U. (2004) Porous silicon for micro-sized fuel cell reformer units. *Mat. Sci. Eng. B*, **108**, 162–165.

127 Zapf, R., Becker-Willinger, C., Berresheim, K., Holz, H., Gnaser, H., Hessel, V., Kolb, G., Löb, P., Pannwitt, A.-K. and Ziogas, A. (2003) Detailed characterization of various porous alumina based catalyst coatings within microchannels and their testing for methanol steam reforming. *Chem. Eng. Res. Design A*, **81**, 721–729.

128 Germani, G., Stefanescu, A., Schuurman, Y. and van Veen, A.C. (2007) Preparation and characterization of porous alumina-based catalyst coatings in microchannels. *Chem. Eng. Sci.*, **62** (18–20) 5084–5091.

129 Agrafiotis, C. and Tsetsekou, A. (2000) The effect of powder characteristics on washcoat quality. Part I: alumina washcoats. *J. Eur. Ceramic Soc.*, **20**, 815–824.

130 Avila, P., Montes, M. and Miro, E.E. (2005) Monolithic reactors for environmental applications. A review on preparation techniques. *Chem. Eng. J.*, **109**, 11–36.

131 Cristiani, C., Valentini, M., Merazzi, M., Neglia, S. and Forzatti, P. (2005) Effect of aging time on chemical and rheological evolution in γ-Al_2O_3 slurries for dip-coating. *Catal. Today*, **105**, 492–498.

132 Meille, V., Pallier, S., Santa, Cruz Bustamante, G.V., Roumanie, M. and Reymond, J.-P. (2005) Deposition of γ-Al_2O_3 layers on structured supports for the design of new catalytic reactors. *Appl. Catal. A*, **286**, 232–238.

133 Stefanescu, A., van Veen, A.C., Mirodatos, C., Bziat, J.C. and Duval-Brunel, E. (2007) Wall coating optimization for microchannel reactors. *Catal. Today*, **125** (1–2), 16–23.

134 Hwang, S.-M., Kwon, O.J. and Kim, J.J. (2007) Method of catalyst coating in micro-reactors for methanol steam reforming. *Appl. Catal. A*, **316**, 83–89.

135 Pfeifer, P., Schubert, K. and Emig, G. (2005) Preparation of copper catalyst washcoats for methanol steam reforming in microchannels based on nanoparticles. *Appl. Catal. A*, **286**, 175–185.

136 Tomasic, V. and Jovic, F. (2006) State-of-the-art in the monolithic catalysts/reactors. *Appl. Catal. A*, **311**, 112–121.

137 Haas-Santo, K., Fichtner, M. and Schubert, K. (2001) Preparation of microstructure compatible porous supports by sol-gel synthesis for catalyst coatings. *Appl. Catal. A*, **220**, 79–92.

138 Bravo, J., Karim, A., Contant, T., Lopez, G.P. and Dayte, A. (2004) Wall coating of a $CuO/ZnO/Al_2O_3$ methanol steam reforming catalyst for microchannel reformers. *Chem. Eng. J.*, **101**, 113–121.

139 Thybo, S., Jensen, S., Johansen, J., Johannesen, T., Hansen, O. and Quaade, U.J. (2004) Flame spray deposition of porous catalysts on surfaces and in microsystems. *J. Catal.*, **223**, 271–277.

140 Födisch, R., Kursawe, A. and Hönicke, D. (2002) Immobilizing heterogeneous catalysts in microchannel reactors, in Proceedings of the 6th International Conference on Microreaction Technology, IMRET 6, pp. 140–146; (11–14 March); AIChE Pub. No. 164, New Orleans, USA.

141 Pfeifer, P., Görke, O. and Schubert, K. (2002) Washcoats and electrophoresis with coated and uncoated nanoparticles on microstructured metal foils and microstructured reactors, in Proceedings of the 6th International Conference on Microreaction Technology, IMRET 6, pp. 281–285; (11–14 March); AIChE Pub. No. 164, New Orleans, USA.

142 Villegas, L., Masset, F. and Guilhaume, N. (2007) Wet impregnation of alumina-washcoated monoliths: Effects of the drying procedure on Ni distribution and on autothermal reforming activity. *Appl. Catal. A*, **320**, 43–55.

143 Kim, T. and Kwon, S. (2006) Design, fabrication. and testing of a catalytic microreactor for hydrogen production. *J. Micromech. Microeng.*, **16**, 1760–1768.

144 Kim, T. and Kwon, K. (2006) Catalyst preparation for fabrication of a MEMS fuel reformer. *Chem. Eng. J.*, **123**, 93–102.

145 Takahashi, T., Tanaka, S. and Esashi, M. (2006) Development of an *in situ* chemical vapour deposition method for an alumina catalyst bed in suspended membrane micro fuel reformer. *J. Micromech. Microeng.*, **16**, 206–210.

146 Xu, J. and Froment, G.F. (1989) *AIChE J.*, **35**, 88.

147 Karim, A., Bravo, J., Gorm, D., Conant, T. and Dayte, A. (2005) Comparison of wall-coated and packed-bed reactors for steam reforming of methanol. *Catal. Today*, **110**, 86–91.

148 Karim, A., Bravo, J. and Datye, A. (2005) Nonisothermality in packed bed reactors for steam reforming of methanol. *Appl. Catal. A*, **282**, 101–109.

149 Purnama, H., Ressler, T., Jentoft, R.E., Soerijanto, H., Schlögl, R. and Schomaecker, R. (2004) CO formation/selectivity for steam reforming of methanol with a commercial $CuO/ZnO/Al_2O_3$ catalyst. *Appl. Catal. A*, **259**, 83–94.

150 Roychoudhury, S., Lyubovski, M., Walsh, D., Chu, D. and Kallio, E. (2006) Design and development of a diesel and JP-8 logistic fuel processor. *J. Power Sources*, **160**, 510–513.

151 Rampe, T., Heinzel, A. and Vogel, B. (2000) Hydrogen generation from biogenic. and fossil fuels by autothermal reforming. *J. Power Sources*, **86**, 536–541.

152 Springmann, S., Bohnet, M., Sommer, M., Himmen, M. and Eigenberger, G. (2003) Steady-state and dynamic simulation of an autothermal gasoline reformer. *Chem. Eng. Technol.*, **26**, 790–796.

153 Flytzani-Stephanopoulos, M. and Voecks, G.E. (1983) Autothermal reforming of aliphatic and aromatic hydrocarbon liquids. *Int. J. Chem. Reactor Enging.*, **8** (7), 539–548.

154 Wiese, W., Emonts, B. and Peters, R. (1999) Methanol steam reforming in a fuel cell drice system. *J. Power Sources*, **84**, 187–193.

155 Agrell, J., Birgersson, H. and Boutonnet, M. (2002) Steam reforming of methanol over a $Cu/ZnO/Al_2O_3$ catalyst: a kinetic analysis and strategies for suppression of CO formation. *J. Power Sources*, **106**, 249–257.

156 Lindstroem, B. and Petterson, L.J. (2001) Deactivation of copper-based catalysts for fuel cell applications. *Catal. Lett.*, **74** (1–2), 27–30.

157 Reitz, T.L., Ahmed, S., Krumpelt, M., Kumar, R. and Kung, H.H. (2000) Methanol reforming over CuO/ZnO under oxidizing conditions. *Stud. Surf. Sci. Catal.*, **130**, 3645–3650.

158 Men, Y., Gnaser, H., Zapf, R., Kolb, G. and Ziegler, C. (2004) Steam reforming of methanol over $Cu/CeO_2/\gamma\text{-}Al_2O_3$ catalysts

in a microchannel reactor. *Appl. Catal. A*, **277**, 83–90.

159 Men, Y., Gnaser, H., Zapf, R., Kolb, G., Hessel, V. and Ziegler, C. (2004) Parallel screening of Cu/CeO$_2$/γ-Al$_2$O$_3$ for steam reforming of methanol in a 10-channel micro-structured reactor. *Catal. Commun.*, **5**, 671–675.

160 Men, Y., Gnaser, H., Ziegler, C., Zapf, R., Hessel, V. and Kolb, G. (2005) Characterization of Cu/CeO$_2$/γ-Al$_2$O$_3$ thin film catalysts by thermal desorption spectroscopy. *Catal. Lett.*, **105** (1–2), 35–40.

161 Gnaser, H., Bock, W., Rowlett, E., Men, Y., Ziegler, C., Zapf, R. and Hessel, V. (2004) Secondary-ion mass spectrometry (SIMS) analysis of catalyst coatings used in microreactors. *Nucl. Instrum. Methods Phys. Res. B*, **219–220**, 880–885.

162 Kolb, G., Cominos, V., Drese, K., Hessel, V., Hofmann, C., Löwe, H., Wörz, O. and Zapf, R. (2002) A novel catalyst testing microreactor for heterogeneous gas phase reactions, in Proceedings of the 6th International Conference on Microreaction Technology, IMRET 6, pp. 61–72; (11–14 March); *AIChE Pub. No. 164*, New Orleans, USA.

163 Men, Y., Kolb, G., Zapf, R., Tiemann, D., Wichert, M., Hessel, V. and Löwe, H. (2006) A complete miniaturised microstructured methanol fuel processor/fuel cell system for low power applications. *Int. J. Hydrogen Energ.*, accepted for publication.

164 Ghenciu, A.F. (2002) Review of fuel processing catalysts for hydrogen production in PEM fuel cell systems. *Curr. Solid State Mater. Sci.*, **6**, 389–399.

165 Reitz, T.L., Ahmed, S., Krumpelt, M., Kumar, R. and Kung, H.H. (2000) Characterization of CuO/ZnO under oxidizing conditions for the oxidative methanol reforming reaction. *J. Mol. Catal. A*, **162**, 275–285.

166 Iwasa, N., Yoshikawa, M., Nomura, W. and Arai, M. (2005) Transformation of methanol in the presence of steam and oxygen over ZnO-supported transition metal catalysts under steam reforming conditions. *Appl. Catal. A*, **292**, 215–222.

167 Jiang, C., Trimm, D.L. and Wainwright, M.S. (1995) New technology for hydrogen production by the catalytic oxidation and steam reforming of methanol at low temperatures. *Chem. Eng. Technol.*, **18**, 1–6.

168 Lindström, B. and Petterson, L.J. (2003) Development of a methanol fuelled reformer for fuel cell applications. *J. Power Sources*, **118**, 71–78.

169 Lindström, B. and Petterson, L.J. (2002) Steam reforming of methanol over copper-based monoliths: the effects of zirconia doping. *J. Power Sources*, **106**, 264–273.

170 Lindström, B., Agrell, J. and Petterson, L.J. (2003) Combined methanol reforming for hydrogen generation over monolithic catalysts. *Chem. Eng. J.*, **93**, 91–101.

171 Jeong, H., Kim, K.I., Kim, T.H., Ko, C.H., Park, H.C. and Song, I.K. (2006) Hydrogen production by steam reforming of methanol in a micro-channel reactor coated with Cu/ZnO/ZrO$_2$/Al$_2$O$_3$ catalyst. *J. Power Sources*, **159**, 1296–1299.

172 Murcia-Mascaros, S., Navarro, R.M., Gomez-Sainero, L., Costantiono, U., Nocchetti, M. and Fierro, J.L.G. (2001) Oxidative methanol reforming reactions on CuZnAl catalysts derived from hydrotalcite-like precursors. *J. Catal.*, **198**, 338–347.

173 Velu, S., Suzuki, K., Okazaki, M., Kapoor, M.P., Osaki, T. and Ohashi, F. (2000) Oxidative steam reforming of methanol over CuZnAl(Zr)-oxide catalysts for the selective production of hydrogen for fuel cells: catalyst characterisation and performance evaluation. *J. Catal.*, **194**, 373–384.

174 Lattner, J.R. and Harold, M.P. (2007) Autothermal reforming of methanol: Experiments and modeling. *Catal. Today*, **120**, 78–89.

175 Koga, H., Fukahori, S., Kitaoka, T., Tomoda, A., Suzuki, R. and Wariishi, H. (2006) Autothermal reforming of methanol using paper-like Cu/ZnO catalyst composites prepared by a papermaking technique. *Appl. Catal. A*, **309**, 263–269.

176 Iwasa, N., Kudo, S., Takahashi, H., Masuda, S. and Takezawa, N. (1993) Highly selective supported Pd catalyst for steam reforming of methanol. *Catal. Lett.*, **19**, 211.

177 Iwasa, N., Mayanagi, T., Nomura, W., Arai, M. and Takezawa, N. (2003) Effect of Zn addition to supported Pd catalyst in the steam reforming of methanol. *Appl. Catal. A*, **248**, 153–160.

178 Ranganathan, E.S., Bej, S.K. and Thompson, L.T. (2005) Methanol reforming over Pd/ZnO and Pd/CeO$_2$ catalysts. *Appl. Catal. A*, **289**, 153–162.

179 Iwasa, N., Masuda, S., Ogawa, N. and Takezawa, N. (1995) Steam reforming of methanol over Pd/ZnO: effect of the formation of PdZn alloys upon the reaction. *Appl. Catal. A*, **125** (1), 145–157.

180 Pfeifer, P., Fichtner, M., Schubert, K., Liauw, M.A. and Emig, G. (2000) Microstructured catalysts for methanol-steam reforming, in Microreaction Technology: 3rd International Conference on Microreaction Technology, Proceedings of IMRET 3, (ed. W. Ehrfeld), Springer–Verlag, Berlin, pp. 372–382.

181 Pfeifer, P., Schubert, K., Fichtner, M., Liauw, M.A. and Emig, G. (2002) Methanol-steam reforming in microstructures: Difference between palladium and copper catalysts and testing of reactors for 200 W fuel cell power, in Proceedings of the 6th International Conference on Microreaction Technology, IMRET 6, pp. 125–130; (11–14 March); *AIChE Pub. No. 164*, New Orleans, USA.

182 Pfeifer, P., Schubert, K., Liauw, M.A. and Emig, G. (2004) PdZn catalysts prepared by washcoating microstructured reactors. *Appl. Catal. A*, **270**, 165–175.

183 Pfeifer, P., Schubert, K., Liauw, M.A. and Emig, G. (2003) Electrically heated microreactors for methanol steam reforming. *Chem. Eng. Res. Dev.* **81** (A7), 711–720.

184 Chin, Y.-H., Dagle, R., Hu, J., Dohnalkova, A.C. and Wang, Y. (2002) Steam reforming of methanol over highly active Pd/ZnO catalyst. *Catal. Today*, **77**, 79–88.

185 Hu, J., Wang, W., VanderWiel, D., Chin, C., Palo, D., Rozmiarek, R., Dagle, R., Cao, J., Holladay, J. and Baker, E. (2003) Fuel processing for portable power applications. *Chem. Eng. J.*, **93**, 55–60.

186 Chin, Y.-H., Wang, Y., Dagle, R.A. and Li, X.S. (2003) Methanol steam reforming over Pd/ZnO: catalyst preparation and pretreatment studies. *Fuel Process. Technol.*, **83**, 193–201.

187 Palo, D., Rozmiarek, R., Steven, P., Holladay, J., Guzman, C., Wang, Y., Hu, J., Dagle, R. and Baker, E. (2001) Fuel processor development for a soldier-portable fuel cell system, in Microreaction Technology 5th International Conference on Microreaction Technology Proceedings of JMRET 5, (ed. M. Matlosz, W. Ehrfeld and J.P. Baselt), Springer-Verlag, Berlin. pp. 359–367.

188 Holladay, J.D., Jones, E.O., Phelps, M. and Hu, J. (2002) Microfuel processors for use in a miniature power supply. *J. Power Sources*, **108**, 21–27.

189 Liu, S., Takahashi, K., Uematsu, K. and Ayabe, M. (2005) Hydrogen production by oxidative methanol reforming on Pd/ZnO. *Appl. Catal. A*, **283**, 125–135.

190 Lyubovsky, M. and Roychoudhury, S. (2004) Novel catalytic reactor for oxydative reforming of methanol. *Appl. Catal. A*, **54**, 203–215.

191 Liu, S., Takahashi, K., Uematsu, K. and Ayabe, M. (2004) Hydrogen production by oxidative methanol reforming on Pd/ZnO catalyst: effect of the addition of a third metal component. *Appl. Catal. A*, **277**, 265–270.

192 Chen, G., Li, S. and Yuan, Q. (2007) Pd-Zn/Cu-Zn-Al catalysts prepared for

methanol oxidation reforming in microchannel reactors. *Catal. Today*, **120**, 63–70.
193 Borup, R.L., Inbody, M.A., Semelsberger, T.A., Tafoya, J.I. and Guidry, D.R. (2005) Fuel composition effects on transporation fuel reforming. *Catal. Today*, **99**, 263–270.
194 Cubeiro, M.L. and Fierro, J.L.G. (1998) Partial oxidation of methanol over supported palladium catalysts. *Appl. Catal. A*, **168**, 307–322.
195 Wanat, E.C., Suman, B. and Schmidt, L.D. (2005) Partial oxidation of alcohols to produce hydrogen and chemicals in millisecond-contact time reactors. *J. Catal.*, **235**, 18–27.
196 Vaidya, P.D. and Rodrigues, A.E. (2006) Insight into steam reforming of ethanol to produce hydrogen for fuel cells. *Chem. Eng. J.*, **117**, 39–49.
197 Klouz, V., Fierro, V., Denton, P., Katz, H., Lisse, J.P., Bouvot-Maudit, S. and Mirodatos, C. (2002) Ethanol reforming for hydrogen production in a hybrid electric vehicle: process optimisation. *J. Power Sources*, **105**, 26–34.
198 Frusteri, F., Freni, S., Chiodo, V., Donato, S., Bonura, G. and Cavallaro, S. (2006) Steam and autothermal reforming of bio-ethanol over MgO and CeO_2 Ni supported catalysts. *Int. J. Hydrogen Energ.*, **31**, 2193–2199.
199 Fatsikostas, A.N. and Verykios, X.E. (2004) Reaction network of steam reforming of ethanol over Ni-based catalysts. *J. Catal.*, **225**, 439–452.
200 Fatsikostas, A.N., Kondarides, D.I. and Verykios, X.E. (2001) Steam reforming of biomass-derived ethanol for the production of hydrogen for fuel cell applications. *Chem. Commun.*, 851–852.
201 Batista, M.S., Santos, R.K.S., Assaf, E.M., Assaf, J.M. and Ticianelli, E.A. (2004) High efficiency steam reforming of ethanol by cobalt-based catalysts. *J. Power Sources*, **134**, 27–32.
202 Batista, M.S., Assaf, E.M., Assaf, J.M. and Ticianelli, E.A. (2006) Double bed reactor for the simultaneous steam reforming of ethanol and water gas shift reactions. *Int. J. Hydrogen Energ.*, **31**, 1204–1209.
203 Sahoo, D.R., Vajpai, S., Patel, S. and Pant, K.K. (2007) Kinetic modelling of steam reforming of ethanol for the production of hydrogen over Co/Al_2O_3 catalyst. *Chem. Eng. J.*, **125**, 139–147.
204 Llorca, J. Ramirez de la Piscina, P., Dalmon, J.A., Sales, J. and Homs, N. (2003) CO-free hydrogen from steam-reforming of bioethanol over ZnO-supported cobalt catalyst. Effect of the metallic precursor. *Appl. Catal. B*, **43**, 355–369.
205 Llorca, J., Homs, N., Sales, J. and Ramirez de la Piscina, P. (2002) Efficient production of hydrogen over supported cobalt catalysts from ethanol steam reforming. *J. Catal.* **209**, 306–317.
206 Homs, N., Llorca, J. and Ramirez de la Piscina, P. (2006) Low-temperature steam reforming of ethanol over ZnO-supported Ni and Cu catalysts. The effect of nickel and copper addition to ZnO-supported cobalt catalysts. *Catal. Today*, **116**, 361–366.
207 Liguras, D.K., Kondarides, D.I. and Verykios, X.E. (2003) Production of hydrogen for fuel cells by steam reforming of ethanol over supported noble metal catalysts. *Appl. Catal. B*, **43**, 345–354.
208 Toth, M., Doemoek, M., Rasko, J., Hancz, A. and Erdohelvi, A. (2004) Reforming of ethanol on different supported Rh catalysts. *Chem. Eng. Trans.*, **4**, 229–234.
209 Wanat, E.C., Venkataraman, K. and Schmidt, L.D. (2004) Steam reforming and water-gas shift of ethanol on Rh and Rh-Ce catalysts in a catalytic wall reactor. *Appl. Catal. A*, **276**, 155–162.
210 Deluga, G.A., Salge, J.R., Schmidt, L.D. and Verykios, X.E. (2004) Renewable hydrogen from ethanol by autothermal reforming. *Science*, **303**, 993–997.
211 Men, Y., Kolb, G., Zapf, R., Hessel, V. and Löwe, H. (2007) Ethanol steam reforming in a microchannel reactor. *Chem. Eng. Res. Dev.*, **85** (B5), 1–6.

212 Liguras, D.K., Goundani, K. and Verykios, X.E. (2004) Production of hydrogen for fuel cells by catalytic partial oxidation of ethanol over structured Ru catalysts. *Int. J. Hydrogen Energ.*, **29**, 419–427.

213 Liguras, D.K., Goundani, K. and Verykios, X.E. (2004) Production of hydrogen for fuel cells by catalytic partial oxidation of ethanol over structured Ni catalysts. *J. Power Sources*, **130**, 30–37.

214 Trimm, D.L. and Önsan, Z.I. (2001) Onboard fuel conversion for hydrogen-fuel-cell-driven vehicles. *Catal. Rev.*, **43** (1&2), 31–84.

215 Newson, E. and Truong, T.B. (2003) Low-temperature catalytic partial oxidation of hydrocarbons (C_1–C_{10}) for hydrogen production. *Int. J. Hydrogen Energ.*, **28**, 1379–1386.

216 Freni, S., Calogero, G. and Cavallaro, S. (2000) Hydrocarbon production from methane through catalytic partial oxidation reactions. *J. Power Sources*, **87**, 28–38.

217 Hoang, D.L. and Chan, S.H. (2007) Experimental investigation on the effect of natural gas composition on performance of autothermal reforming. *Int. J. Hydrogen Energ.*, **32**, 548–556.

218 Find, J., Lercher, J.A., Cremers, C., Stimming, U., Kurtz, O. and Crämer, K. Characterisation of supported methane steam reforming catalyst for micro-reactor systems, in Proceedings of the 6th International Conference on Microreaction Technology, IMRET 6, pp. 99–104; (11–14 March 2002); *AIChE Pub. No. 164*, New Orleans, USA.

219 Nurunnabi, M., Mukainakano, Y., Kado, S., Miyazawa, T., Okumura, K., Miyao, T., Naito, S., Suzuki, K., Fujimoto, K.-I., Kunimori, K. and Tomishige, K. (2006) Oxidative steam reforming of methane under atmospheric pressurized conditions over Pd/NiO-MgO solid solution catalysts. *Appl. Catal. A*, **308**, 1–12.

220 Berman, A., Karn, R.K. and Epstein, M. (2005) Kinetics of steam reforming of methane on Ru/Al_2O_3 catalyst promoted with Mn oxides. *Appl. Catal. A*, **282**, 73–83.

221 Wang, Y., Chin, Y.H., Rozmiarek, R.T., Johnson, B.R., Gao, Y., Watson, J., Tonkovich, A.Y.L. and Vander Wiel, D.P. (2004) Highly active and stable Rh/MgO-Al_2O_3 catalysts for methane steam reforming. *Catal. Today*, **98**, 575–581.

222 Johnson, B.R., Canfield, N.L., Tran, D.N., Dagle, R.A., Li, X.S.L., Holladay, J.D. and Wang, Y. (2007) Engineered SMR catalysts based on hydrothermally stable, porous, ceramic supports for microchannel reactors. *Catal. Today*, **120**, 54–62.

223 Hohn, K.L. and Schmidt, L.D. (2001) Partial oxidation of methane to syngas at high space velocities over Rh-coated spheres. *Appl. Catal. A*, **53**, 211.

224 Bizzi, M., Basini, L., Saracco, G. and Specchia, V. (2003) Modeling a transport phenomena limited reactivity in short contact time catalytic partial oxidation reactors. *Ind. Eng. Chem. Res.*, **42**, 62–71.

225 Hickman, D.A. and Schmidt, L.D. (1993) Production of syngas by direct catalytic oxidation of methane. *Science*, **259**, 343–347.

226 Huff, M., Torniainen, P.M., Hickman, D.A. and Schmidt, L.D. (1994) Partial oxidation of CH_4, C_2H_6 and C_3H_8 on monoliths at short contact times, in *Natural Gas Conversion II* (eds H.E. Curry-Hyde and R.F. Howe), Elsevier, Amsterdam, pp. 315–320.

227 Bodke, A.S., Bharadwaj, S.S. and Schmidt, L.D. (1998) The effect of ceramic supports on partial oxidation of hydrocarbons over noble metal coated monoliths. *J. Catal.*, **179**, 138–149.

228 Li, D., Shigara, M., Atake, I., ShiShido, T., Oumi, Y., Sano, T. and Takehira, K. (2007) Partial oxidation of propane over Ru promoted Ni/Mg(Al)O catalysts. Self-activation and prominent effect of reduction-oxidation treatment of the catalyst. *Appl. Catal. A*, **321**, 155–164.

229 Beretta, A. and Forzatti, P. (2004) Partial oxidation of light paraffins in short contact time reactors. *Chem. Eng. J.*, **99**, 219–226.

230 Kolb, G., Zapf, R., Hessel, V. and Löwe, H. (2004) Propane steam reforming in microchannels. *Appl. Catal. A*, **277**, 155–166.

231 Silberova, B., Venvik, H.J., Walmsley, J.C. and Holmen, A. (2005) Small-scale hydrogen production from propane. *Catal. Today*, **100**, 457–462.

232 Silberova, B., Venvik, H.J. and Holmen, A. (2005) Production of hydrogen by short contact time partial oxidation and oxidative steam reforming of propane. *Catal. Today*, **99**, 69–76.

233 Pino, L., Vita, A., Cipiti, F., Lagana, M. and Recupero, V. (2006) Performance of Pt/CeO_2 catalyst for propane oxidative steam reforming. *Appl. Catal. A*, **306**, 68–77.

234 Laosiripojana, N. and Assabumrungrat, S. (2006) Hydrogen production from steam and autothermal reforming of LPG over high surface area ceria. *J. Power Sources*, **158**, 1348–1357.

235 Sperle, T., Chen, D., Loedeng, R. and Holmen, A. (2005) Pre-refroming of natural gas in a Ni catalyst. Criteria for carbon-free operation. *Appl. Catal. A*, **282**, 195–204.

236 Chen, Y., Xu, H., Jin, X. and Xiong, G. (2006) Integration of gasoline prereforming into autothermal reforming for hydrogen production. *Catal. Today*, **116**, 334–340.

237 Schwank, J., Tadd, A. and Gould, B. (2004) Autothermal reforming of simulated gasoline in compact fuel processor. *Chem. Eng. Trans.*, **4**, 75–80.

238 Tadd, A.R., Gould, B.D. and Schwank, J.W. (2005) Packed bed versus microreactor performance in autothermal reforming of isooctane. *Catal. Today*, **110** (1–2), 68–75.

239 Zhang, J., Xu, H., Jin, X., Ge, Q. and Li, W. (2005) Characterizations and activities of the nanosized Ni/Al_2O_3 and $Ni/La\ Al_2O_3$ catalysts for NH_3 decomposition. *Appl. Cat A*, **290**, 87–96.

240 Bartholomew, C.H. (2001) Mechanisms of catalyst deactivation. *Appl. Cat. A*, **212**, p. 17–60.

241 Kalia, R.K. and Krause, A.O.I. (2006) Autothermal reforming of simulated gasoline and diesel fuels. *Int. J. Hydrogen Energy*, **31**, 1934–1941.

242 Qi, A., Wang, S., Fu, G., Ni, C. and Wu, D. (2005) La-Ce-Ni-O monolithic perovskite catalysts potential for gasoline autothermal reforming system. *Appl. Catal. A*, **281**, 233–246.

243 Murata, K., Wang, L., Saito, M., Inaba, M., Takahara, I. and Mimura, N. (2004) Hydrogen production from steam reforming of hydrocarbons over alkaline-earth metal-modified Fe- or Ni-based catalysts. *Energy Fuels*, **18**, 122–126.

244 Trimm, D.L. and Adesina, A.A. Praharso and Cant, N.W. (2004) The conversion of gasoline to hydrogen for on-board vehicle applications. *Catal. Today*, **93–95**, 17–22.

245 Ferrandon, M. and Krause, T. (2006) Role of the oxide support on the performance of rhodium catalysts for the autothermal reforming of gasoline and gasoline surrogates to hydrogen. *Appl. Catal. A*, **311**, 135–145.

246 Krumpelt, M., Krause, T.R., Carter, J.D., Kopasz, J.P. and Ahmed, S. (2002) Fuel Processing for fuel cell systems in transportation and portable power applications. *Catal. Today*, **77**, 3–16.

247 Krumpelt, M., Ahmed, S., Kumar, R. and Doshi, Partial Oxidation Catalyst, US 6,110,861.

248 O'Connor, R.P., Klein, E.J. and Schmidt, L.D. (2000) High yields of synthesis gas by millisecond partial oxidation of higher hydrocarbons. *Catal. Lett.* **70**, 99–107.

249 Qi, A., Wang, S., Fu, G. and Wu, D. (2005) Autothermal reforming of n-octane on Ru-based catalysts. *Appl. Catal. A*, **293**, 71–82.

250 Gould, B.D., Tadd, A.R. and Schwank, J.W. (2007) Nickel-catalyzed autothermal reforming of jet fuel surrogates: n-dodecane, tetralin and their mixture. *J. Power Sources*, **164**, 344–350.

251 Rosa, F., Lopez, E., Briceno, Y., Sopena, D., Navarro, R.M., Alvarez-Galvan, M.C.,

Fierro, J.L.G. and Bordons, C. (2006) Design of a diesel reformer coupled to a PEMFC. *Catal. Today*, **116**, 324–333.

252 Dinka, P. and Mukasyan, A.S. (2007) Perovskite catalysts for the auto-reforming of sulfur containing fuels. *J. Power Sources*, **167**, 472–481.

253 Liu, D.-J. and Krumpelt, M. (2005) Activity and structure of perovskites as diesel-reforming catalysts for solid oxide fuel cell. *Int. J. Appl. Ceramic Technol.*, **2** (4), 301–307.

254 Suzuki, T., Iwanami, H. and Yoshinari, T. (2000) Steam reforming of kerosene on Ru/Al_2O_3 catalyst to yield hydrogen. *Int. J. Hydrogen Energ.*, **25**, 119–126.

255 Fukunaga, T., Katsuno, H., Matsumoto, H., Takahashi, O. and Akai, Y. (2003) Development of kerosene fuel processing system for PEMFC. *Catal. Today*, **84**, 197–200.

256 Kang, I. and Bae, J. (2006) Autothermal reforming study of diesel for fuel cell application. *J. Power Sources*, **159**, 1283–1290.

257 Cheekatamarla, P.K. and Lane, M.A. (2005) Efficient bimetallic catalysts for hydrogen generation from diesel fuel. *Int. J. Hydrogen Energ.*, **30**, 1277–1285.

258 Cheekatamarla, P.K. and Lane, A.M. (2006) Efficient sulfur-tolerant bimetallic catalysts for hydrogen generation from diesel fuel. *J. Power Sources*, **153**, 157–164.

259 Palm, C., Cremer, P., Peters, R. and Stolten, D. (2002) Small-scale testing of precious metal catalyst in the autothermal reforming of various hydrocarbon fuels. *J. Power Sources*, **106**, 231–237.

260 Kopasz, J.P., Applegate, D., Miller, L., Liao, H.K. and Ahmed, S. (2005) Unravelling the maze: Understanding diesel reforming through the use of simplified fuel blends. *Int. J. Hydrogen Energ.*, **30**, 1243–1250.

261 Krummenacher, J.J., West, K.N. and Schmidt, L.D. (2003) Catalytic partial oxidation of higher hydrocarbons at millisecond contact times: decane, hexadecane and diesel fuel. *J. Catal.*, **215**, 332–343.

262 Subramanian, R., Panuccio, G.J., Krummenacher, J.J., Lee, I.C. and Schmidt, L.D. (2004) Catalytic partial oxidation of higher hydrocarbon mixtures: reactivities and selectivities of mixtures. *Chem. Eng. Sci.*, **59**, 5501–5507.

263 Krummenacher, J.J. and Schmidt, L.D. (2004) High yield of olefins and hydrogen from decane in short contact time reactors: rhodium versus platinum. *J. Catal.*, **222**, 429–438.

264 Dreyer, B.J., Lee, I.C., Krummenacher, J.J. and Schmidt, L.D. (2006) Autothermal steam reforming of higher hydrocarbons: n-decane, n-hexadecane and JP-8. *Appl. Catal. A*, **307**, 184–194.

265 Choudhary, T.V., Sivadinarayana, C., Chusuei, C.C., Klinghoffer, A. and Goodman, D.W. (2001) Hydrogen production via catalytic decomposition of methane. *J. Catal.*, **199**, 9–18.

266 Takenaka, S., Kawashima, K., Matsune, H. and Kishida, M. (2007) Production of CO-free hydrogen through the decomposition of LPG and kerosene over Ni-based catalysts. *Appl. Catal. A*, **321**, 165–174.

267 Iojoiu, E.E., Domine, M.E., Davidian, T., Guilhaume, N. and Mirodatos, C. (2007) Hydrogen production by sequential cracking of biomass-derived pyrolysis oil over noble metal catalysts supported on ceria-zirconia. *Appl. Catal. A*, **323**, 147–161.

268 Cheekatamarla, P.K. and Lane, A.M. (2006) Catalytic autothermal reforming of diesel fuel for hydrogen generation in fuel cells II. Catalyst poisoning and characterization results. *J. Power Sources*, **154**, 223–231.

269 Kolb, G., Pennemann, H. and Zapf, R. (2005) Water-gas shift reaction in micro-channels—Results from catalyst screening and optimisation. *Catal. Today*, **110** (1–2), 121–131.

270 Rostrup-Nielson, J.R., Christensen, T.S. and Dybkaer, I. (1998) Steam reforming of

liquid hydrocarbons. *Stud. Surf. Sci. Catal.*, **113**, 81–95.

271 Trimm, D.L. (1999) Catalysts for the control of coking during steam reforming. *Catal. Today*, **49**, 3–10.

272 Bridger, G.W. (1980) Steam reforming of hydrocarbons in catalysis (eds. C. Kamball, D.A. Dowden) *Chem. Soc.*, London pp. 39–49.

273 Shamsi, A., Baltrus, J.P. and Spivey, J.J. (2005) Characerization of coke deposited on Pt/alumina catalyst during reforming of liquid hydrocarbons. *Appl. Catal. A*, **293**, 145–152.

274 Oudar, J.H., Pracher, C.-M., Vaasilakis, D., and Bertheir, Y. (1998) A reaction between co-adsorbed sulfer and butadiene on the Pt (111) surface. *Catal. Lett.* 1(11), 339–343.

275 Ming, Q., Healey, T., Allen, L. and Irving, P. (2002) Steam reforming of hydrocarbon fuels. *Catal. Today*, **77**, 51–64.

276 Lombard, C., Le Doze, S., Marcenak, E., Marquaire, P.-M., Le Noc, D., Bertrand, G. and Lapicque, F. (2006) *in situ* regeneration of the Ni-based catalytic reformer of a 5 kW PEMFC system. *Int. J. Hydrogen Energy*, **31**, 437–440.

277 Semelsberger, T.A., Ott, K.C., Borup, R.L. and Greene, H.L. (2006) Generating hydrogen-rich fuel-cell feeds from dimethyl ether (DME) using Cu/Zn supported on various solid-acid substrates. *Appl. Catal. A*, **309**, 210–223.

278 Zhang, Q., Li, X., Fujimoto, K. and Asami, K. (2005) Hydrogen production by partial oxidation and reforming of DME. *Appl. Catal. A*, **288**, 169–174.

279 Kariya, N., Fukuoka, A. and Ichikawa, M. (2002) Efficient evolution of hydrogen from liquid cyclohexanes over Pt-containing catalysts supported on active carbons under 'wet-dry multiphase conditions'. *Appl. Catal. A*, **233**, 91–102.

280 Roumanie, M., Meille, V., Pijolat, C., Tournier, G., de Bellefont, C., Pouteau, P. and Delattre, C. (2005) Design and fabrication of a structured catalytic reactor at micrometer scale: example of methylcyclohexane dehydrogenation. *Catal. Today*, **110**, 164–170.

281 Dong, H., Yang, H.X., Ai, X.P. and Cha, C. (2003) Hydrogen production from catalytic hydrolysis of sodium borohydride solution using nickel boride catalyst. *Int. J. Hydrogen Energ.*, **28**, 1095–1100.

282 Pinto, A.M.F.R., Falcao, D.S., Silva, R.A. and Rangel, C.M. (2006) Hydrogen generation and storage from hydrolysis of sodium borohydride in batch reactors. *Int. J. Hydrogen Energ.*, **31**, 1341–1347.

283 Choudhary, T.V., Sivadinarayana, C. and Goodman, D.W. (2001) Catalytic ammonia decomposition: CO_x-free hydrogen production for fuel cell applications. *Catal. Lett.*, **72**, 3–4. 197–201.

284 Ganley, J.C., Seebauer, E.G. and Masel, R.I. (2004) Development of a microreactor for the production of hydrogen from ammonia. *J. Power Sources*, **137**, 53–61.

285 Chowdhury, R., Pedernera, E. and Reimert, R. (2002) Trickle-bed reactor model for desulfurization and dearomatization of diesel. *AIChE J.*, **48** (1), 126–135.

286 Zaki, T., Riad, M., Saad, L. and Mikhail, S. (2005) Selected oxide materials for sulphur removal. *Chem. Eng. J.*, **113**, 41–46.

287 Richard, F., Boita, T. and Perot, G. (2007) Reaction mechanism of 4,6 dimethyldibenzothiophene desulfurization over sulfided NiMoP/Al_2O_3-zeolite catalysts. *Appl. Catal. A*, **320**, 69–79.

288 Khare, G.P., Delzer, G.A., Kubicek, D.H. and Greenwood, G.J. (1995) Hot gas desulfurization with Phillips Z-Sorb sorbent im moving bed and fluidized bed reactors. *Environ Prog.*, **14** (3), 146–150.

289 Jayne, D., Zhang, Y., Haij, S. and Erkey, C. (2005) Dynamics of removal of organosulfur compounds from diesel by adsorption on carbon aerogels for fuel cell applications. *Int. J. Hydrogen Energ.*, **30**, 1287–1293.

290 Haji, S. and Erkey, C. (2003) Removal of dibenzothiophenes from model diesel by adsorption on carbon aerogels for fuel cell applications. *Ind. Eng. Chem. Res.*, **42** (26), 6933–6937.

291 Zhang, J.C., Song, L.F., Hu, J.Y., Ong, S.L., Ng, L.Y., Lee, L.Y., Wang, Y.H., Zhao, J.G. and Ma, R.Y. (2005) Investigation on gasoline deep desulfurisation for fuel cell applications. *Energy Conversion Management*, **46**, 1–9.

292 Xue, M., Chitrakar, R., Sakane, K., Hirotsu, T., Ooi, K., Yoshimura, Y., Toba, M. and Feng, Q. (2006) Preparation of cerium-loaded Y-zeolites for removal of organic compounds from hydro-desulfurized gasoline and diesel oil. *J. Colloid Interface Sci.*, **298**, 535–542.

293 Rosso, I., Galletti, C., Bizzi, M., Saracco, G. and Specchia, V. (2003) Zinc oxide sorbents for the removal of hydrogen sulphide from syngas. *Ind. Eng. Chem. Res.*, **42**, 1688–1697.

294 Li, L. and King, D.L. (2006) H_2S removal with ZnO during fuel processing for PEM fuel cell applications. *Catal. Today*, **116**, 537–541.

295 Wails, D. Johnson Matthey Fuel Cells, personal communication.

296 Lampert, J. (2004) Selective catalytic oxidation: a new catalytic approach to the desulfurization of natural gas and liquid petroleum gas for fuel cell reformer applications. *J. Power Sources*, **131**, 27–34.

297 Gardner, T.H., Berry, D.A., Lyons, K.D., Beer, S.K. and Freed, A.D. (2002) Fuel processor integrated H_2S catalytic partial oxidation technology for sulfur removal in fuel cell power plant. *Fuel*, **81**, 2157–2166.

298 Meeyoo, V. and Trimm, D.L. (1997) Adsorption-reaction processes for the removal of hydrogen sulphide from gas streams. *J. Chem. Technol. Biotechnol.*, **68**, 411–416.

299 Flytzani-Stephanopoulos, M., Sakbodin, M. and Wang, Z. (2006) Regenerative adsorption and removal of H_2S from hot fuel gas streams by rare earth oxides. *Science*, **312**, 1508–1510.

300 Utaka, T., Sekizawa, K. and Eguchi, K. (2000) CO removal by oxygen-assisted water-gas shift reaction over supported Cu catalysts. *Appl. Catal. A*, **194–195**, 21–26.

301 Patt, J., Moon, D.J., Phillips, C. and Thompson, L. (2000) Molybdenum carbide catalysts for water-gas shift. *Catal. Lett.*, **65**, 193–195.

302 Ruettinger, W., Ilinich, O. and Farrauto, R.J. (2003) A new generation of water-gas shift catalysts for fuel cell applications. *J. Power Sources*, **118**, 61–65.

303 Hilaire, S., Wang, X., Luo, T., Gorte, R.J. and Wagner, J. (2001) A comparative study of water-gas shift reaction over ceria supported metallic catalysts. *Appl. Catal. A*, **215**, 271–278.

304 Koryabkina, N.A., Phatak, A.A., Ruettinger, W.F., Farrauto, R.J. and Ribeiro, F.H. (2003) Determination of kinetic parameters for the water-gas shift reaction on copper-based catalysts under realistic conditions for fuel cell applications. *J. Catal.*, **217**, 233–239.

305 Amenomiya, Y. and Pleizier, G. (1982) Alkali-promoted alumina catalysts: II. Water-gas shift reaction. *J. Catal.*, **76** (2), 345–353.

306 Phatak, A.A., Koryabkina, N., Rai, S., Ratts, J.L., Ruettinger, W., Farrauto, R.J., Blau, G.E., Delgass, W.N. and Ribeiro, F.H. (2007) Kinetics of the water-gas shift reaction on Pt catalysts supported on alumina and ceria. *Catal. Today*, **123**, 224–234.

307 Gorte, R.J. and Zhao, S. (2005) Studies of the water-gas shift reaction with ceria-supported precious metals. *Catal. Today*, **104**, 18–24.

308 Germani, G., Alphonse, P., Courty, M., Schuurman, Y. and Mirodatos, C. (2005) Platinum/ceria/alumina catalysts on microstructures for carbon monoxide conversion. *Catal. Today*, **110**, 1–2. 114–120.

309 Pasel, J., Cremer, P., Wegner, B., Peters, R. and Stolten, D. (2004) Combination of autothermal reforming with water-gas shift reaction- small-scale testing of

different water-gas shift catalysts. *J. Power Sources*, **126**, 112–118.

310 Liu, X., Ruettinger, W., Xu, X. and Farrauto, R. (2005) Deactivation of Pt/CeO_2 water-gas shift catalysts due to shutdown/startup modes of fuel cell applications. *Appl. Catal. B.*, **56**, 69–75.

311 Zalc, J.M., Sokolovskii, V. and Löffler, D.G. (2002) Are noble metal-based water-gas shift catalysts practical for automotive fuel processing? *J. Catal.*, **206**, 169–171.

312 Kolb, G., Schürer, J., Tiemann, D., Wichert, M., Zapf, R., Hessel, V. and Löwe, H. (2007) Fuel processing in integrated microstructured heat-exchanger reactors. *J. Power Sources*, **171**, 198–204.

313 Choung, S.Y., Ferrando, M. and Krause, T. (2005) Pt-Re bimetallic supported on CeO_2-ZrO_2 mixed oxides as water-gas shift catalysts. *Catal. Today*, **99**, 257–262.

314 Wang, X., Gorte, R.J. and Wagner, J.P. (2002) Deactivation mechanisms for Pd/ceria during the water-gas shift reactios. *J. Catal.*, **212**, 225–230.

315 Wheeler, C., Jhalani, A., Klein, E.J., Tummala, S. and Schmidt, L.D. (2004) The water-gas shift reaction at short contact times. *J. Catal.*, **223**, 191–199.

316 Utaka, T., Okanishi, T., Takeguchi, T., Kikuchi, R. and Eguchi, K. (2003) Water gas shift reaction of reformed fuel over supported Ru catalyst. *Appl. Catal. A*, **245**, 343–351.

317 Tonkovich, A.L., Zilka, J.L., LaMont, M.J., Wang, Y. and Wegeng, R. (1999) Microchannel chemical reactor for fuel processing applications.- I. Water gas shift reactor. *Chem. Eng. Sci.*, **54**, 2947–2951.

318 Basiniska, A., Kepinski, L. and Domka, F. (1999) The effect of support on WGSR activity of ruthenium catalyst. *Appl. Catal. A*, **183**, 143–153.

319 Jacobs, G., Ricote, S., Patterson, P.M., Graham, U.M., Dozier, A., Khalid, S., Rhodus, E. and Davis, B.H. (2005) Low temperature water-gas shift: Examining the efficiency of Au as a promoter for ceria-based catalysts prepared by CVD of a Au precursor. *Appl. Catal. A*, **292**, 229–243.

320 Fu, Q., Weber, A. and Flytzani-Stephanopoulos, M. (2001) Nanostructured Au-CeO_2 catalysts for low-temperature water-gas shift. *Catal. Lett.*, **77** (1–3), 87–95.

321 Luengnaruemitchai, A., Osuwan, S. and Gulari, E. (2003) Comparative studies of low-temperature water-gas shift reaction over Pt/CeO_2, Au/CeO_2 and Au/Fe_2O_3 catalysts. *Catal. Commun.*, **4**, 215–221.

322 Zhou, S., Yuan, Z. and Wang, S. (2006) Selective CO oxidation with real methanol reformate over monolithic Pt group catalysts: PEMFC applications. *Int. J. Hydrogen Energ.*, **31**, 924–933.

323 Manaslip, A. and Gulari, E. (2002) Selective oxidation over Pt/alumina catalysts for fuel cell applications. *Appl. Catal. B.*, **37**, 17–25.

324 Cominos, V., Hessel, V., Hofmann, C., Kolb, G., Zapf, R., Ziogas, A., Delsman, E. and Schouten, J. (2005) Selective oxidation of carbon monoxide in a hydrogen-rich fuel cell feed using a catalyst coated microstrucred reactor. *Catal. Today*, **110** (1–2), 140–153.

325 Castaldi, M.J., LaPierre, R., Lyubovski, M., Pfefferle, W. and Roychoudhury, S. (2005) Effect of water on performance. and sizing of fuel-processing reactors. *Catal. Today*, **99**, 339–346.

326 Son, I.H., Lane, A.M. and Johnson, D.T. (2003) The study of the deactivation of water-pretreated Pt/γ-Al_2O_3 for low-temperature selective CO oxidation in hydrogen. *J. Power Sources*, **124**, 415–419.

327 Oh, S.H. and Sinkevitch, R.M. (1993) Carbon monoxide removal from hydrogen-rich fuel cell feed streams by selective catalytic oxidation, *J. Catal.*, **142**, 254–262.

328 Dudfield, C.D., Chen, R. and Adock, P.L. (2001) A carbon monoxide PROX reactor for PEM fuel cell automotive application. *Int. J. Hydrogen Energ.*, **26**, 763–775.

329 Trimm, D.L. (2005) Minimisation of carbon monoxide in a hydrogen stream

for fuel cell application. *Appl. Catal. A,* **296**, 1–11.

330 Ayastuy, J.L., Gil-Rodriguez, A., Gonzales-Marcos, M.P. and Gutierrez-Ortiz, M.A. (2006) Effect of process variables on Pt/CeO_2 catalyst behaviour for the PROX reaction. *Int. J. Hydrogen Energ.,* **31**, 2231–2242.

331 Schubert, M.M., Kahlich, M.J., Feldmeyer, G., Hüttner, M., Hackenberg, S., Gasteiger, H.A. and Behm, R.J. (2001) Bimetallic PtSn catalyst for selective CO oxidation in H_2-rich gases at low temperatures. *Phys. Chem. Chem. Phys.,* **3**, 1123–1131.

332 Delsman, R.E., de Croon, M.H.J.M., Pierik, A., Kramer, G.J., Cobden, P.D., Hofmann, C., Cominos, V. and Schouten, J.C. (2004) Design. and operation of a preferential oxidation microdevice for a portable fuel processor. *Chem. Eng. Sci.,* **59**, 4795–4802.

333 Maier, W.F. and Saalfrank, J. (2004) Discovery, combinatorial chemistry and a new selective CO-oxidation catalyst. *Chem. Eng. Sci.,* **59**, 4673–4678.

334 Chen, G., Yuan, Q., Li, H. and Li, S. (2004) CO selective oxidation in a microchannel reactor for PEM fuel cell. *Chem. Eng. J.,* **101** (1–3), 101–106.

335 Haruta, M. and Masakazu, D. (2001) Advances in the catalysis of Au nanoparticles. *Appl. Catal. A,* **222**, 427–437.

336 Luengnaruemitchai, A., Osuwan, S. and Gulari, E. (2004) Selective catalytic oxidation of CO in the presence of H_2 over gold catalysts. *Int. J. Hydrogen Energ.,* **29**, 429–435.

337 Panzera, G., Modafferi, V., Candamano, S., Donato, A., Frusteri, F. and Antonucci, P.L. (2004) CO selective oxidation on ceria-supported Au catalysts for fuel cell application. *J. Power Sources,* **135**, 177–183.

338 Wakita, H., Kani, Y., Ukai, K., Tomizawa, T., Takeguchi, T. and Ueda, W. (2005) Effect of SO_2 and H_2S in CO preferential oxidation in H_2-rich gas over Ru/Al_2O_3 and Pt/Al_2O_3 catalyst. *Appl. Catal. A,* **283**, 53–61.

339 Wang, J.B., Lin, S.-C. and Huang, T.-J. (2002) Selective CO-oxidation in rich hydrogen over CuO/samaria-doped ceria. *Appl. Catal. A,* **232**, 107–120.

340 Park, J.W., Jeong, J.H., Yoon, W.L. and Rhee, Y.W. (2004) Selective oxidation of carbon monoxide in hydrogen-rich stream over $Cu/Ce/\gamma$-Al_2O_3 catalysts promoted with cobalt in a fuel processor for proton exchange membrane fuel cells. *J. Power Sources,* **132**, 18–28.

341 Snytnikov, P.V., Stadnichenko, A.I., Semin, G.L., Belyaev, V.D., Boronin, A.I. and Sobyanin, V.A. (2007) Copper-cerium oxide catalysts for the selective oxidation of carbon monoxide in hydrogen-containing mixtures: I. Catalytic activity. *Kinet. Catal.,* **48** (3), 439–447.

342 Takenaka, S., Shimizu, T. and Otsuka, K. (2004) Complete removal of carbon monoxide in hydrogen-rich gas stream through methanation over supported metal catalysts. *Int. J. Hydrogen Energ.,* **29**, 1065–1073.

343 Görke, O., Pfeifer, P. and Schubert, K. (2005) Highly selective methanation by the use of a microchannel reactor. *Catal. Today,* **110** (1–2), 132–139.

344 Men, Y., Kolb, G., Zapf, R., Hessel, V. and Löwe, H. (2007) Selective methanation of carbon oxides in a microchannel reactor – primary screening and impact of oxygen additive. *Catal. Today,* **125**, 81–87.

345 Kuijpers, E.G.M., Tjepkma, R.B. and Geus, J.W. (1984) Elimination of the water-gas shift reaction by direct processing of $CO/H_2/H_2O$ over Ni/SiO_2 catalysts. *J. Mol. Catal.,* **25**, 241–251.

346 Rinnemo, M., Fasshi, M. and Kasemo, B. (1993) The critical condition for catalytic ignition. H_2/O_2 on Pt. *Chem. Phys. Lett.,* **211** (1), 60–64.

347 Rinnemo, M., Deutschmann, O., Behrendt, F. and Kasemo, B. (1997) Experimental and numerical investigation of the catalytic ignition of mixtures of hydrogen and oxygen on platinum. *Combust. Flame,* **111**, 312–326.

348 Fernandes, N.E., Park, Y.K. and Vlachos, D.G. (1999) The autothermal behaviour of platinum catalyzed hydrogen oxidation: Experiments and modelling. *Combust. Flame*, 118, 164–178.

349 Lindström, B. and Petterson, L.J. (2003) Catalytic oxidation of liquid methanol as a heat source for an automotive reformer. *Chem. Eng. Technol.*, 26 (4), 473–478.

350 Choudhary, T.V., Banerjee, S. and Choudhary, V.R. (2002) Catalysts for combustion of methane and lower alkanes. *Appl. Catal. A*, 234, 1–23.

351 Gelin, P. and Primet, M. (2002) Complete oxidation of methane at low temperature over noble metal based catalysts: a review. *Appl. Catal. B*, 39, 1–37.

352 Ma, L., Trimm, D.L. and Jiang, C. (1996) The design and testing of an autothermal reactor for the conversion of light hydrocarbons to hydrogen I. The kinetics of the catalytic oxidation of light hydrocarbons. *Appl. Catal. A*, 138, 275–283.

353 Opoku-Gyamfi, K. and Adesina, A.A. (1999) Kinetic Studies of CH_4 oxidation over Pt-NiO/δ-Al_2O_3 in a fluidized bed reactor. *Appl. Catal. A*, 180, 113–122.

354 Kiwi-Minsker, L., Yuranov, I., Slavinskaia, E., Zaikovskii, V. and Renken, A. (2000) Pt and Pd on glass fibres as effective combustion catalysts. *Catal. Today*, 59, 61–68.

355 Saint-Just, J.J. and der Kinderen, J. (1996) Catalytic combustion: from reaction mechanisms to commercial applications. *Catal. Today*, 29, 387–395.

356 Corro, G., Montiel, R. and Vazquez, L.C. (2002) Promoting and inhibiting effect of SO_2 on propane oxidation over Pt/Al_2O_3. *Catal. Commun.*, 3, 533–539.

357 Corro, G., Fierro, J.L.G. and Vazquez, C. (2005) Strong improvement on CH_4 oxidation over Pt/γ-Al_2O_3 catalysts. *Catal. Commun.*, 6, 287–295.

358 Yazawa, Y., Yoshida, H., Komai, S.-I. and Hattori, T. (2002) The additive effect on propane combustion over platinum catalysts. Control of oxidation-resistance of platinum by the electronegativity of additives. *Appl. Catal. A*, 233, 113–124.

359 Yoshida, H., Yazawa, Y. and Hattori, T. (2003) Effects of support and additive on oxidation state and activity of Pt catalyst in propane combustion. *Catal. Today*, 87, 19–28.

360 Yazawa, Y., Takagi, N., Yoshida, H., Komai, S.-I., Satsuma, A., Tanaka, T., Yoshida, S. and Hattori, T. (2002) The support effect on propane combustion over platinum catalyst: control of the oxidation resistance of platinum by the acid strength of support materials. *Appl. Catal. A*, 233, 103–112.

361 Guan, G., Zapf, R., Kolb, G., Men, Y., Hessel, V., Löwe, H., Ye, J. and Zentel, R. (2007) Low temperatue catalytic combustion of propane over MoO_3-promoted Pt/Al_2O_3 catalyst with inverse opal microstructure in microchannel reactor. *Chem. Commun.*, 3, 260–262.

362 Mouaddib, N., Feumi-Jantou, C., Garbowski, E. and Primet, M. (1992) Catalytic oxidation of methane over palladium supported on alumina. *Appl. Catal. A*, 87, 129–144.

363 Guerrero, S., Araya, P. and Wolf, E.E. (2006) Methane oxidation on Pd supported on high area zirconia catalysts. *Appl. Catal. A*, 298, 243–253.

364 Deng, Y. and Nevell, T.G. (1993) Sulfur poisoning, recovery and related phenomena over supported palladium, rhodium and iridium catalysts for methane oxidation. *Appl. Catal. A*, 101, 51–62.

365 Yue, B., Zhou, R., Wang, Y. and Zheng, X. (2005) Effect of rare earths (La, Pr, Nd, Sm and Y) on the methane combustion over Pd/Ce-Zr-Al_2O_3 catalysts. *Appl. Catal. A*, 295, 31–39.

366 Yuranov, I., Dunand, N., Kiwi-Minsker, L. and Renken, A. (2002) Metal grids with high-porous surface as structured catalysts: preparation, characterization. and activity in propane total oxidation. *Appl. Catal. B*, 36, 183–191.

367 Yuranov, I., Kiwi-Minsker, L. and Renken, A. (2003) Structured combustion catalysts based on sintered metal fibres. *Appl. Catal. B*, **43**, 217–227.

368 Zavyalova, U., Scholz, P. and Ondruschka, B. (2007) Infuence of cobalt precursor and fuels on the performance of combustion synthesized $Co_3O_4/\gamma\text{-}Al_2O_3$ catalysts for total oxidation of methane. *Appl. Catal. A*, **323**, 226–233.

369 Zhao, S. and Gorte, R.J. (2004) A comparison of ceria and Sm-doped ceria for hydrocarbon oxidation reactions. *Appl. Catal. A*, **277**, 129–136.

370 Brown, L.F. (2001) A comparative study of fuels for on-board hydrogen production for fuel-cell-powered automobiles. *Int. J. Hydrogen Energ.*, **26**, 381–397.

371 Specchia, S., Cutillo, A., Saracco, G. and Specchia, V. (2006) Concept study on ATR and STR fuel processors for liquid hydrocarbons. *Ind. Eng. Chem. Res.*, **45**, 5298–5307.

372 Ma, L., Jiang, C., Adesina, A.A., Trimm, D.L. and Wainwright, M.S. (1996) Simulation studies of autothermal reactor system for H_2 production from methanol steam reforming. *Chem. Eng. J.*, **62**, 103–111.

373 Avci, A.K., Trimm, D.L. and Önsan, Z.I. (2001) Heterogeneous reactor modelling for simulation of catalytic oxidation and steam reforming of methane. *Chem. Eng. Sci.*, **56**, 641–649.

374 Johnston, A.M. and Haynes, B.S., Heatric steam reforming technology, in Proceedings of the AIChE Spring National Meeting - 2nd Topical Conference on Natural Gas Utilization, pp. 112–118; (10–12 March 2001); (ed. AIChE), New Orleans, USA.

375 Frauhammer, J., Eigenberger, G., von Hippel, L. and Arntz, D. (1999) A new reactor concept for endothermic high-temperature reactions. *Chem. Eng. Sci.*, **54** (15/16), 3661–3670.

376 Kolios, G., Frauhammer, J. and Eigenberger, G. (2002) Efficient reactor concepts for coupling of endothermic and exothermic reactions. *Chem. Eng. Sci.*, **57**, 1505–1510.

377 Gritsch, A., Kolios, G. and Eigenberger, G. (2004) Reaktorkonzepte zur autothermen Führung endothermer Hochtemperaturreaktion. *Chem. Ing. Tech.*, **76** (6), 722–725.

378 Zalc, J.M. and Löffler, D.G. (2002) Fuel processing for PEM fuel cells: transport and kinetic issues of system design. *J. Power Sources*, **111**, 58–64.

379 Zanfir, M. and Gavriilidis, A. (2003) Catalytic combustion assisted methane steam reforming in a catalytic plate reactor. *Chem. Eng. Sci.*, **58**, 3947–3960.

380 Robbins, F.A., Zhu, H. and Jackson, G.S. (2003) Transient modeling of combined catalytic combustion/CH_4 steam reforming. *Catal. Today*, **83**, 141–156.

381 Kirillov, V.A., Fadeev, S.I., Kuzin, N.A. and Shigarov, A.B. (2007) Modeling of a heat-coupled catalytic reactor with co-current oxidation and conversion flow. *Chem. Eng. J.*, doi: 10.1016/j.cej.2007.03.050.

382 Pan, L. and Wang, S. (2005) Modeling of a compact plate-fin reformer for methanol steam reforming in fuel cell systems. *Chem. Eng. J.*, **108**, 51–58.

383 Petrachi, G.A., Negro, G., Specchia, S., Saracco, G., Maffetone, P.L. and Specchia, V. (2005) Combining catalytic combustion and steam reforming in a novel multifunctional reactor for on-board hydrogen production from middle distillates. *Ind. Eng. Chem. Res.*, **44**, 9422–9430.

384 Cao, C., Wang, Y. and Rozmiarek, R.T. (2005) Heterogeneous reactor model for steam reforming of methane in a microchannel reactor with microstructured catalyst. *Catal. Today*, **110**, 92–97.

385 Tonkovich, A.L., Yang, B., Perry, S.T., Fitzgerald, S.P. and Wang, Y. (2007) From seconds to milliseconds to microseconds through tailored microchannel reactor design of a steam methane reformer. *Catal. Today*, **120**, 21–29.

386 Delsman, E.R., Laarhoven, B.J.P.F., de Croon, M.H.J.M., Kramer, G.J. and Schouten, J.C. (2005) Comparison between conventional fixed-bed and microreactor technology for a portable hydrogen production case. *Chem. Eng. Res. Design*, **83** (A9), 1063–1075.

387 Stutz, M.J., Hotz, N. and Poulikakos, D. (2006) Optimisation of methane reforming in a microreactor - effects of catalyst loading and geometry. *Chem. Eng. Sci.*, **61**, 4027–4040.

388 Dybkjaer, I. (1995) Tubular reforming and autothermal reforming of natural gas - an overview of available processes. *Fuel Process. Technol.*, **42**, 85–107.

389 Springmann, S., Bohnet, M., Docter, A., Lamm, A. and Eigenberger, G. (2004) Cold start simulations of a gasoline based fuel processor for mobile fuel cell applications. *J. Power Sources*, **128**, 13–24.

390 Choudhary, V.R., Banerjee, S. and Rajput, A.M. (2001) Continuous production of H_2 at low temperature from methane decomposition over Ni-containing catalyst followed by gasification by steam of the carbon on the catalyst in two parallel reactors operated in cyclic manner. *J. Catal.*, **198**, 136–141.

391 Levent, M. (2001) Water-gas shift reaction over porous catalyst: temperature and reactant concentration distribution. *Int. J. Hydrogen Energ.*, **26**, 551–558.

392 Giunta, P., Amadeo, N. and Laborde, M. (2006) Simulation of a low temperature water-gas shift reactor using the heterogeneous model/application to a PEM fuel cell. *J. Power Sources*, **156**, 489–496.

393 TeGrotenhuis, W.E., King, D.L., Brooks, K.P., Holladay, B.J. and Wegeng, R.S. Optimizing microchannel reactors by trading-off equilibrium and reaction kinetics through temperature management, in Proceedings of the 6th International Conference on Microreaction Technology, IMRET 6, pp. 18–28; (11–14 March 2002); *AIChE Pub. No. 164*, New Orleans, USA.

394 Kim, G.-Y., Mayor, J.R. and Ni, J. (2005) Parametric study of microreactor design for water gas shift reactor using an integrated reaction and heat exchange model. *Chem. Eng. J.*, **110**, 1–10.

395 Baier, T. and Kolb, G. (2007) Temperature control of the water-gas shift reaction in microstructured reactors. *Chem. Eng. Sci.*, **62**, 4602–4611.

396 Pasel, J., Cremer, P., Stalling, J., Wegner, B., Peters, R. and Stolten, D. Comparison of two different reactor concepts for the water-gas shift reaction, in Proceedings of the Fuel Cell Seminar, pp. 607–610 (November 18–21 2002); Palm Springs, CA, USA.

397 Ouyang, X. and Besser, R.S. (2005) Effect of reactor heat transfer limitations on CO preferential oxidation. *J. Power Sources*, **141**, 39–46.

398 Schuessler, M., Portscher, M. and Limbeck, U. (2003) Monolithic integrated fuel processor for the conversion of liquid methanol. *Catal. Today*, **79–80**, 511–520.

399 Xu, G., Chen, X. and Zhang, Z.-G. (2006) Temperature-staged methanation: An alternative method to purify hydrogen-rich fuel gas for PEFC. *Chem. Eng. J.*, **121**, 97–107.

400 Gepert, V., Kilgus, M., Schiestel, T., Brunner, H., Eigenberger, G. and Merten, C. (2006) Ceramic supported capillary Pd membranes for hydrogen separation: Potential and present limitations. *Fuel Cells*, **6**, 472–481.

401 Ledjeff-Hey, K., Formanski, V., Kalk, T. and Roes, J. (1998) Compact hydrogen production systems for solid polymer fuel cells. *J. Power Sources*, **71**, 199–207.

402 Han, J., Kim, I.-S. and Choi, K.-S. (2002) High purity hydrogen generator for on-site hydrogen production. *Int. J. Hydrogen Energ.*, **27**, 1043–1047.

403 Wieland, S., Melin, T. and Lamm, A. (2002) Membrane reactors for hydrogen production. *Chem. Eng. Sci.*, **57**, 1571–1576.

404 Qi, A., Peppley, B. and Kunal, K. (2007) Integrated fuel processors for fuel cell applications: A review. *Fuel Process Technol.*, **88**, 3–22.

405 Lattner, J.R. and Harold, M.P. (2004) Comparison of conventional and membrane reactor fuel processors for hydrocarbon-based PEM fuel cell systems. *Int. J. Hydrogen Energ.*, **29**, 393–417.

406 Oklany, J.S., Hou, K. and Hughes, R. (1998) A simulative comparison of dense and microporous membrane reactors for the steam reforming of methane. *Appl. Catal. A*, **170** (1), 13–22.

407 Amphlett, J.C., Baumert, R.M., Mann, R.F. and Peppley, B.A., System analysis of an integrated methanol steam reformer/PEM fuel cell power generating system, in Proceedings of the 27th Intersociety Energy Conversion Engineering Conference, pp. 3343–3348; (03–07 August 1992); San Diego, CA,USA.

408 Li, A., Liang, W. and Hughes, R. (2000) Fabrication of dense palladium composite membranes for hydrogen separation. *Catal. Today*, **56**, 45–51.

409 Liang, W. and Hughes, R. (2005) The effect of diffusion direction on the permeation rate of hydrogen in palladium composite membranes. *Chem. Eng. J.*, **112**, 81–86.

410 Joergensen, S.L., Nielsen, P.E.H. and Lehrmann, P. (1995) Steam reforming of methane in a membrane reactor. *Catal. Today*, **25**, 303–307.

411 Barbieri, G. and Di Maio, F.P. (1997) Simulation of the methane steam reforming process in a catalytic Pd-membrane reactor. *Ind. Eng. Chem. Res.*, **36**, 2121–2127.

412 Gallucci, F. and Basile, A. (2006) Co-current and counter-current modes for methanol steam reforming membrane reactor. *Int. J. Hydrogen Energ.*, **31**, 2243–2249.

413 Marigliano, G., Barbieri, G. and Drioli, E. (2001) Effect of energy transport on a palladium-based membrane reactor for methane steam reforming process. *Catal. Today*, **67**, 85–99.

414 Barbieri, G., Marigliano, G., Perri, G. and Drioli, E. (2001) Conversion-temperature diagram for a palladium membrane reactor. Analysis of an endothermic reaction: methane steam reforming. *Ind. Eng. Chem. Res.*, **40**, 2017–2026.

415 Marigliano, G., Barbieri, G. and Drioli, E. (2003) Equilibrium conversion for a Pd-based membrane reactor. Dependence on the temperature and pressure. *Chem. Eng. Process*, **42**, 231–236.

416 Lin, Y.-M., Lee, G.-L. and Rei, M.-H. (1998) An integrated purification and production of hydrogen with a palladium membrane-catalytic reactor. *Catal. Today*, **44**, 343–349.

417 Gallucci, F., Patzuro, L. and Basile, A. (2004) A simulation study of the steam reforming of methane in a dense tubular membrane reactor. *Int. J. Hydrogen Energ.*, **29**, 611–617.

418 Vogiatzis, E., Koukou, M.K., Papayannakos, N. and Markatos, N.C. (2004) Heat dispersion effects on the functional characteristics of industrial-scale adiabatic membrane reactors. *Chem. Eng. Technol.*, **27** (8), 857–865.

419 Edlund, D.J. (1996) Steam reformer with internal hydrogen purification US, 5,861,137.

420 Iyuke, S.E., Daud, W.R.W., Mohamad, A.B., Kadhum, A.A.H., Fisal, Z. and Shariff, A.M. (2000) Application of Sn-activated carbon in pressure swing adsorption for purification of H_2. *Chem. Eng. Sci.*, **55**, 4745–4755.

421 Peramanu, S., Cox, B.G. and Pruden, B.B. (1999) Economics of hydrogen recovery processes for the purification of hydroprocessor purge and off-gases. *Int. J. Hydrogen Energ.*, **24**, 405–424.

422 Sircar, S., Waldron, W.E., Rao, M.B. and Anand, M. (1999) Hydrogen production by hybrid SMR-PSA-SSF membrane system. *Sep. Purif. Technol.*, **17**, 11–20.

423 Dalle Nogare, D., Baggio, P., Tomasi, C., Mutri, L. and Canu, P. (2007) A

thermodynamic analysis of natural gas reforming processes for fuel cell application. *Chem. Eng. Sci.*, **62**, 5418–5424.

424 Veser, G., Friedrich, G., Freygang, M. and Zengerle, R. (2000) A modular microreactor design for high-temperature catalytic oxidation reactions, in Microreaction Technology: 3rd International Conference on Microreaction Technology, Proceedings of IMRET 3, (ed. W. Ehrfeld), Springer-Verlag, Berlin, pp. 674–686.

425 Veser, G. (2001) Experimental and theoretical investigation of H_2 oxidation in a high-temperature catalytic microreactor. *Chem. Eng. Sci.*, **56**, 1265–1273.

426 Chattopadhyay, S. and Veser, G. (2006) Heterogeneous-homogeneous interactions in catalytic microchannel reactors. *AIChE J.*, **52** (6), 2217–2220.

427 Wierzba, I. and Wang, Q. (2006) The flammability limits of H_2-CO-CH_4 mixtures in air at elevated temperatures. *Int. J. Hydrogen Energ.*, **31**, 485–489.

428 Veser, G. and Schmidt, L.D. (1996) Ignition and extinction in the catalytic oxidation of hydrocarbons over platinum. *AIChE J.*, **42** (9), 1077–1087.

429 Raimondeau, S., Norton, D., Vlachos, D.G. and Masel, R.I. (2002) Modeling of high-temperature microburners. *Proc. Combustion Inst.*, **29**, 901–907.

430 Wang, S. and Wang, S. (2007) Distribution optimization for plate-fin catalytic combustion heat exchanger. *Chem. Eng. J.*, **131**, 171–179.

431 Peters, R., Riensche, E. and Cremer, P. (2000) Pre-reforming of natural gas in solid oxide fuel cell systems. *J. Power Sources*, **86**, 432–441.

432 Riensche, E., Meusinger, J., Stimming, U. and Unverzagt, G. (1998) Optimization of a 200 kW SOFC cogeneration power plant Part II: Variation of the flowsheet. *J. Power Sources*, **71**, 306–314.

433 Mathiak, J., Heinzel, A., Roes, J., Kalk, T.H.K. and Brandt, H. (2004) Coupling a 2.5 kW steam reformer with a 1 kW_{el} PEM fuel cell. *J. Power Sources*, **131**, 112–119.

434 Doss, E.D., Kumar, R., Ahluwalia, R.K. and Krumpelt, M. (2001) Fuel processors for automotive fuel cells: a parametric analysis. *J. Power Sources*, **102**, 1–15.

435 Ahmed, S., Kopasz, J., Kumar, R. and Krumpelt, M. (2002) Water balance in a polymer electrolyte fuel cell system. *J. Power Sources*, **112**, 519–530.

436 Shah, K. and Besser, R.S. (2007) Key issues in the microchemical systems-based methanol fuel processor: Energy density, thermal integration and heat loss mechanisms. *J. Power Sources*, **166**, 177–193.

437 Semelsberger, T.A. and Borup, R.L. (2005) Fuel effects on start-up energy and efficiency for automotive PEM fuel cell systems. *Int. J. Hydrogen Energ.*, **30**, 425–435.

438 Wang, S. and Wang, S. (2006) Exergy analysis and optimization of methanol generating hydrogen system for PEMFC. *Int. J. Hydrogen Energ.*, **31**, 1747–1755.

439 Hotz, N., Lee, M.-T., Grigoropoulos, C.P., Senn, S.M. and Poulikakos, D. (2006) Exergetic analysis of fuel cell micropowerplants fed by methanol. *Int. J. Heat Mass Transfer*, **49**, 2397–2411.

440 Ioannides, T. (2001) Thermydynamic analysis of ethanol processors for fuel cell applications. *J. Power Sources*, **92**, 17–25.

441 Francesconi, J.A., Mussati, M.C., Mato, R.O. and Aguirre, P.A. (2007) Analysis of the energy efficiency of an integrated ethanol processor for PEM fuel cell systems. *J. Power Sources*, **167**, 151–161.

442 Specchia, S., Tillemans, F.W.A., van den Oosterkamp, P.F. and Saracco, G. (2005) Conceptual design and selection of a biodiesel fuel processor for a vehicle fuel cell auxiliary power unit. *J. Power Sources*, **145**, 683–690.

443 Cutillo, A., Specchia, S.A.M., Saracco, G. and Specchia, V. (2006) Diesel fuel processor for PEM fuel cells: Two possible alternatives (ATR versus STR). *J. Power Sources*, **154** (2), 379–385.

444 Hoelleck, G.L. (1970) Diffusion and solubility of hydrogen in palladium and

palladium silver alloys. *J. Phys. Chem.*, **74** (5), 503–511.

445 Liu, Z., Mao, Z., Xu, J., Hess-Mohr, N. and Schmidt, V.M. (2006) Modelling of a PEM fuel cell system with propane ATR reforming. *Fuel Cells*, **6** (5), 376–386.

446 Ahluwalia, R.K., Doss, E.D. and Kumar, R. (2003) Performance of high-temperature polymer electrolyte fuel cell systems. *J. Power Sources*, **117**, 45–60.

447 Aoki, H., Mitsui, H., Shimazu, T., Kimura, K. and Masui, T. (2006) A numerical study on thermal efficiency of fuel cell systems combined with reformer, in Proceedings of the 16th World Hydrogen Energy Conference, Lyon, France, pp. 1–10.

448 Ahmed, S., Ahluwalia, R., Lee, S.H.D. and Lottes, S. (2006) A gasoline fuel processor designed to study quick-start performance. *J. Power Sources*, **154**, 214–222.

449 Seo, Y.-S., Seo, D.-J., Seo, Y.-T. and Yoon, W.-L. (2006) Investigation of the characteristics of a compact steam reformer integrated with water-gas shift reactor. *J. Power Sources*, **161**, 1208–1216.

450 Seo, Y.T., Seo, D.J., Jeong, J.H. and Yoon, W.L. (2006) Design of an integrated fuel processor for residential PEMFCs application. *J. Power Sources*, **160**, 505–509.

451 Qi, A., Wang, S., Fu, G. and Wu, D. (2006) Integrated fuel processor built on autothermal reforming of gasoline: A proof of principle study. *J. Power Sources*, **162**, 1254–1264.

452 Chen, Y.-H., Yu, C.-C., Liu, Y.-C. and Lee, C.-H. (2006) Start-up strategies of an experimental fuel processor. *J. Power Sources*, **160**, 1275–1286.

453 Shin, W.C. and Besser, R.S. (2007) Towards autonomous control of microreactor system for steam reforming of methanol. *J. Power Sources*, **164**, 328–335.

454 Görgün, H., Arcak, M., Varigonda, S. and Bortoff, S.A. (2005) Observer designs for fuel processing reactors in fuel cell power systems. *Int. J. Hydrogen Energ.*, **30**, 447–457.

455 Capobianco, L., Del Prete, Z., Schiavetti, P. and Violante, V. (2006) ,Theoretical analysis of a pure hydrogen production separation plant for fuel cells dynamical applications. *Int. J. Hydrogen Energ.*, **31**, 1079–1090.

456 Honda (2007) http://www.hondapowerequipment.com/ModelDetail.asp?ModelName= eu1000i.

457 Groppi, G. and Tronconi, E. (2000) Design of novel monolith catalyst supports for gas/solid reactions with heat exchange. *Chem. Eng. Sci.*, **55**, 2161–2171.

458 Castaldi, M., Lyubovski, M., LaPierre, R., Pfefferle, W.C. and Roychoudhury, S. (2003) Performance of microlith based catalytic reactors for an isooctane reforming system. *Soc. Automot. Eng. SAE*, 2003-01-1366, 1707–1714.

459 Bae, J.-M., Ahmed, S., Kumar, R. and Doss, E. (2005) Microchennel development for autothermal reforming of hydrocarbon fuels. *J. Power Sources*, **139**, 91–95.

460 Mitchell, M.M. and Kenis, P.J.A. (2006) Ceramic microreactors for on-site hydrogen production from high temperature steam reforming of propane. *Lab Chip*, **6**, 1328–1337.

461 Twigg, M.V. and Richardson, J.T. (1995) in *Scientific bases for the preparation of heterogeneous catalysts* (ed. G. Poncelet), Elsevier, Amsterdam. pp. 345–359.

462 Reay, A.D. (1993) Catalytic combustion: current status and implications for energy efficiency in the process industries. *Heat Recov. Syst. CHP*, **13**, 383–390.

463 Frauhammer, J., Friedrich, G., Kolios, G., Klingel, T., Eigenberger, G., von Hippel, L. and Arntz, D. (1999) Flow distribution concepts for new type monolithic co- or countercurrent reactors. *Chem. Eng. Technol.*, **22**, 1012–1016.

464 Minjolle, L (1981) Method of manufacturing a ceramic unit for indirect heat exchange and a heat exchanger unit obtained thereby, US 4271110.

465 von Hippel, L., Arntz, D., Frauhammer, J., Eigenberger, G. and Friedrich, G. (1998) Reaktorkopf für einen

monolitischen Gleich-oder Gegenstrom reaktor, DE 19653989.
466 Ramshaw, C. (1985) Process intensification - a game for N-players. *Chem. Eng.*, **415**, 30–33.
467 Kolb, G. and Hessel, V. (2004) Microstructured reactors for gas phase reactions. *Chem. Eng. J.*, **98** (1–2), 1–38.
468 Kiwi-Minsker, L. and Renken, A. (2005) Microstructured reactors for catalytic reactions. *Catal. Today*, **110** (1–2), 2–14.
469 Hessel, V., Hardt, S. and Löwe, H. (2004) *Chemical micro process engineering - Fundamentals, modelling and reactions*, Wiley-VCH, Weinheim.
470 Hessel, V., Löwe, H., Müller, A. and Kolb, G. (2005) *Chemical micro process engineering - Processing and plants*, Wiley-VCH, Weinheim.
471 Hessel, V., Schouten, J.C. and Renken, A.I. (2007) *Handbook of micro-reactor engineering and micro-reactor chemistry*, in preparation, Wiley-VCH, Weinheim.
472 Jenkins, J.W. (1990) Catalytic generation of hydrogen from hydrocarbons. US 4,897,253, Priority.
473 Peters, R., Düsterwald, H.G. and Höhlein, B. (2000) Investigation of a methanol reformer concept considering the particular impact of dynamics and long-term stability for use in a fuel-cell powered passenger car. *J. Power Sources*, **86**, 507–514.
474 Düsterwald, H.G., Höhlein, B., Kraut, H., Meusinger, J., Peters, R. and Stimming, U. (1997) Methanol steam-reforming in a catalytic fixed bed reactor. *Chem. Eng. Technol.*, **20**, 617–623.
475 Mitchell, W.E., Thijssen, H.J., Bentley, J.M. and Marek, N.M. (1995) Development of a catalytic partial oxidation ethanol reformer for fuel cell applications. *Soc. Automotive Eng.*, SAE **952761**, 209–217.
476 Patel, K.S. and Sunol, A.K. (2006) Dynamic behaviour of methane heat exchange reformer for residential fuel cell power generation systems. *J. Power Sources*, **161**, 503–512.

477 Recupero, V., Pino, L., Di Leonardo, R., Lagana, M. and Maggio, G. (1998) Hydrogen generator via catalytic partial oxidation of methane for fuel cells. *J. Power Sources*, **71**, 208–214.
478 Lee, D., Lee, H.C., Lee, K.H. and Kim, S. (2007) A compact and highly efficient natural gas fuel processor for 1-kW residential polymer electrolyte membrane fuel cells. *J. Power Sources*, **165**, 337–341.
479 Horng, R.-F. (2005) Transient behaviour of a small methanol reformer for fuel cell during hydrogen production after cold starts. *Energy Conv. Management*, **46**, 1193–1207.
480 Horng, R.-F., Chen, C.-R., Wu, T.-S. and Chan, C.-H. (2006) Cold start response of a small methanol reformer by partial oxidation reforming of hydrogen for fuel cell. *Appl. Therm. Eng.*, **26**, 1115–1124.
481 Horny, C., Kiwi-Minsker, L. and Renken, A. (2004) Micro-structured string-reactor for autothermal production of hydrogen. *Chem. Eng. J.*, **101** (1–3), 3–9.
482 Horny, C., Renken, A. and Kiwi-Minsker, L. (2007) Compact string reactor for autothermal hydrogen production. *Catal. Today*, **120**, 45–53.
483 Catillon, S., Louis, C., Topin, F., Vicente, J. and Rouget, R. (2004) Improvement of methanol steam reformer for H_2 production by addition of copper foam in both the evaporator and the catalytic reactors. *Chem. Eng. Trans.*, **4**, 111–116.
484 Jung, H., Yoon, W.L., Lee, H., Park, J.S., Shin, J.S., La, H. and Lee, J.D. (2003) Fast start-up reactor for partial oxidation of methane with electrically heated metallic monolith catalyst. *J. Power Sources*, **124**, 76–80.
485 Fichtner, M., Mayer, J., Wolf, D. and Schubert, K. (2001) Microstructured rhodium catalysts for the partial oxidation of methane to syngas under pressure. *Ind. Eng. Chem. Res.*, **40** (16), 3475–3483.
486 Aartun, I., Gjervan, T., Venvik, H., Görke, O., Pfeifer, P., Fathi, M., Holmen, A. and Schubert, K. (2004) Catalytic conversion of propane to hydrogen in micro-

structured reactors. *Chem. Eng. J.*, **101** (1–3), 93–99.

487 Aartun, I., Venvik, H.J., Holmen, A., Pfeifer, P., Görke, O. and Schubert, K. (2005) Temperature profiles and residence time effects during catalytic partial oxidation and oxidative steam reforming of propane in metallic microchannel reactors. *Catal. Today*, **110** (1–2), 98–107.

488 Aartun, I., Silberova, B., Venvik, H., Pfeifer, P., Görke, O., Schubert, K. and Holmen, A. (2005) Hydrogen production from propane in Rh-impregnated metallic microchannel reactors and alumina foams. *Catal. Today*, **105**, 469–478.

489 Docter, A., Konrad, G. and Lamm, A. (2000) *Reformer für Benzin und benzinähnliche Kraftstoffe*, VDI Berichte No. 1565, pp. 399–411.

490 Roychoudhury, S., Lyubovski, M. and Shabbir, A. (2005) Microlith catalytic reactors for reforming iso-octane-based fuels into hydrogen. *J. Power Sources*, **152**, 75–86.

491 Liu, D.-J., Kaun, T.D., Liao, H.-K. and Ahmed, S. (2004) Characterization of kilowatt-scale autothermal reformer for production of hydrogen from heavy hydrocarbons. *Int. J. Hydrogen Energ.*, **29**, 1035–1046.

492 Lenz, B., Full, J. and Siewek, C. Reforming of jet fuel for fuel cell APU's in commercial aircraft, in Proceedings of the Hydrogen Expo Conference, published on CD; (August 31st–September 1st 2005); Hamburg, Germany.

493 de Wild, P.J. and Verhaak, M.J.F.M. (2000) Catalytic production of hydrogen from methanol. *Catal. Today*, **60**, 3–10.

494 Pan, L. and Wang, S. (2005) Methanol steam reforming in a compact plate-fin reformer for fuel cell systems. *Int. J. Hydrogen Energ.*, **30**, 973–979.

495 Riensche, E., Stimming, U. and Unverzagt, G. (1998) Optimization of a 200 kW SOFC cogeneration power plant. Part I: Variation of process parameters. *J. Power Sources*, **73**, 251–256.

496 Polman, E.A., Der Kinderen, J.M. and Thuis, F.M.A. (1999) Novel compact steam reformer for fuel cells with heat generation by catalytic combustion augmented by induction heating. *Catal. Today*, **47**, 347–351.

497 Reuse, P., Tribolet, P., Kiwi-Minsker, L. and Renken, A. (2001) Catalyst coating in microreactors for methanol steam reforming: Kinetics, in Proceedings of the 5th International Conference on Microreaction Technology IMRET 5, (eds. M. Matlosz, W. Ehrfeld and J.P. Baselt), Springer-Verlag, Berlin, pp. 322–331.

498 Cominos, V., Hardt, S., Hessel, V., Kolb, G., Löwe, H., Wichert, M. and Zapf, R. A methanol steam micro-reformer for low power fuel cell applications, in Proceedings of the 6th International Conference on Microreaction Technology, IMRET 6, pp. 113–124; (11–14 March 2002); AIChE Pub. No. 164, New Orleans, USA.

499 Reuse, P., Renken, A., Haas-Santo, K., Görke, O. and Schubert, K. (2004) Hydrogen production for fuel cell application in an autothermal microchannel reactor. *Chem. Eng. J.*, **101** (1–3), 133–141.

500 Park, G.-G., Seo, D.J., Park, S.-H., Yoon, Y.-G., Kim, C.-S. and Yoon, W.-L. (2004) Development of microchannel methanol steam reformer. *Chem. Eng. J.*, **101** (1–3), 87–92.

501 Seo, D.J., Yoon, W.-L., Yoon, Y.-G., Park, S.-H., Park, G.-G. and Kim, C.-S. (2004) Development of a micro fuel processor for PEMFCs. *Electrochim. Acta*, **50**, 719–723.

502 Yu, X., Tu, S.-T., Wang, Z. and Qi, Y. (2006) Development of a microchannel reactor concerning steam reforming of methanol. *Chem. Eng. J.*, **116**, 123–132.

503 Kundu, A., Ahn, J.E., Park, S.-S., Shul, Y.G. and Han, H.S. (2008) Process intensification by micro-channel reactor for steam reforming of methanol. *Chem. Eng. J.*, **135**, 113–119.

504 Delsman, E.R., de Croon, M.H.J.M., Kramer, G.J., Cobden, P.D., Hofmann, C., Cominos, V. and Schouten, J.C. (2004) Experiments and modelling of an integrated preferential oxidation-heat exchanger microdevice. *Chem. Eng. J.*, **101** (1–3), 123–131.

505 Cremers, C., Dehlsen, J., Stimming, U., Reuse, P., Renken, A. and Haas-Santo, K. Görke, O. and Schubert, K. Micro-structured-reactor-system for the steam reforming of methanol, in Proceedings of the 7th International Conference on Micro-reaction Technology, IMRET 7, pp. 56; (7–10 September 2003); (ed. DECHEMA, Frankfurt), Lausanne, Switzerland.

506 Kolb, G., Hessel, V., Cominos, V., Pennemann, H., Schürer, J., Zapf, R. and Löwe, H. (2006) Microstructured fuel processors for fuel-cell applications. *J. Mater. Eng. Perform.*, **15** (4), 389–393.

507 Park, G.-G., Yim, S.-D., Yoon, Y.-G., Lee, W.-Y., Kim, C.-S., Seo, D.-J. and Eguchi, K. (2005) Hydrogen production with integrated microchannel fuel processor for portable fuel cell systems. *J. Power Sources*, **145**, 702–706.

508 Park, G.-G., Yima, S.-D., Yoon, Y.-G., Kim, C.-S., Seo, D.-J. and Eguchi, K. (2005) Hydrogen production with integrated microchannel fuel processor using methanol for portable fuel cell systems. *Catal. Today*, **110** (1–2), 108–113.

509 Tonkovich, A.L.Y., Zilka, J.L., Powell, M.R. and Call, C.J. (1998) The catalytic partial oxidation of methane in a microchannel chemical reactor, in Process Miniaturization: 2nd International Conference on Microreaction Technology, IMRET 2; Topical Conf. Preprints, AIChE, (eds W. Ehrfeld, I.H. Rinard and R.S. Wegeng), New Orleans, USA, pp. 45–53.

510 Cremers, C., Stummer, M., Stimming, U., Find, J., Lercher, J.A., Kurtz, O., Crämer, K., Haas-Santo, K., Görke, O. and Schubert, K. Micro-structured reactors for coupled steam-reforming and catalytic combustion of methane, in Proceedings of the 7th International Conference on Microreaction Technology, IMRET 7, p. 100; (7–10 September 2003); (ed. DECHEMA, Frankfurt), Lausanne, Switzerland.

511 Cremers, C., Pelz, A., Stimming, U., Haas-Santo, K., Görke, O., Pfeifer, P. and Schubert, K. (2007) Micro-structured methane steam reformer with integrated catalytic combustor. *Fuel Cells*, **2**, 91–98.

512 Pfeifer, P., Bohn, L., Görke, O., Haas-Santo, K. and Schubert, K. (2004) Mikrostrukturkomponenten für die Wasserstofferzeugung aus unterschiedlichen Kohlenwasserstoffen. *Chem. Ing. Tech.*, **76** (5), 618–620.

513 Ryi, S.-K., Park, J.-S., Choi, S.-H., Cho, S.-H. and Kim, S.-H. (2005) Novel micro fuel processor for PEMFCs with heat generation by catalytic combustion. *Chem. Eng. J.*, **113**, 47–53.

514 Ryi, S.K., Park, J.S., Cho, S.H. and Kim, S.H. (2006) Fast start-up of microchannel fuel processor integrated with an igniter for hydrogen combustion. *J. Power Sources*, **161**, 1234–1240.

515 Hermann, I., Lindner, M., Winkelmann, H. and Düsterwald, H.G. Microreaction technology in fuel processing for fuel cell vehicles, in Proceedings of the VDE World Microtechnologies Congress, MICRO. tec 2000, pp. 447–453; (25–27 September 2000); VDE Verlag, Berlin, EXPO Hannover.

516 Whyatt, G.A., TeGrotenhuis, W.E., Wegeng, R.S. and Pederson, L.R. Demonstration of energy efficient steam reforming in microchannels for automotive fuel processing, in Proceedings of the 5th International Conference on Microreaction Technology IMRET 5, (27–30 May 2001); Strasbourg, France, pp. 302–312.

517 Whyatt, G.A., Fischer, C.M. and Davis, J.M. Progress on the development of a microchannel steam reformer for automotive applications, in Proceedings of the 6th International Conference on Microreaction Technology, IMRET 6, pp. 85–98; (11–14 March 2002); *AIChE Pub. No. 164*, New Orleans, USA.

518 Whyatt, A., Fischer, C.M. and Davis, J.M. Development of a rapid-start on-board automotive steam reformer, in Proceedings of the AIChE Spring Meeting, paper 120C. published on CD (25–29 April 2004); New Orleans, USA.

519 Uemyia, S., Sato, N., Ando, H., Matsuda, T. and Kikuchi, E. (1991) Steam reforming of methane in a hydrogen-permeable membrane reactor. *Appl. Catal. A*, **67**, 223–230.

520 Kikuchi, E., Nemoto, Y., Kajiwara, M., Uemyia, S. and Kojima, T. (2000) Steam reforming of methane in membrane reactors: comparison of electroless-plating and CVD membranes and catalyst packing modes. *Catal. Today*, **56**, 75–81.

521 Itoh, N., Kaneko, Y. and Igarashi, A. (2002) Efficient hydrogen production via methanol steam reforming by preventing back-permeation of hydrogen in a palladium membrane reactor. *Ind. Eng. Chem. Res.*, **41**, 4702–4706.

522 Tong, J. and Matsumura, Y. (2005) Effect of catalytic activity on methane steam reforming in hydrogen-permeable membrane reactor. *Appl. Catal. A*, **286**, 226–231.

523 Barbieri, G., Violante, V., Di Maio, F.P., Criscuoli, A. and Drioli, E. (1997) Methane steam reforming analysis in a palladium-based catalytic membrane reactor. *Ind. Eng. Chem. Res.*, **36**, 3369–3374.

524 Kikuchi, E. (2000) Membrane reactor application to hydrogen production. *Catal. Today*, **56**, 97–101.

525 Basile, A., Tereschenko, G.F., Orekhova, N.V., Ermilova, M.M., Gallucci, F. and Iulianelli, A. (2006) An experimental investigation on methanol steam reforming with oxygen addition in a flat Pd-Ag membrane reactor. *Int. J. Hydrogen Energ.*, **31**, 1615–1622.

526 Wilhite, B.A., Weiss, S.E., Ying, J.Y., Schmidt, M.A. and Jensen, K.F. (2006) High-purity hydrogen generation in a microfabricated 23 wt.% Ag-Pd membrane device integrated with 8:1 $LaNi_{0.95}Co_{0.05}O_3/Al_2O_3$ catalyst. *Adv. Mater.*, **18**, 1701–1704.

527 Ferreira-Aparicio, P., Rodriguez-Ramos, I. and Guerrero-Ruiz, A. (2002) Pure hydrogen production from methylcyclohexane using a new high performance membrane reactor. *Chem. Commun.*, 2082–2083.

528 Kurungot, S., Yamaguchi, T. and Nakao, S.-I. (2003) Rh/γ-Al_2O_3 catalytic layer integrated with sol-gel synthesized microporous silica membrane for compact membrane reactor applications. *Catal. Lett.*, **86** (4), 273–278.

529 Ikeguchi, M., Mimura, T., Sekine, Y., Kikuchi, E. and Matsukata, M. (2005) Reaction and oxygen permeation studies in $Sm_{0.4}Ba_{0.6}Fe_{0.8}Co_{0.2}O_{3-\delta}$ membrane reactor for partial oxidation of methane to syngas. *Appl. Catal. A*, **290**, 212–220.

530 Kusakabe, K., Fumio, S., Eda, T., Oda, M. and Sotowa, K.-I. (2005) Hydrogen production in zirconia membrane reactors for use in PEM fuel cells. *Int. J. Hydrogen Energ.*, **30**, 989–994.

531 Pattekar, A.V. and Kothare, M.V. (2004) A microreactor for hydrogen production in micro fuel cells. *J. Microelectromech. Syst.*, **13** (1), 7–18.

532 Pattekar, V.P. and Kothare, M.V. (2005) A radial microfluidic fuel processor *J. Power Sources*, **147**, 116–127.

533 Kundu, A., Jang, J.H., Lee, H.R., Kim, S.-H., Gil, J.H., Jung, C.R. and Oh, Y.S. (2006) MEMS-based micro-fuel processor for application in a cell phone. *J. Power Sources*, **162**, 572–578.

534 Yoshida, K., Tanaka, S., Hiraki, H. and Esashi, M. (2006) A micro fuel reformer intergrated with a combustor and a microchannel evaporator. *J. Micromechan. Microeng.*, **16**, 191–197.

535 Park, D.-E., Kim, T., Kwon, S., Kim, C.-K. and Yoon, E. (2007) Micromachined methanol steam reforming system as hydrogen supplier for portable proton exchange membrane fuel cells. *Sens. Actuators A*, **135**, 58–66.

536 Shah, K., Ouyang, X. and Besser, R.S. (2003) Microreaction for microfuel processing: challenges and prospects. *Chem. Eng. Technol.*, **28** (3), 303–313.

537 Chang, K.-S., Tanaka, S. and Esashi, M. (2005) A micro-fuel processor with trench-refilled thick silicon dioxide for thermal isolation fabricated by water-immersion contact photolithography. *J. Micromech. Microeng.*, **15**, S171–S178.

538 Kim, T., Hwang, J.S. and Kwon, S. (2007) A MEMS methanol reformer heated by decomposition of hydrogen peroxide. *Lab Chip*, **7**, 835–841.

539 Bromberg, L., Cohn, D.R., Rabinovich, A. and Alexeev, N. (1999) Plasma catalytic reforming of methane. *Int. J. Hydrogen Energ.*, **24**, 1131–1137.

540 Sobacchi, M.G., Saveliev, A.V., Fridman, A.A., Kennedy, L.A., Ahmed, S. and Krause, S. (2002) Experimental assessment of a combined plasma/catalytic system for hydrogen production via partial oxidation of hydrocarbon fuels. *Int. J. Hydrogen Energ.*, **27**, 635–642.

541 Czernichowski, A., Czernichowski, J., Czernichowski, P. and Wesolowska, K., Reformate gas production from various liquid fuels, Proceedings of the International Symposium and workshop on fuel cells and hydrogen for aerospace and maritime applications, pp. 121–130; (September 16–17 2004); Shaker, Hamburg, Germany.

542 Dokupil, M., Spitta, C., Mathiak, J., Beckhaus, P. and Heinzel, A. (2006) Compact propane fuel processor for auxiliary power unit application. *J. Power Sources*, **157**, 906–913.

543 Kolb, G., Baier, T., Schürer, J., Tiemann, D., Ziogas, A., Specchia, S., Galetti, E., Germani, G. and Schuurman, Y. (2007) A micro-structured 5 kW complete fuel processor for iso-octane as hydrogen supply system for mobile auxiliary power units. Part II - Development of water-gas shift and preferential oxidation reactors and assembly of the fuel processor. *Chem. Eng. J.*, in press.

544 Barbieri, G., Bernardo, P., Mattia, R., Drioli, E., Bredesen, R. and Klette, H. (2004) Pd-based membrane reactor for water-gas shift reaction. *Chem. Eng. Trans.*, **4**, 55–60.

545 Lee, S.H., Han, J. and Lee, K.-Y. (2002) Development of 10 kWe preferential oxidation system for fuel cell vehicles. *J. Power Sources*, **109**, 394–402.

546 Pan, L. and Wang, S. (2006) A compact integrated fuel-processing system for proton exchange membrane fuel cells. *Int. J. Hydrogen Energ.*, **31**, 447–454.

547 Chin, P., Sun, X., Roberts, G.W. and Spivey, J.J. (2006) Preferential oxidation of carbon monoxide with iron-promoted platinum catalysts supported on metal foams. *Appl. Catal. A*, **302**, 22–31.

548 Ahluwalia, R.K., Zhang, Q., Chmielewski, D.J., Lauzze, K.C. and Inbody, M.A. (2005) Performance of CO preferential oxidation reactor with noble-metal catalyst coated on ceramic monolith for on-board fuel processing applications. *Catal. Today*, **99**, 271–283.

549 Dudfield, C.D., Chen, R. and Adcock, P.L. (2000) Evaluation and modelling of a selective oxidation reactor for solid polymer fuel cell automotive applications. *J. Power Sources*, **85**, 237–244.

550 Dudfield, C.D., Chen, R. and Adcock, P.L. (2000) A compact CO selective oxidation reactor for solid polymer fuel cells powered vehicle applications. *J. Power Sources*, **86**, 214–222.

551 Srinivas, S., Dhingra, A., Im, H. and Gulari, E. (2004) A scaleable silicon microreactor for preferential CO oxidation: Performance comparison with a tubular packed-bed microreactor. *Appl. Catal. A*, **274**, 285–293.

552 Sotowa, K.-I., Hasegawa, Y., Kusakabe, K. and Morooka, S. (2002) Enhancement of CO oxidation by use of H_2-selective membranes impregnated with noble-metal catalysts. *Int. J. Hydrogen Energ.*, **27**, 339–346.

553 Hansen, J.B., Fastrup, B., Joergensen, S.L., Boe, M., Emonts, B., Grube, T., Hoehlein,

B., Peters, R., Tschauder, A., Schmidt, H. and Preidel, W. (2000) Compact methanol reformer test - Design, construction and operation of a 25 kW unit, *Final Publishable Report to European Commission*, Haldor Topsoe, Forschungszentrum Jülich and Siemens AG.

554 Kusakabe, K., Miyagawa, D., Gu, Y., Maeda, H. and Morooka, S. (2001) Preparation of microchannel palladium membranes by electrolysis, in Proceedings of the 5th International Conference on Microreaction Technology IMRET 5, (eds M. Matlosz, W. Ehrfeld and J.P. Baselt), Springer-Verlag, Berlin, pp. 78–85.

555 Emonts, B. (1999) Catalytic radiant burner for stationary and mobile applications. *Catal. Today*, **47**, 407–414.

556 Höhlein, B., Boe, M., Bögild-Hansen, J., Bröckerhoff, P., Colsman, G., Emonts, B., Menzer, R. and Riedel, E. (1996) Hydrogen from methanol for fuel cells in mobile systems: development of a compact reformer. *J. Power Sources*, **61**, 143–147.

557 Pasel, J., Emonts, B. and Peters, R. and Stolten, D. (2001) A structured test reactor for the evaporation of methanol on the basis of a catalytic combustion. *Catal. Today*, **69**, 193–200.

558 Arana, L.R., Schaevitz, S.B., Franz, A.J., Schmidt, M.A. and Jensen, K.F. (2002) A microfabricated suspended-tube chemical reactor for thermally efficient fuel processing. *J. Microelectromechan. Syst.*, **12** (5), 147–158.

559 Haas-Santo, K., Görke, O., Schubert, K., Fiedler, J. and Funke, H. (2001) A microstructure reactor system for the controlled oxidation of hydrogen for possible application in space, in Proceedings of the 5th International Conference on Microreaction Technology IMRET 5, (eds M. Matlosz, W. Ehrfeld and J.P. Baselt), Springer-Verlag, Berlin, pp. 313–320.

560 Görke, O., Pfeifer, P. and Schubert, K., Determination of kinetic data in the isothermal microstructure reactor based on the example of catalyzed oxidation of hydrogen, in Proceedings of the 6th International Conference on Microreaction Technology, IMRET 6, pp. 262–274; (11–14 March 2002); AIChE publ. no. 164, New Orleans, USA.

561 Janicke, M.T., Kestenbaum, H., Hagendorf, U., Schüth, F., Fichtner, M. and Schubert, K. (2000) The controlled oxidation of hydrogen from an explosive mixture of gases using a microstructured reactor/heat exchanger and Pt/Al_2O_3 catalyst. *J. Catal.*, **191**, 282–293.

562 Zaheed, L. and Jachuck, R.J.J. (2005) Performance of a square, cross-corrugated, polymer film, compact heat-exchanger with potential application in fuel cell systems. *J. Power Sources*, **140**, 304–310.

563 Clark, T. and Arner, M. (2002) Motor blower technologies for fuel cell automotive power systems, Progress report hydrogen, fuel cells and infrastructure technologies, US Department of Energy.

564 Selecman, G.E. and McTaggart, P.E. (2002) Hybrid compressor/expander module, Progress report hydrogen, fuel cells and infrastructure technologies, US Department of Energy.

565 Gee, M.K. (2002) Turbocompressor for PEM fuel cells, Progress report hydrogen, fuel cells and infrastructure technologies, US Department of Energy.

566 Gee, M.K. (2005) Cost and performance enhancements for a PEM fuel cell system turbocompressor, Hydrogen Program Report, US Department of Energy.

567 Sgroi, M., Bollito, G., Saracco, G. and Specchia, S. (2005) BIOFEAT: Biodiesel fuel processor for a vehicle fuel cell auxiliary power unit. Study of the feed system. *J. Power Sources*, **149**, 8–14.

568 Kawannura, Y. (2007) A microfuel processor with microreactor for a small fuel cell system in proceedings of the Small Fuel Cell Conference (March 8–9) (ed. Knowledge Foundation, Brooklyn, MA) Miami, FL, US.

569 Peters, R., Düsterwald, H.-G., Höhlein, B., Meusinger, J. and Stimming, U. (1997) Scouting study about the use of microreactors for gas supply in a PEM fuel cell system for traction, in 1st International Conference on Microreaction Technology, IMRET 1, (ed. W. Ehrfeld), Springer-Verlag, Berlin, pp. 27–34.

570 Emonts, B., Hansen, J.B., Grube, T., Höhlein, B., Peters, R., Schmidt, H., Stolten, D. and Tschauder, A. (2002) Operational experience with the fuel processing system for fuel cell drives. *J. Power Sources*, **106**, 333–337.

571 Dams, R.A.J., Hayter, P.R., Moore, S.C., Verhaak, M., DeWilde, P., Adcock, P., Dudfield, C., Chen, R., Lillie, K., Bourne, C., Azevedo, J. and Cunha, J. (1999) Development and evaluation of an integrated methanol reformer and catalytic gas clean-up system for an SPFC electric vehicle, *Final Publishable report of MERCATOX project to European Commission*, Wellman CJB Limited, ECN, Loughborough University, Rover Car Company, Instituto Spuerior Technico.

572 Jenkins, J.W. and Shutt, E. (1989) The Hot Spot Reactor. Hydrogen generation using a novel concept. *Platinum Met. Rev.*, **33** (3), 118–127.

573 Golunski, S. (1998) HotSpot fuel processor. Advancing the case for fuel cell powered cars. *Platinum Met. Rev.*, **42** (1), 2–7.

574 Edwards, N., Ellis, S.R., Frost, J.C., Golunski, S.E., van Keulen, A.N.J., Lindewald, N.G. and Reinkingh, N.G. (1998) On-board hydrogen generation for transport applications: the HotSpot™ methanol processor. *J. Power Sources*, **71**, 123–128.

575 Gray, P.G. and Petch, M.I. (2000) Advances with HotSpot fuel processing. Efficient hydrogen production for use with solid polymer fuel cells. *Platinum Met. Rev.*, **44** (3), 108–111.

576 Reinkingh, J., Carpenter, I., Edwards, N., Ellis, S., Frost, J., Golunski, S., van Keulen, N. and Petch, M.J.P. (1999) On-board hydrogen generation for PEM fuel cells in automotive applications, Society of Automotive Engineers, *SAE Paper* 1999-01-1320, pp. 173–178.

577 Ralph, T.R. and Hards, G.A. (1998) Powering the cars and homes of tomorrow. *Chem. Ind.*, **69**, 337–342.

578 Schuessler, M., Lamla, O., Stefanovski, T., Klein, C. and zur Megede, S. (2001) Autothermal reforming of methanol in an isothermal reactor - concept and realisation. *Chem. Eng. Technol.*, **24** (11), 1141–1145.

579 Panik, F. (1998) Fuel cells for vehicle applications in cars - bringing the future closer. *J. Power Sources*, **71**, 36–38.

580 Cooper, R. and Feasey, G. Summary of dbb's fuel cell engine development activities, in Proceedings of the Fuel Cell Reformer Conference, pp. 116–125 (1999); Diamond Bar, CA, USA.

581 Cooper, R. Summary of dbb's MeOH fuel cell engine development activities, in Proceedings of the Fuel Cell Reformer Conference, pp. 116–126; (November 20 1998); B/T Books, Diamond Bar, CA, USA.

582 Podolski, W.F. Reformer technology challenges, in Proceedings of the Fuel Cell Reformer Conference, pp. 70–76; (1999); Diamond Bar, CA, USA.

583 Birch, S. (1999) Hard Cell. *Automot. Eng. Int.* April, 42–49.

584 Yan, X., Wang, S., Li, X., Hou, M., Yuan, Z., Li, D., Pan, L., Zhang, C., Liu, J., Ming, P. and Yi, B. (2006) 75 kW methanol reforming fuel cell system. *J. Power Sources*, **162**, 1265–1269.

585 Sattler, G. (2000) Fuel Cells going on-board. *J. Power Sources*, **86**, 61–67.

586 Edlund, D. (2005) Development of a compact 250w portable power supply for commercial and military markets, presentation at the Fuel Cell Conference (November 15–17), Palm Springs, California, US.

587 Han, J., Lee, S.-M. and Chang, H. (2002) Metal membrane-type 25 kW methanol

fuel processor for fuel-cell hybride vehicle. *J. Power Sources*, **112**, 484–490.

588 kerchner, G.A. (2006) Recent regulatory developments affecting the transport of Small Fuel Cells in proceedings of the Small Fuel Cells Conference, (ed. Knowledge Foundation, Brooklyn, MA) April 2–4, Washington DC, US.

589 Terazaki, T., Nomura, M., Takeyama, K., Nakamura, O. and Yamamoto, T. (2005) Development of multi-layered microreactor with methanol reformer for small PEMFC. *J. Power Sources*, **145**, 691–696.

590 Ogura, N., Kawamura, Y., Yahata, T., Yamamoto, K., Terazaki, T., Nomura, M., Takeyama, K., Nakamura, O., Namai, T., Terada, T., Bitoh, H., Yamamoto, T. and Igarashi, A. Small PEMFC system with multi-layered microreactor, in Proceedings of the PowerMEMS, pp. 142–145; (28–30 Nov 2004); Kyoto, Japan.

591 Kawamura, Y., Ogura, N., Yamamoto, T. and Igarashi, A. (2006) A miniaturised methanol reformer with Si-based microreactor for a small PEMFC. *Chem. Eng. Sci.*, **61**, 1092–1101.

592 http://www.motorola.com/mediacenter/news/detail/0,3099_2539_23,00.html (29 October 2003 Motorola, Engelhard, University of Michigan win ATP award to develop miniature fuel-cell power source; *Press Release* (5 August 2003).

593 Koripella, R.C., Dyer, C.K., Gervasio, F.D., Rogers, S.P., Wilconx, D. and Ooms, W.J. (2003) Fuel processor with integrated fuel cell utilizing ceramic technology, US 6,569,553.

594 Hallmark, J., Samms, S.H., Casellano, C. and Liu Y. (2006) Development of a ceramic wall-coated methanol fuel processor using novel co-firable steam reforming catalysts. Presention at the Fuel Cell Seminar, November 17, Honolulu, Hawaii, US.

595 Palo, D.R., Holladay, J.D., Rozmiarek, R.T., Guzman-leong, C.E., Wang, Y., Hu, J., Chin, Y.-H., Dagle, R.A. and Baker, E.G. (2002) Development of a soldier-portable fuel cell power system. Part I: A breadboard methanol fuel processor. *J. Power Sources*, **108**, 28–34.

596 Kwon, O.J., Hwang, S.-M., Chae, J.H., Kang, M.S. and Kim, J.J. (2007) Performance of a miniaturized silicon reformer-PrOx-fuel cell system. *J. Power Sources*, **165**, 342–346.

597 Jones, E., Holladay, J., Perry, S., Orth, R., Rozmiarek, B., Hu, J., Phelps, M. and Guzman, C. (2001) Sub-watt power using an integrated fuel processor and fuel cell, in Proceedings of the 5th International Conference on Microreaction Technology IMRET 5, (eds M. Matlosz, W. Ehrfeld and J.P. Baselt), Springer–Verlag, Berlin, pp. 277–285.

598 Cao, C.C., Wang, Y., Holladay, J.D., Jones, E.O. and Palo, D.R. (2005) Design of micro-scale fuel processors assisted by numerical modelling. *AIChE J.*, **51** (3), 982–988.

599 Holladay, J., Jones, E., Palo, D.R., Phelps, M., Chin, Y.-H., Dagle, R., Hu, J., Wang, Y. and Baker, E. (2003) Miniature fuel processors for portable fuel cell power supplies. *Mater. Res. Soc. Symp. Proc.*, **756**, FF9.2.1–FF9.2.6.

600 Holladay, J.D., Wainright, J.S., Jones, E.O. and Gano, S.R. (2004) Power generation using a mesoscale fuel cell integrated with a microscale fuel processor. *J. Power Sources*, **130**, 111–118.

601 Holladay, J.D., Jones, E.O., Dagle, R.A., Xia, G.G., Cao, C. and Wang, Y. (2004) High efficiency and low carbon monoxide micro-scale methanol processors. *J. Power Sources*, **131**, 69–72.

602 Seris, E.L.C., Abramowitz, G., Johnston, A.M. and Haynes, B.S. (2005) Demonstration plant for distributed production of hydrogen from steam reforming of methane. *Chem. Eng. Res. Design*, **83** (A6), 619–625.

603 Lee, S.H.D., Applegate, D.V., Ahmed, S., Calderone, S.G. and Harvey, T.L. (2005) Hydrogen from natural gas: Part I - autothermal reforming in an integrated fuel processor. *Int. J. Hydrogen Energ.*, **30**, 829–842.

604 INNOTECH B. (2007) Beta 1.5 PLUS product flyer, Baxi Innotech GmbH.

605 Vaillant (2007) The future: fuel cells, fuel cell heating appliances, www.vaillant.com.

606 Osaka, N., Nishisaki, K., Kawamura, M., Ito, K., Fujiwara, N., Nishisaki, Y. and Kitazawa, H. (2005) Development of residential PEMFC Co-generation system. Presentation at the Fuel Cell Seminar (November 15–17), Palm Springs, California, US.

607 Echigo, M., Shinke, N., Takami, S. and Tabata, T. (2004) Performance of a natural gas fuel processor for residential PEFC system using novel CO preferential oxidation catalyst. *J. Power Sources*, **132**, 29–35.

608 Wärtsila (2007) WFC20 product flyer Wärtsila Fuel Cells.

609 Seo, H.-K., Eom, Y.-C., Kim, Y.-C., Lee, S.-. and Gu, J.-H. (2007) Operation results of a 100 kW class reformer for molten carbonate fuel cell. *J. Power Sources*, **166**, 165–171.

610 Beckhaus, P., Dokupil, M., Heinzel, A., Souzani, S. and Spitta, C. (2005) On-board fuel cell power supply for sailing yachts. *J. Power Sources*, **145**, 639–643.

611 Spitta, C., Mathiak, J., Dokupil, M. and Heinzel, A. (2007) Coupling of a small scale hydrogen generator and a PEM fuel cell. *Fuel Cells*, **07** (3), 197–203.

612 Recupero, V., Pino, L., Vita, A., Cipiti, F., Cordaro, M. and Lagana, M. (2005) Development of a LPG fuel processor for PEFC systems: Laboratory scale evaluation of autothermal reforming and preferential oxidation units. *Int. J. Hydrogen Energ.*, **30**, 963–971.

613 Cipiti, F., Recupero, V., Pino, L., Vita, A. and Lagana, M. (2006) Experimental analysis of a 2 kWe LPG-based fuel processor for polymer electrolyte fuel cells. *J. Power Sources*, **157**, 914–920.

614 Bosch, R. (2005) Perspective on Fuel Cells vs incumbent technologies. Presentation at the Fuel Cell Seminar, (November 15–17), Palm Springs, California, US.

615 Flynn, T.J., Privette, R.M. Pema, M.A., Kneidel, K.E., King, D.L. and Cooper, M. (1999) Compact fuel processor for fuel cell powered vehicles. *Soc. Automotive Eng.*, SAE paper 1999-01-0536, 47–53.

616 King, J.M. and O'day, M.J. (2000) Applying fuel cell experience to sustainable power products. *J. Power Sources*, **86**, 16–22.

617 Bosco, M., Hajbolouri, F., Truong, T.B., De Boni, E., Vogel, F. and Scherer, G.G. (2006) Link-up of a bench-scale 'shift-less' gasoline fuel processor to a polymer electrolyte fuel cell. *J. Power Sources*, **159**, 1034–1041.

618 Severin, C., Pischinger, S. and Ogrzewalla, J. (2005) Compact gasoline fuel processor for passenger vehicle APU. *J. Power Sources*, **145**, 675–682.

619 Goebel, S.G., Miller, D.P., Pettit, W.H. and Cartwright, M.D. (2005) Fast starting fuel processor for automotive fuel cell systems. *Int. J. Hydrogen Energ.*, **30**, 953–962.

620 Bowers, B.J., Zhao, J.L., Ruffo, M., Khan, R., Sweetland, V., Beziat, J. and Boudjemaa, F. (2006) Advanced onboard fuel processor for PEM fuel cell vehicles. *Soc. Automot. Eng.*, SAE paper 2006-01-0216.

621 Besiat, J., Boudyemaa, F., Bowers, B.J., Zhao, J.C., Dattatraya, D., Ruffo, M., Quet, P., Khan, R., Sweetland, V., Darba, E., Shi,Y., Dorfman, Y., Dushman, N., Foro, A., Alberti, J., and Conti, A. (2007) Multi fuel processor and PEM fuel cell system for vehicles. Presentation at the SOC. of Automot. Eng. World Congress, April (16–19), Detroit, Michigan, US

622 Baratto, F., Dikewar, U.M. and Manca, D. (2005) Impacts assessment and trade-offs of fuel cell-based auxiliary power units Part I: System performance and cost modelling. *J. Power Sources*, **139**, 205–213.

623 Piwetz, M.M., Larsen, J.S. and Christensen, T.S. in Proceedings of the Fuel Cell Seminar, pp. 780–783; (1996); Orlando, FL, USA.

624 Pasel, J. Autothermal reforming of kerosene for application in aviation, in

Proceedings of the Hydrogen Expo Conference, (August 31st–September 1st 2005); Hamburg, Germany.

625 Daggett, D.L., Lowery, N. and Wittmann, J. Fuel cell APU for commercial aircraft, in Proceedings of the Hydrogen Expo Conference, published on CD; (August 31st–September 1st 2005); Hamburg.

626 Krummrich, S., Tuinstra, B., Kraaij, G., Roes, J. and Olgun, H. (2006) Diesel fuel processing for fuel cells - DESIRE. *J. Power Sources*, **160**, 500–504.

627 Meyer, A.P., Schroll, C.R. and Lesieur, R. (2000) Development and evaluation of multi - fuel fuel cell power plant for transport application. *Soc. Automot. Eng.*, SAE paper 2000-01-0008, 45–50.

628 Irving, P.M. Lloyd Allen, W., Healey, T. and Thomson, W.J. (2001) Catalytic micro reactor systems for hydrogen generation, in Proceedings of the 5th International Conference on Microreaction Technology IMRET 5, (eds M. Matlosz, W. Ehrfeld and J.P. Baselt), Springer–Verlag, Berlin, pp. 286–294.

629 Jruing, P.M. and Pideles, J.S. (2006) Operational requirments for a multi-fuel processor that generates hydrogen from bio- and petroleum bassed fuels. Presentation at the Fuel Cell Seminar (November 15–17), Palm Springs, California, US.

630 Metkemeijer, R. and Achard, P. (1994) Ammonia as feedstock for a hydrogen fuel cell; reformer and fuel cell behaviour. *J. Power Sources*, **49**, 271–282.

631 Zhang, Q., Smith, G., Wu, Y. and Mohring, R. (2006) Catalytic hydrolysis of sodium borohydride in an autothermal fixed-bed reactor. *Int. J. Hydrogen Energ.*, **31**, 961–965.

632 Molning, R.M. (2007) Chemical hydride technology for portable PEM FC applications. Proceedings of the Small Fuel Cells Conference (March 8–9), (ed. Knowledge Foundation, Brooklyn, MA) Miami, Florida, US.

633 Teagan, W.P., Bentley, J. and Barnett, B. (1998) Cost reductions of fuel cells for transport applications: fuel processing options. *J. Power Sources*, **71**, 80–85.

634 Jochem, E. and Schirrmeister, E. (2003) Potential economic impact of fuel cell technologies, in *Handbook of Fuel Cells* (eds W. Vielstich, A. Lamm and H.A. Gasteiger), Wiley, Chichester. vol. 4, pp. 1330.

635 Garche, J. and Jörissen, L. (2003) PEMFC fuel cell systems, in *Handbook of Fuel Cells* (eds W. Vielstich and A.A.G.H. Lamm), Wiley, Chichester. vol. 4, p. 1244.

636 Beziat, J., Boudjemaa, F., Bowers, B.J., Zhao, J.L., Ruffo, M., Khan, R. and Sweetland, V. Onboard fuel reforming for fuel cell powered vehicles, in Proceedings of the World Hydrogen Energy Conference, (2006); Lyon, France.

637 Wang, X., Zhu, J., Bau, H. and Gorte, R.J. (2001) Fabrication of micro-reactors using tape-casting methods. *Catal. Lett.*, **77** (4), 173–177.

638 Takeda, T., Kunitomi, K., Horie, T. and Iwata, K. (1997) Feasibility study on the applicability of a diffusion-welded compact intermediate heat exchanger to next-generation high temperature gas-cooled reactor. *Nucl. Eng. Design*, **168**, 11–21.

639 Bier, W., Keller, W., Linder, G., Seidel, D., Schubert, K. and Martin, H. (1993) Gas-to-gas heat transfer in micro heat exchangers. *Chem. Eng. Process.*, **32** (1), 33–43.

640 Schubert, K., Brandner, J., Fichtner, M., Linder, G., Schygulla, U. and Wenka, A. (2001) Microstructure devices for applications in thermal and chemical process engineering. *Microscale Therm. Eng.*, **5**, 17–39.

641 Shin, Y., Kim, O., Hong, J.-C., Oh, J.-H., Kim, W.-J., Haam, S. and Chung, C.-H. (2006) The development of micro-fuel processor using low temperature co-fired ceramic (LTCC). *Int. J. Hydrogen Energ.*, **31**, 1925–1933.

642 Ehrfeld, W., Hessel, V. and Löwe, H. (2000) *Microreactors*, Wiley-VCH, Weinheim.

643 Tosti, S., Bettinali, L. and Violante, V. (2000) Rolled thin Pd and Pd-Ag membranes for hydrogen separation and production. *Int. J. Hydrogen Energ.*, **25**, 319–325.

644 Franz, A.J., Jensen, K.J. and Schmidt, M.A. (2000) Palladium membrane microreactors, in Proceedings of the 3rd International Conference on Microreaction Technology, IMRET 3, (ed. W. Ehrfeld), Springer-Verlag, Berlin, pp. 267–276.

645 Gielens, E.C., Tong, H.D., van Rijn, C.J.M., Vorstman, M.A.G. and Keurentjes, J.T.F. (2002) High flux palladium-silver alloy membranes fabricated by microsystem technology. *Desalination*, **147**, 417–423.

646 Adomaitis, J.R., Galligan, M.P., Kubsh, J.E. and Whittenberger, W.A. (1996) Metal converter technology using precoated metal foil. Soc. Automot. Eng., *SAE Paper*, 962080 p. 185.

647 Lylykanas, R. and Lappi, P. (1991) Soc. Automot. Eng., *SAE Paper*, 910614.

Index

a
acetaldehyde 20, 79
acid leaching 233
activated carbon 108f., 174
activation
– effect 126
– energy 113
active carbon 5
– impregnating 109
active species 57, 60, 66, 74, 78
– dispersion 57f.
– mass 59
adhesion 67, 88
– layer 61, 64f.
adiabatic
– bed 111
– conditions 151
– operation 156
– pre-reforming 39, 88
– quasi- 158
– temperature rise 26f., 48, 269, 275, 338
adsorbent 48, 108ff.
– bed 174
– lifetime 174
– material 200
adsorption 108ff.
– capacity 109
– carbon oxide 54
– chemical 103
– materials 108
– methanol 73
– permanent 60
– pressure swing (PSA) 174ff.
– reversible 60
– selective 108
– techniques 108
– temperature 109
afterburner (AFB) 193, 196, 199, 204

– catalytic 195, 204, 305, 311, 315, 324f.
– methanol steam reformer/catalytic 246
– off-gas 191, 193, 195
air
– bleed 29
– cooler 196
– excess 152
– ratio 31, 41f.
air/fuel mixture 36, 42
alcohol 3, 8, 16f., 81, 98
– fuels 20, 30
alkanes 5, 81, 89
alkylbenzenes 6
alumina 57, 60ff.
– amorphous hydrated 61
– layer 61f., 64, 360, 370
– micro-channel reactors 61
– sol 64
aluminium 61f., 361
– alloys 61f.
– microchannels 64
– plate heat-exchangers 61
– substrates 61
ammonia 5, 46, 107, 350, 359
– cracking 15
– decomposition 46, 107
– formation 43
– getter 352
– hydride generators 352
anode 12ff.
– fuel utilisation 182
– gas flow 13
– materials 15
– off-gas 16, 31, 46, 52, 143, 155, 181f., 191, 280
anodic
– etching 62f.
– oxidation 61f.

Fuel Processing for Fuel Cells. Gunther Kolb
Copyright © 2008 WILEY-VCH Verlag GmbH & Co. KGaA, Weinheim
ISBN: 978-3-527-31581-9

– spark deposition   62
anodisation
– potential   62
– treatment   62
aqueous
– phase reforming   44
– solution   44f.
aromatic diesel compounds   94
aromatics   5, 94f., 98f.
Arrhenius plot   168
auto-ignition   40, 42, 81
automated catalyst coating   370f.
automotive applications   332
automotive drive train application   340, 356
automotive exhaust gas
– systems   68, 135
– treatment   217f., 370
automotive style control components   343
autothermal equilibrium effluent composition   186
autothermal reformer (ATR)   194
Auxiliary Power Unit (APU)   2, 332, 345f., 349
– commercial aircraft   346
– efficiency   197f.
– fuel cell   346
aviation fuel   108

b

balance-of-plant   13, 312, 324, 343
– components   2, 190, 205, 289, 318, 322
bench-scale tests   46
benzene   91
benzothiophene   97
binder   65f.
bio-diesel   4, 193
bio-ethanol   44
bio-fuel   44, 46, 350
– processor   44
blowers   290f.
– high pressure   290
– low power   290
boiling point   4
– distribution   5
boiling range   4, 6
borohydrites   3
Boudouart reaction   22, 54, 229, 239
burner
– external   140
– heterogeneous   52
– homogeneous   52, 228
– integrated   140, 142
– laboratory radiant   52f.
– light-off   124

– methanol   298f.
– microchannel   140
– off-gas   191, 249
– optimisation   187
– permeate   286
– start-up   155, 228
butane   3
– conversion   83
– partial oxidation   83
butylene   23, 95
by-products   8, 19ff.

c

calcination   57, 64, 66
carbon
– aero gels   108
– deposition   54f.
– deposition mechanism   54
– dioxide   3, 13f., 18, 21, 47f., 51
– filament growth   54
carbon formation   54, 129, 131
– rate   87
carbon monoxide   6f., 10, 13ff.
– clean-up devices   47, 155, 190, 199f., 275, 300
– content   279
– converting   47, 50f., 112f., 117
– desorption   51
– output   277f.
– partial pressure   167
– preferential adsorption   16
– preferential oxidation   49ff.
– removal   51
– selective methanation   123, 164
– selectivity   74ff.
carbonaceous species, see coke
carrier material   60ff.
catalyst   6, 12ff.
– accelerated aging   61
– acidic   44, 105
– activity   58, 69, 71ff.
– aging   115
– bed configurations   133
– bimetallic   13, 78, 85, 96f., 119
– blend   133
– carbon-based   155
– carrier   63ff.
– clean-up   109, 183
– coating techniques   59, 61ff.
– cobalt/alumina   78
– cobalt/molybdenum   108
– cobalt/silica   78
– cobalt/zinc oxide   78
– cold light-off   124

– combustion   124f.
– copper-based   73
– copper/ceria   122f.
– copper/ceria/alumina   73, 122
– copper/chrome oxide   74
– copper/zinc oxide   68, 71ff.
– copper/zinc oxide/alumina   72, 74
– copper/zinc oxide/alumina/palladium-zinc oxide   77
– copper/zinc oxide/titania   75
– copper/zirconia   74
– cracking   96
– deactivation reaction   20, 49, 58, 352
– degradation   80, 93
– dual beds   74
– durability   60f.
– effectiveness   68, 155
– fixed bed   69, 71, 82, 155
– formulations   68ff.
– geometries   143
– gold/α-alumina   161
– gold/ceria   121
– gold/iron   161f.
– honeycomb   70
– hydrogenation   108
– impregnated   57
– incorporated   66
– iron oxide/chromium oxide water-gas shift   78
– layer   14, 138ff.
– magnetite ammonia synthesis   350
– mass   58f., 124
– methanation   123f.
– monometallic   79
– nickel-based   39f., 69, 77ff.
– nickel/ceria   78
– nickel/ceria/zirconia   92
– nickel/lanthana   79
– nickel/magnesia   78
– nickel/molybdenum   108
– nickel/molybdenum/alumina   108
– nickel/nickel oxide   80
– nickel/silica   124
– nickel steam reforming catalyst   68
– over-heating   101, 118
– oxidation   132
– palladium/palladium-zinc/palladium/zinc oxide   75
– palladium-zinc alloy   75f.
– palladium-zinc oxide   75f.
– palladium-zirconia   77
– paper-like   75
– partial oxidation   147
– particle size   82
– performance   57ff.
– perovskite   81, 88f., 93
– platinum/alumina   74, 122, 161f.
– platinum/ceria   103
– platinum/palladium partial oxidation   80
– platinum/ruthenium   14f.
– porous structure   158, 162
– powder   75f.
– preferential oxidation   116f., 119
– pre-reformer   86
– Raschig-ring steam reforming   242
– reforming   69, 71
– regeneration   96, 98, 108, 116
– reliability   61
– rhodium   79f., 82, 85, 91
– rhodium/alumina   77, 91
– rhodium/ceria   77, 79
– rhodium/cobalt   77
– rhodium/ruthenium   77
– ruthenium   79, 82, 91f.
– ruthenium/alumina   121
– ruthenium/titania   123
– screeing   118
– selectivity   20, 84
– serial   133
– silver/ceria   121
– sintering   60, 92, 106
– stability   44, 60, 79, 82, 84, 114f.
– steam reforming   132
– surface   50, 52, 60, 98, 115f.
– technology   57
– utilisation   68, 111
– volume   58f.
– wash-coated   61, 73, 77, 138
– water-gas shift   110ff.
catalytic
– activity   49, 59, 78
– afterburner   10, 130
– burners   124f., 143, 285f.
– carbon monoxide fine clean-up   272ff.
– cracking   38f., 96, 154
– decomposition   46
– reaction   22
cathode   12, 15f.
– air   195
– electrolyte   209
– gas flow   13
– off-gas   191, 195, 209f.
cathodic sputtering   67
cells per square inch (cpsi)   242
ceramic
– fibres   75, 360
– foams   221
– honeycomb monolith   239

– jell-roll monolithic catalyst   221
– membrane   259f.
– membrane supports   285
– monolithic structures   221, 236f.
– monoliths   62, 66, 74, 77, 79, 82, 91, 95, 133ff.
– oxide layers   62
– seals   237
– tubes   88
ceria   77, 103
– high surface   86
– sizes   138
channel   137f.
– combustion   139
– diameters   150
– half-height   137
– steam reforming   139
chemical properties   10f.
chemical vapour deposition   67
chemisorption   168
chip-like microreactor   367
– fabrication   367ff.
clean-up
– equipment   22, 338
– reactor   14
– steps   181
– systems   21
coating   61ff.
– dip-   64, 67, 371
– mechanical stability of   68
– precipitation   66f.
– profiles   63
– quality   65
– silica sol   65
– spin-   65, 67
– thickness   62ff.
– wash-   61, 63ff.
co-current
– flow arrangement   135f., 158, 170, 199
– heat-exchanger   223f.
– mode   139, 158, 170, 245
– operation   134, 159, 170f.
– plate heat-exchanger   138, 242
– pressurisation   174
coke   60, 64
– deposit   52, 54, 101
– filamentous   89
– formation   15, 21f., 26f., 29, 33f., 39f., 52, 60, 78, 81, 84ff.
– gasification   154
– molecules   54
– precursor   20
– stable   100f.
coking tendency   99

combinatorial chemistry   120
combined heat and power systems (CHP)   323f., 358f.
combustion   8
– anode off-gas   181f., 187, 240
– catalyst bed   133
– catalyst layer   136
– catalytic   10, 22, 27, 52, 90, 126, 131, 135, 182
– channel   136, 143
– controlled   12
– exothermic   132
– heterogeneous   52
– homogeneous   10, 52, 149, 182
– process   22
– reaction   132, 134, 286
– total   27
– zone   131, 169
compression   202f.
compressor   131, 190, 199, 202, 290ff.
– energy demand   201
– isothermal   291
– membrane   292
– micro-vane   292
– piston   292
– scroll   291
– twin-screw   292
compressor/expander module   291, 308
concentration   57f.
– gradient   52
– profiles   142
– species   58
– vectors   58
condenser   8, 192
conductivity
– effective   219
– heat   219f.
– proton   14
– wall   159
contact time   82ff.
control strategy
– feedback   215
– observer-based   214
control valve   131, 201, 213
conventional
– burner technology (CBT)   358
– electric power generators   215
– energy supply   215
– power production   359
conversion   6, 8, 57, 79
– chemical   6
– equilibrium   18f., 131, 145
– full   52, 71, 79, 83. 84f.
– incomplete   14

– isooctane   141
– outlet   137
– rate of   112
– stable   58
cool flame   39ff.
– evaporator   155f.
– oxidation   41
– stability range of   41
– technology   39f., 42
– temperature   41
cooling   13, 158
– counter-flow   270
– cross-flow   335
– gas flow   158
– inter-stage   162
– loop   201
co-precipitation method   121
cordierite   61, 66, 74, 89, 359
– monoliths   274
corona discharge   43
corrugations   289
cost   355ff.
– breakdown   356f.
– catalyst   355f.
– estimation   357
– fabrication   355, 357
– fuel processor   355, 357
– future   355
– production quantity   355ff.
– type of fuel   355f.
cost-capacity factor   358
counter-current
– concept   133f.
– cooling   156
– depressurisation   174
– design   289
– flow   158, 169f., 200
– heat-exchanger   223f., 363
– mode   135, 170
– operation   158f., 170f.
– pressure equalisation   174
– purge   174
cracking
– cycle   193
– intervals   98
– process   155, 192
– reactions   22, 96, 155, 193
– step   154
cracking/regeneration sequences   98
crystallisation   45f., 66
cycloalkanes   106
cyclohexane   90f.
– cracking   97
cycloparaffins   6

d

deactivation   75f., 78f., 82, 86, 89f., 98, 101ff.
– reversible   103
decaline   101
n-decane   95f.
decomposition reaction   20
Deep Reactive Ion Etching (DRIE)   367
dehydrogenation   106f.
density   3f.
– energy   5
– gravimetric power   3, 5, 248
– volumetric power   3, 5, 248
deposition   65f.
depressurisation   174
desulfurisation   46f., 73, 88, 108f., 333, 348
– adsorbents   108
– catalysts   108
– feed   349
– step   348
dielectric barrier discharge   43
diesel   40f., 92ff.
– bio-   4
– compounds   94f.
– conversion   93
– feedstocks   95
– fuel processor   195
– fuel processor/fuel cell systems   195
– fuelled passenger cars   344
– heavy   108
– low sulfur   109
– partial oxidation   101
– reforming   42
– steam reformer/anode off-gas burner   253
– sulfur free synthetic   102
– surrogates   88, 92, 96
– synthetic   101ff.
– synthetic low sulfur   98
diesel-powered vehicles   13
diethyl sulfide   5
diffusion
– bonding   62, 363f.
– bulk phase   168
– coefficient   167
– effects   158
– intraparticle   68
– intrapellet   68
– limitations   160
– molecular   68
dimethyl ether   44, 105
– conversion   44
– formation   20
– partial oxidation   105
– reforming   45

discharge tank   46
distillation step   190
dodecane   4, 93ff.
n-dodecane   89
downstream   8, 13, 22, 26, 47, 108f., 120f., 149
– processes   19
dry-gas composition   95

e

efficiency   21, 29, 194f.
– autothermal reformer   196
– auxiliary power unit   197f.
– conversion   238
– factor   9
– fuel processor   197f.
– gross   203
– hydrogen separation   203
– losses   194
– maximum   29ff.
– overall   13, 21, 31, 43, 215
– processor   187f., 190
– reformer   197
– reforming   187f.
– system   196f., 203ff.
– target   340
– thermal   215
elastomer   169
electric vehicle   332
electrical
– efficiency   215
– energy   6, 43
– fuel cell stack power   9
– potential   12
– power equivalent   204, 229, 245, 335
– power generation   12f.
– power output   13, 215, 338
– pre-heating   124
electricity generator   215, 353
Electro Discharge Machining (EDM)   365
– die-sinking   365
– wire-cutting   365
electrochemical
– deposition   65
– reactions   9
electrode   12, 14
electroless plating   66
electrolyte   15f.
– materials   15
– solid   15
electron beam evaporation   67
electrophoretic
– deposition   65
– precipitation   65

electroplating, see electrochemical deposition
emission   10, 285, 316, 320
– control regulations   10
endothermic   17
– cracking   38
– reactions   23, 39
energy
– diagram   184, 264
– efficiency   151
– losses   185
energy balance   8, 183f.
– overall   181
energy density
– gravimetric   3
– volumetric   3
energy requirements   129f.
– process   130
– start-up   186, 189
– steam reforming   129
energy supply   129, 131
– conventional   215
enthalpy   9f.
– reaction   10
equilibrium
– concentration   38
– constant   47
– conversion   143
– feed composition   168
– gas composition   19ff.
– reformate composition   151
– water-gas shift   151f.
equivalence ratio   178
etching   62f.
– photochemical   63
ethanol   4f., 20
– bio-   44
– conversion   78, 80
– decompositon   20
– partial oxidation   80
– steam reforming   20
ethylene   20, 23f., 79, 89, 95
European driving circle   10
evaporation   140, 180
– feed   187
– fuel   181f.
– processes   22
– system   42
– water   147, 190, 351
evaporator   155, 181, 187, 305
– diameter   42
exergetic efficiency   187
exergy
– analysis   187
– losses   187

exhaust combustor 200
exothermic 18, 23, 26
– reactions 23, 39, 77
expander 291
explosion limits 176f.
extrusion 66
– techniques 218

f

fabrication
– costs 355ff.
– techniques 355
Fecralloy 360ff.
– metal foils 370
– plate coated 139
feed 12f., 15, 22
– alkane 148
– compostion 17, 23, 158, 163
– compression 202
– concentration 45f., 277, 352
– conditions 60
– flow path 136
– flow rate 59, 129, 147, 186, 213
– impurities 60
– injection 269, 292
– methanol/air/steam 232
– methanol/water 295
– pre-heating 352
– reformate 275
– species 57f.
– temperature 27f., 30, 34ff.
– volume flow 59
feedstock 6, 31, 39, 87f., 90, 230
– blends 39
– compositions 94
– pre-reformed 89
fixed bed 58f., 68f.
– overheating 161
– technology 147
flame
– arrester 42, 176, 242, 292
– guide 267
flammability 4
– limits 36, 124, 178
flow
– distribution 180, 223, 282
– equi-partition 237
– molar 7
– nitrogen 268
– radial 262
– rate 58, 186
– velocity 69
foam 25, 59, 65
– α-alumina 83

– ceramic 81, 83, 85
– coated 88
– pore size 83
– precursor 361
formaldehyde 14
formic acid 14, 19f.
fossil fuels 3, 5, 46
– common 3
fuel cell 6ff.
– air-cooled medium temperature 209
– alkaline 15, 46, 296, 349f.
– ammonia cracker/alkaline 350
– Andromena 349
– autothermal propane reformer/low temperature PEM 203f.
– bio-fuel processor 353
– ceramic membrane 202
– diesel steam reformer/PEM 202f.
– direct borohydride 46
– direct ethanol 14
– direct methanol 1, 14
– efficiency 8f., 202
– electric power output 319
– electric vehicle (FCEV) 306
– fuel processor system 322, 327
– heat-exchanger applications 289
– high temperature 7f., 14f., 43, 47, 147
– hydrocarbon fuelled 349
– low temperature 7, 13
– membrane 14
– mobile 108
– molten carbonate 16, 326, 345, 349
– optimisation 187
– performance 13f., 334
– phosphoric acid 15, 129, 349
– plasmatron 43
– power generation 13
– Proton Exchange Membrane (PEM) 1, 7ff.
– solid oxide 15f., 47, 155, 181, 325, 345f.
– stability 14
– stack 12ff.
– submarine 297
– system 8ff.
– technology 3, 332
– type 6, 12f.
– vehicles 305f., 341
– water-cooled 208
fuel feed rate 204
fuel processing
– applications 124
– chemistry 17ff.
– gasoline 212

– reactions 8
– steps 188
fuel processor 7ff.
– alcohol 169
– autothermal diesel 193
– autothermal gasoline 182
– autothermal methane 183ff.
– basic engineering 192
– bio-diesel 193
– breadboard 280, 333f.
– configurations 182
– control strategies 213
– design concepts 129
– diesel 193f., 200f., 344f.
– diffusion layer 303
– dynamic simulations 205
– efficiency 7ff.
– ethanol 189f., 316
– fabrication 366
– feed 190
– future 337f.
– gasoline 204, 210, 251, 332ff.
– HotSpot 227, 300ff.
– integrated 302
– kerosine 344
– Liquified Petroleum Gas (LPG) 327ff.
– membrane 173, 202
– methane 317ff.
– methane steam reforming 182, 212
– methanol 10, 14, 169, 185, 189, 215, 295ff.
– methanol micro 213
– micro 310
– microstructured gasoline 252
– multi- 348ff.
– natural gas 317
– natural gas/liquified petroleum gas 331
– on-board 296, 341
– overall 181, 191
– product 347
– production techniques 355, 359
– reactor temperatures 339
– requirements 12
– second generation 343f.
– self-sustaining 314
– single plate 302f.
– steam reformer 196
– systems 295, 348
– thermal power output 319
– third generation 340, 342ff.
fuel processor/fuel cell plant 342
fuel processor/fuel cell system 207ff.
fullerene 5
furfurals 98

g

gas
– analysis 347
– composition 37, 71, 155
– consumption (GU) 358
– distribution layers 12
– dump 174
– exhaust 135, 188
– hot combustion 203
– inert 13, 42, 171
– mixture 6, 40, 98
– natural 3f., 6, 10, 16, 34, 39, 46f.
– phase 17, 47, 50
– production 39
– purification 8, 301
– purification process 174
– secondary process 174
– sweep 166, 169, 199, 202
– synthesis 6, 39
Gas Hourly Space Velocity (GHSV) 58f.
gas-liquid separator 351
gasification 44, 101
gasoline 4f., 36
– engine 215
– feed 27
– fuel processor/fuel cell system 333
– hydro processing 108
– low sulfur 109
– pre-reforming 87f.
– reforming 28
– station 332
– sulfur-free 340
– surrogate 88ff.
– system 342f.
gasoline-powered cars 13
gelation 64f.
glas micro-sphere 5
gliding arc
– plasmatron 265ff.
– technology 43
global warming potential 44
glow plugs 230, 347
glucose solution 44, 353
graphite 12

h

H/C ratio 191
– liquids 5
heat
– capacity 4, 9f., 133, 186
– conductivity 149, 152, 158
– exchange capabilities 194
– generation 40f., 134f.
– losses 8f., 40f., 152, 185ff.

– management   164
– of vaporization   4
– recovery   281
– removal problem   13
– requirement   134
– supply   17
– transfer efficiency   212
– transfer limitations   136, 199
– transport limitations   51
heat exchangers   9, 43, 118, 131, 133ff.
heating
– cartridges   244
– system   10
– value   147
n-heptane   21, 89
hexadecane   4, 95f.
hexane   90
– cracking   97
n-hexane   91
hexene   90
homogeneous gas phase chemistry   26
hopcalite   119
– carrier   119
hot-gas blowers   181
hot spot   69f., 86, 149
– formation   199, 233, 247
– temperatures   84, 88
humidification   15, 166, 195
hydrites   3
– MBH4   5
hydrocarbon   3, 17, 38
– activation   80
– cracking   38, 98
– fossil   3
– higher   3, 8, 21, 32, 39, 43
– ignation   81
– light   26, 33, 39
– liquid   3, 5, 14, 44f.
– mixtures   127
– non-volatile   6
– oxidation   125
– pre-reforming   86f.
– reforming   80ff.
– saturated, *see* alkanes
– unconverted   10
– unsaturated, *see* olefins
hydrodesulfurisation   108
hydrogen   3ff.
– back-diffusion   300
– carriers   3
– diffusion   168
– dilution   13, 103, 147
– dissociation   12
– efficiency   129

– filling station   332
– flow rate   129
– flux   167, 285
– generation   105
– liquified   5
– mass flow   129
– oxidation   67
– partial pressure   166, 169f.
– permeator   208
– peroxide conversion   264
– production   155
– ratio (HR)   8
– recovery   171, 285
– selectivity   44, 75, 82, 85
– separation   66, 168, 170
– source   3
– sulfide   3, 14, 47
– unconverted   13
– utilisation   13, 182, 188, 191
– yield   32, 37f., 72, 94f., 102, 106, 149, 151, 189
hydrolysis   107
hydrotalcite   84
hydrotreater   345

i

ignition   178
– auto   36
– butane   179f.
– ethane   179f.
– homogeneous   178ff.
– methane   179f.
– propane   179f.
– surface   179
– temperature   36, 41f., 179f.
impregnation   66, 73, 76
– post   76
– pre   76
impurities   174
industrial processes   68f., 80f., 110, 169
inhibiting effect   126
injection
– nozzle   292, 338
– systems   42, 267
insulating materials   180, 293
insulation   185
– inner thermal   267
– material   280f., 286
integrated
– heat-exchangers   158
– reformer/combustor   246
– water-gas shift reactor/heat-exchanger   271
internal combustion engine   10, 45, 52, 148
– technology   215

ion exchange resin 107
isothermal conditions 156, 201

j
jet diesel fuel 347f.
– A 6
– A1 6
– hydro processing 108
– Jet-A 347
– JP-8 6, 102, 347, 350

k
kerosene 81, 94, 97f., 108
– compound cracking 97
– hydrodesulfurised 93
ketones 98
kinetics 136, 149
– firts order 136
Knudsen diffusion 168, 172

l
laser ablation 366
legislation 10
lifetime analysis 108
light-off 124, 126f.
– partial oxidation 152
liquid
– phase 44
– pump 290, 308
Liquified Petroleum Gas (LPG) 4f., 10, 44, 47, 52, 84, 86, 327ff.
– cracking 97
– low sulfur 109
– storage equipment 44
– surrogate 97
lithium aluminium hydride 352
lithium ion battery 332
load changes 153, 248
– transient changes 248
Low Temperature Co-fired Ceramic tapes (LTCC) 366
Lower Heating Value (LHV) 4, 6f., 9, 29, 38, 44

m
mass
– flow rate 130, 197f.
– fraction 53
– production 355, 365f.
– ratio 42
– transfer coefficients 221
– transfer limitations 51, 133, 136, 138, 140, 155, 163, 199, 229, 234
membrane 12, 164ff.
– alloying 165
– dense 172
– dry-out 8f.
– fabrication 369f.
– imperfections 166
– lifetime 165
– material 12, 15
– metal 166, 209
– microporous ceramic 172f.
– Nafion 12, 15
– palladium 18, 165ff.
– polybenzimidazole 15
– polymeric 164
– porous 172f.
– porous glas 254
– purification unit 202
– quality 168
– separation 18f., 164ff.
– separation devices 283f.
– separation modules 284
– surface area 168, 202
– thickness 166ff.
– tubular 172
– types 172
– vanadium 165f.
Membrame Electrode Assembly (MEA) 12
mercaptanes 5
– ethyl 5
– tertiary butyl 5
metal 12f., 54, 60
– clusters 57
– foams 65, 88, 274
– noble 78, 82, 84, 88, 93
– non-precious 68
– passivating 62
– porous 279
– precious 68, 75, 77, 117
– promoter 71
– salt 57
– surface 52, 54
metallic
– gauze technology 220
– ions 14
– membranes 66
methanation 21, 49, 51, 101, 112, 120, 123f.
– activity 116, 121
– catalysts 51
– reaction 18, 51, 106
– selective 109, 123, 164
– thermodynamic equilibrium of 78, 85
methane 3, 5, 7ff.
– combustion 133ff.
– conversion 25f., 82f., 137, 155
– cracker 192f.

– cracking   96f.
– decomposition   154
– fuel processing   8
– partial oxidation   27, 83, 100
– reformer/combustor   247
– selectivity   85, 122
– steam reforming   47, 133ff.
methanol   4f.
– combustion   299
– conversion   72f., 75, 77, 256ff.
– dehydration   20
– fuel processor/fuel cell system   312
– patial oxidation   77, 80
– reformer/evaporator/burner   263f.
– steam reformer/PEM fuel cell system   299
– steam reforming   68f., 73, 75f., 313f.
methane-fuelled bus   305
methane-fuelled portable fuel cell   308
methyl formate   19
methylcyclohexane   45, 89, 106f.
– dehydrogenation reaction   45
methylnaphthalene   94f.
microchannel   50f., 81f., 84, 114f., 119, 149
– coupled reformer/afterburner   209
– fabrication   365f.
– heat-exchanger   158, 211
– plates   82
– wash-coated   85
microelectromechanical system (MEMS)   367f.
Microlith   220, 231f., 347
– gauzes   232
– packed metal gauze technology   237, 270, 275
micro-membrane pump   290
micro-organisms   44
micro reaction technology   226
microreactor   225, 243, 262
– chip-like   260
– technology   298
microstructure fabrication   365f.
microstructured
– evaporators   249
– heat-exchanger   187, 249, 305
– plates   81
– reformers   249, 151
microwaves   43
mobile applications   345
model
– CAD (Computer-Aided Design)   346, 349
– chemical reaction network   149
– homogeneous (HOM)   160
– kinetic   139, 158, 199
– one-dimensional dynamic   152

– one-dimensional isobaric   139
– transient   138
– two-dimensional   149, 158, 281
modulation range   42
molar
– flow rate   17
– oxygen flow rate   17
– steam flow rate   17
monolith   59
– ceramic   61, 82, 218ff
– cylindrical   219f., 231
– fabrication   359
– Fecralloy   234, 360ff.
– first   95
– metallic   61, 82, 138, 152, 218ff.
– porous   66
– production technique   360f.
– second   94
– silicon carbide   221f.
– stability   361
– structure   361
– third   94
– void fraction   59, 220
multi-fuel conversion capability   348

n

nanoparticle   65, 75
– dispersion   65
– gold   120
nanotubes   5
naphthalenes   6, 96, 99
nickel
– steam reforming catalyst   68
nickel-based alloys   361
nitrogen   3
nitrous oxides (NO_x)   10
– emissions   52, 124

o

O/CO ratio   194, 197, 213
octane
– isomers   23
– number   5
iso-octane   6, 25f., 28, 34, 36f., 81, 89ff.
n-octane   5, 24f., 89ff.
odorants   44, 47
– sulfur-free   47
off-heat   318
olefins   5, 86, 99
– light   95
on-board power
– demand   332
– plant   349
organic

– compounds 10, 108
– materials 65f.
– volatile 10
osmotic drag 15
oxidation
– flameless 320
– partial 7, 22ff.
– preferential 60, 161, 274ff.
– reactions 133
– total 22f., 34ff.
oxidative steam reforming 69
oxygen 12
– conversion 118
– deficient atmosphere 22
– mole fraction 150
– stoichiometry 191
– storage capability 80
– treatment 101
– unconverted 12
oxygen/carbon ratio (O/C) 17, 23f., 26ff.

p

palladium 75f.
palladium-zinc alloy 75
paraffinic diesel compounds 94
paraffins 99
– partial oxidation of light paraffins 84
iso-paraffins 6
n-paraffins 6
parallel-flow 169f.
parasitic power losses 190, 199, 202
partial
– oxidation mode 148
– recirculation 345
– system load 9
peak burner (PB) 358
peptisation, *see* gelation
performance
– experimental 50
– losses 13ff.
permeability 167f., 100, 283f.
– coefficient 168
– hydrogen 172
– specific 172
permeate 166f., 169ff.
– hydrogen/steam mixture 202
– partial pressure 170
– pressure 170
permeation 283
– hydrogen 199, 254f.
– rate 166, 285
– reduction 166f.
perovskite structures 70, 81, 89
phase diagram

– C-H-O 54f.
phase transition 10, 60, 64
phenanthrenes 6
physical vapour deposition 67
plasma generation 267
plate heat-exchanger 136, 180, 195, 217
– catalytic afterburner 147
– design 156
– fabrication 361ff.
– microchannel 158
– technology 139, 207
platinum 12ff.
platinum/ruthenium 13f.
poisoning
– chlorine 73
– effect 13f.
– partial 104
– reversible 104
– sulfur 73, 88, 91, 101ff.
polymer 362
– corrugated 289
– cross-flow 289
– fluorocarbon 12
polymerisation 40, 98f.
pore
– blockage 80
– sizes 25f.
– volume distribution 65
power
– density 348
– output 340, 345
precipitation 66f.
preferential oxidation
– two-stage 162
– water-cooled 193
pre-heated 181
– air stream 40, 155
– catalytic combustion 206
– reformer 152, 200
pre-heating 227
– electrical 206
– feed 192, 206, 212
– fuel/air/steam mixture 200, 232
– temperature 27, 147
pressure swing adsorption (PSA) 15, 19, 45
Printed Circuit Heat-Exchangers (PCHE) 133
product mass flow rate 197f.
promoter 80
propane 3ff.
– conversion 185
– cracking 38
– partial oxidation 84f.
– undiluted 97
propulsion of naval systems 348

propylene 20, 23f., 95
prototype 311f.
purification
– hydrogen 174
– process 19
pulse
– duration 270
– frequency 270
pulsed laser deposition 67
pyrenes 6
Pyrex glas 368
– cover 315
– layer 311
– plate 281
pyrolysis 39, 99
pyrophoric 72, 80, 111

r

radiation
– external 53
– internal 53
– losses 293
– reflection property 293
– shield 293
radical 180
– reactions 41
reaction
– chemical 57
– endothermic 134f., 222f., 350
– exothermic 134f., 222f.
– gas phase 225
– heterogeneous 57, 177
– homogeneous 177f., 180, 200
– light-off 88
– network 50
– patial oxidation 152
– preferential oxidation 161, 189
– pressure 131, 170ff.
– rate 41, 50f., 58, 156, 158, 271
– reforming 186
– steam reforming 131, 133, 169
– temperature 18ff.
– trimolecular 177
– water-gas shift 155, 159, 162f.
reactor
– adiabatic 24, 70, 100, 133, 158, 160, 247
– alumina micro-channel 61
– aluminium 107
– autothermal membrane reformer 201
– carbon monoxide purification 327
– catalyst coated reformer/burner heat-
  exchanger 144
– Catalytic Plate (CPR) 223
– catalytic wall 134, 223

– clean-up 188, 213
– coaxial cylindrical 132
– concentric spherical 131f.
– concentric tubular 131
– configurations 131
– copper 244
– counter-flow 132
– cracking 193, 350
– decomposition 46
– dimensions 177
– dual fixed bed 131f.
– dual stage preferential oxidation
  tubular 274
– electrical heated 244
– exit 57f.
– fixed bed 48, 51, 69, 146, 211, 217, 227
– furnace temperature 86
– heat-exchanger 159
– high temperature water-gas shift (HTWGS)
  194, 207
– housing 23
– inlet 58, 69, 88, 118, 149, 156, 232
– integrated heat-exchangers 131
– length 137f., 142, 149, 152
– lithium aluminium hydride 352
– load 163
– low temperature water-gas shift (LTWGS)
  194, 203
– medium temperature 269
– membrane 168ff.
– metal surface 85
– methanation 164, 199
– methane steam/burner 229
– methanol reformer/burner 145
– methanol steam reforming
  membrane 170ff.
– microchannel 22, 91, 143, 159, 209
– microchannel honeycomb 233
– Microlith 231
– microstructured 288
– microstructured heat-exchanger 187, 225, 244
– microstructured testing 280
– monolithic 48, 69, 116, 209, 217, 230, 269, 360
– non-steady state 193
– outlet composition 196
– partial oxidation 26f., 70, 81, 147
– performance 158, 169, 237
– plasmatron 268
– plate heat-exchanger 221ff.
– porous 162
– preferential oxidation (CO PROX) 120f.,
  162, 187, 190, 194, 201, 203, 211

- quasi-monolithic 232
- reformer/burner heat-exchanger 146
- silicon 281f.
- single heat-exchangers 133
- specific chemical 57
- spray-pulse type catalytic 269
- stainless steel 244f.
- suspended tube 286ff.
- temperature 26, 118
- ten-fold screening 73
- tubular fixed bed 136, 350
- tubular membrane 170
- tubular steam reforming 23, 131, 169, 229
- volume 186, 244
- wall 134f., 223
- water-cooled 207f.
- water-gas shift 121, 155, 160, 187ff.
- water-gas shift membrane 172
reformate 6f., 10, 13f., 170ff.
- composition 129, 148
- hydrocarbon 15
- flow constant 13
- flow rate 170, 251, 279
- pressure 170
- purification 181
- surrogate 270f., 274, 285
- synthetic 277f.
reformation 44
reformer 7f., 16
- adiabatic temperature 28
- alcohol steam 285
- autothermal diesel 238
- autothermal gasoline 206f.
- autothermal methanol 77
- catalytic 266
- chip-like methanol steam 261f.
- diesel autothermal (ATR) 193, 196, 347
- diesel steam (STR) 196
- efficiency 31, 130, 238, 299
- electrical heated 244
- ethanol 49
- feed evaporation system 299
- fixed bed plate 136
- FLOX 320f.
- gasoline 238
- Glid/Arc plasma 268
- HotSpot 300f.
- methanol 14, 298, 304ff.
- methanol steam 310f.
- microchannel oxidative steam 209
- microchannel steam 142
- microstructured methanol 249
- modular autothermal 300
- monolithic autothermal methanol 230

- monolithic autothermal propane 235f.
- monolithic propane 70
- octane 158
- outlet 109
- partial oxidation 148
- plasma 43
- plasmatron 43, 264f.
- plate 136
- plate-fin (PFR) 139f., 240f.
- pre- 16, 181, 325, 331
- propane autothermal (ATR) 347
- tubular fixed bed steam 136, 331
- tubular natural gas steam 326
- volume 202
reformer/heater/evaporator system 321
reforming
- aqueous 44
- autothermal 7, 16, 21, 26, 29f., 32ff.
- butane steam 84f.
- catalyst layer 136
- catalytic steam 149
- channel 136f.
- diesel 70, 92ff.
- dimethyl ether steam 105
- ethanol steam 77ff.
- gasoline 88, 94
- homogeneous plasma 43
- hydrocarbon 80, 91
- internal 15f.
- isooctane 32
- kerosene 92f.
- layer 135
- liquified petroleum gas 84
- methane steam 18f., 68f., 80f., 100, 129ff.
- methanol steam 71, 73, 75f., 80, 105, 131f., 232f.
- natural gas/methane 81
- non-thermal plasma 43
- oxidative steam 29, 69, 82, 149
- partial oxidation 81
- platinum/rhodium steam 80
- pre- 39f., 86, 147, 155, 269
- process design 129, 181
- propane steam 84
- quasi-autothermal 132
- reactions 101
- reactors 227
- spray-pulse 270
- steam 7, 15ff.
- temperature 44
re-hydrogenation 45
residence time 50, 79, 84, 86, 105, 248
- modified 59, 157
retenate 166, 171

– partial pressure   170
Reverse Water Gas-Shift Reaction (RWGS)   49

S
sealing techniques   362f.
Selective Adsorption for Removing Sulfur (SARS)   108
selectivity   25f., 51, 57f.
– product   25
self-hydrolysis   45
side reaction   20, 22, 51
Sievert's law   168
Sievert's solubility constant   167
silica   57, 60, 77f., 96
– membrane   283
– mesoporous   247
– sol   64f.
silicon   67
– black   107
– chip   367
– microchannels   67
– microreactors   67
– reaction zone   287f.
silicon nitride   286
simulation   70, 133, 152
– CFD   281f., 287
– dynamic   205ff.
– numerical   50, 136, 170, 281
– reformate surrogate   276
– static   193
– three-step   152
sintering process   98
slurry
– catalyst   81f.
– viscosity   64f.
sodium borohydride   45f., 107
– concentration   352
– hydrogen generation system   351
sol-gel
– coating   64f.
– method   64f., 82
solid plate material   293
soot formation   40, 340
specific surface area   57f.
spray-pulse techniques   43
stage
– dual   277
– second   339
– small series   289
– three   280
– water-gas shift   339
stainless steel   54f., 61, 65
– heat-exchanger   242
– metal foils   237

– radicals   180
– surface   54
– tube   178
– vessel   217, 267
standard enthalpy of formation   6
start-up
– behaviour   206f.
– burner   228, 338
– energy   339
– gas supply   205
– gasoline fuel processor   337
– power   339f.
– procedure   205f., 339, 343
– rapid   217
– strategy   81, 205, 213
– time   205ff.
start-up/shut-down procedures   61, 111
stationary conditions   244
steam
– condensation   9
– cracking   39
– gas turbines (SGT)   358f.
– generation   181f., 199f.
– generators   338
– jackets   292
– partial pressure   47
– reformer/catalytic combustor   82, 249
– reformer/heat exchanger   250
– reforming   68f., 241f.
– superheated   227
– supported partial oxidation   26f.
– unconverted   17
steam/carbon ratio (S/C)   17ff.
stoichiometric
– factor   58
– ratio (SR)   8, 12f.
storage density
– gravimetric   5, 45
– volumetric   5, 45
submarine applications   348
sulfur   3, 5f., 46f., 102f.
– compounds   3, 14, 88, 91, 101
– free-   47
– polycyclic   47
– species   103
– tolerance   101, 104
sulfur dioxide   102f., 125
Super Ultra Low Emissions vehicle regulation   10
suprastoichiometric values   125
surface area   60, 62, 65f., 88, 103
– specific   98
surfactant   64
syngas   26

system
– load 343
– pressure 18, 41

t
tank to wheel 340
temperature
– adiabatic 157
– coolant 159
– cycling 60
– distribution 351f.
– feed 110
– flame 149
– fluctuations 73f., 92, 133
– gradient 52, 68, 134, 136f., 351
– isothermal 157
– light-off 81
– mal-distribution 139, 338
– operating 109, 155, 167, 190
– optimum 156ff.
– pre-heating 100
– profile 70, 139f., 142, 152, 156, 271
– reaction 73, 76ff.
– reduction 75f.
– resistance 81, 111
– shift stage 111
– wall 137, 161
tetra-decane 101
tetralin 93
thermal
– conductivity 158, 233
– expansion coefficients 285
– explosion 177
– oil cycle 305
– output 343
– partial oxidation 148
– plasmas 43
– power output 323, 340
– resistance 66
– shock resistance 218
– stress 134
thermo neutral conditions 30, 191
thermocouples 268, 278, 280f.
thermodynamic equilibrium 18, 22f., 27, 156
– calculations 27, 32f., 147
– gas composition 28, 38, 71, 239
thermophoresis 65
thiophenes 88
toxic 44
toluene 45, 89ff.
– conversion 89
– unconverted 338
transition test 340
transport limitations 68, 310
– mass 83
tube-bundle cooler 285
turbine expander 204f.
turn-down ratios 72, 343
Turn-Over Frequency (TOF) 59, 84

u
upstream 8, 108f., 120f.

v
vaporisation 43
vaporizer 187
velocity
– flow 137
– gas hourly space 69, 71, 87, 137f., 143, 160, 228ff.
– inlet 137
– weight hourly space 69ff.
viscosity 64f.

w
wafer substrates 65
water
– balance 8, 183, 190f., 195ff.
– condensation 289
– cooling 13, 210
– excess 8
– feed flow rate 194
– grey 346
– injection 148, 158, 213
– migration 15
– recovery 8, 191, 199
– separation 195
– service 346
– supercritical 44
water-gas shift reaction (WGS) 8, 14, 17f., 22, 30, 32, 44, 48f., 60, 114ff.
Weight-Hourly Space Velocity (WHSV) 58ff.
welding procedure 244
wet chemical etching 365

x
XPS (X-ray Photon Spectroscopy) 73

y
yield 57f.

z
zeolite 57, 88, 96, 105
– membranes 282
zinc oxide 73, 77
– matrix 76
zinc sulfide 73
zirconia 105